Experiments in Applied Chemistry

Dr. Sunita Rattan

EXPERIMENTS IN APPLIED CHEMISTRY

(For Engineering Students)

By

Dr. Sunita Rattan

M.Sc, M.Phil, Ph.D

Department of Chemistry

Amity School of Engineering and Technology, New Delhi

S.K. KATARIA & SONS®

Publisher of Engineering Books

4885/109, Prakash Mahal, Dr. Subhash Bhargav Lane,

Opposite Delhi Medical Association, Daryaganj, New Delhi–110002 (INDIA)

Phone: +91-11-43551243; *Mobile:* +91-9871775858, 7678425209

e-mail: katariabook@gmail.com, katariabooks@yahoo.com

Website: www.skkatariaandsons.com

(e-books, subscriptions and bulk buying also available on our website)

Published by:

S.K. Kataria & Sons®

4885/109, Prakash Mahal, Dr. Subhash Bhargav Lane,
Opposite Delhi Medical Association, Daryaganj, New Delhi-110002 (INDIA)
Phone: +91-11-43551243; *Mobile:* +91-9871775858, 7678425209
e-mail: katariabook@gmail.com, katariabooks@yahoo.com
Website: www.skkatariaandsons.com

First Edition: 2002; Second Edition: 2003; Third Edition: 2011; Reprint: 2025

***Price :* ₹ 495/-**

Laser Type Setting by: **SARA ASSIGNMENTS**, Patparganj, Delhi-110092.

Printed at: **A.S. OFFSET**, Delhi

Preface to the Third Edition

"The only important thing in a book is the meaning it has for you"

—Somerset Maugham.

I express my profound gratitude to the faculty and students of various Engineering colleges, who have given a splendid response to my first edition. Their tremendous appreciation has boosted my confidence and has encouraged me to present the third revised edition.

The objective of the third edition remains the same i.e. to provide conceptual understanding of the principles involved in experimental work. Although, the organization of third edition is essentially unchanged from the second edition, some chapters have been revised and expanded considerably. To meet the specific demands of a segment of readers, a number of new experiments are incorporated in various sections. A new section on 'Environmental Chemistry' has been added and section on 'Distribution Law' has been removed.

I sincerely hope that the revised third edition will prove more useful and would provide a better and explicit understanding of the concepts involved in experimental chemistry.

I would be highly grateful to readers for sending comments and suggestion for further improvement of the book and would be pleased to incorporate their suggestions in my subsequent editions.

Sunita Rattan

Dedicated to

My mother

(Mrs. Sharda Lakhanpal)

Acknowledgements

Gratitude unlocks the fullness of life;
It turns what we have into enough, and more;
It turns denial into acceptance; Chaos into order;
Confusion into clearity;
Gratitude makes sense of our past,
Brings peace for today,
And creates vision for tomorrow.

"Melody Beattie"

I take this opportunity to acknowledge the assistance and contributions of a large number of individual to this effort.

My special thanks are due to my revered teacher, Professor B.N. Mishra, Emeritus Fellow, Department of Chemistry, H.P. University, Shimla. The legacy of scholarly and scientific inquiry and the temperament which I have inherited from him, will remain with me as invaluable treasure to be invested in everything, I do and seek in life.

I will be failing in my duties if I do not acknowledge Dr (Mrs) I. K. Mehta (Himachal Pradesh University) whose affectionate advice and moral support have left a deep impression on my life.

It gives me great pleasure to place on record my profound gratitude and indebtness to our esteemed president Dr. Ashok Kumar Chauhan, Founder President RBEF, whose inspiring words, stimulating encouragement and uniflinching support have been a source of great strength in accomplishing this assignment.

I am deeply grateful to Prof. B.P. Singh, Director, Amity School of Engineering and Technology, for his excellent professional guidance and sincere advice for encouraging this venture.

It is also a pleasure to acknowledge the enthusiastic support and assistance given to me by my colleagues at other institutions. Their constructive comments and suggestions during informal discussions are responsible for smooth completion of the book. I am grateful to Dr Puja Singh and Dr Kusum Sharma (Maharaja Agrasen Institute of Technology) whose professional skills have helped me in covering the various topics upto the required depth. My heartfelt thanks are due to Dr. Jyoti Singh and Dr Sushmita Mahapatra (Bharati Vidyapeeth Institute of Technology) for their appreciation and encouragement which has given a great impetus to my long cherished desire. I also express my thanks to Mrs Ritu Malik (Guru Prem Sukh Memorial College of Engineering) for their invaluable and significant inputs. I profoundly thank Miss Meetu Walia (Guru Tegh Bahadur Institute of Technology) for extending me splendid cooperation and envincing keen interest in this venture. I wish to record my sincere thanks for the concrete suggestions received from Dr. Mona Siddhu and Dr. Satbir Singh (Maharaja Surajmal Institute of Technology). The encouragement received from Dr Manasi Singhi (Ambedkar Institute of Technology) worked as a catalyst in accomplishing this work. Dr Gulshan Kumar (Indira Gandhi Institute of Technology) deserves special thanks for his useful suggestions. My special thanks are due to Dr. Ranu Gadi, Indra Gandhi Institute of Technology, for her valuable contribution in the Chapter on Environmental Pollution.

My sincere thanks are due to all my colleagues in Amity University, Uttar Pradesh, for their suggestions and valuable inputs.

I also record my sense of appreciation to my students whose inquisitive questions during my laboratory classes provided valuable input into this effort.

Words elude me in expressing my sincere thanks to Praneesh who offered every possible help and supported the long hours of preparation that went into this effort. I feel greatly indebted to my kids, Tanishq and Tanvi without whose patience and sacrifice it was rather impossible to bring out the book in its present form.

My immeasurable gratitude and thanks are due to my family members who have been a driving force behind all my achievements. Their affection and benign blessings will ever remain to be cherished.

Finally, I am thankful to Mr. S.K. Kataria Publications and his entire staff for their efforts without which the book could not have seen the light of the day.

Mr. Dinesh, Mr. Pawan deserve a special thanks for typing this text as error free as possible.

Sunita Rattan

Contents

1

Volumetric Analysis

CHEMICAL ANALYSIS IS CARRIED OUT IN TWO WAYS:

- **Qualitative analysis.**
- **Quantitative analysis**

Qualitative analysis is the detection of ions or radicals in an inorganic salt whereas, **quantitative analysis** involves the determination of quantity of a particular component present in a substance. Quantitative analysis can be carried out in two ways:

(i) Gravimetric analysis

(ii) Volumetric analysis.

When the analysis is based on the *measurement of weight* then it is called **gravimetric analysis**. On the other hand when it is based on *measuring the volume of* the solution of a substance then it is called **volumetric analysis**.

Thus volumetric analysis is a technique of finding the volume of one solution which will exactly react with a definite volume of another solution of known concentration.

In this method, a known volume of one solution is taken in any glass vessel (conical flask) and the solution of other substance is added very slowly to it till the reaction is complete. The completion of the reaction is indicated by certain substances known as **indicators**, which change colour on completion of the reaction. This process is known as **titration**. The stage during which the reaction is just complete is known as **end-point**. (Since the indicator changes colour when the reaction is very *close* to equivalence point and not at the equivalence point, it is called *end point*).

From the volume of unknown solution which is required to completely react with definite volume of the other solution of known strength, the strength of the unknown solution can be calculated using the *Law of Equivalence* or *Normality Equation.*

NORMALITY EQUATION

According to the law of equivalence, the substances react with each other in a definite ratio by mass. For example, if an acid is titrated against a base, then at the end point according to normality equation,

Number of gram equivalents of the acid = Number of gram equivalents of the base

or $$N_1V_1 = N_2V_2$$

where N_1 and N_2 are normalities of acid and base and V_1 and V_2 are the volumes of acid and base respectively. This is known as the normality equation and is most fundamental principle of chemistry.

If N_1, V_1 and V_2 are known then N_2, the normality of the base and hence its strength can be calculated

$$N_1V_1 = N_2V_2$$

$$N_2 = \frac{N_1V_1}{V_2} = x \text{ (say)}$$

$\therefore$ Strength= Normality × Equivalent weight = x × Equivalent weight.

TERMS USED IN VOLUMETRIC ANALYSIS

1. Titration: The process of finding out the volume of one of the solution required to react completely with the definite volume of the other solution is known as titration.

2. Titrant: The solution of known strength is called the titrant.

3. Titrate: The solution which contains a substance to be estimated or whose strength is to be estimated.

4. * End point and Equivalence point: The point at which the completion of reaction occurs is called equivalence point whereas an end point is an observable physical change that occurs near the equivalence point.

5. Indicator: The substance which indicates the end point of titration is called indicator. The indicator indicates the completion of reaction by change in colour at the end point.

6. Strength: The amount of substance dissolved in one litre of a solution when expressed in grams is termed as strength of a solution. Strength of the solution can also be expressed in any of the following ways:

---

* **Note.** The end point differs very slightly from the equivalence point. The stage at which the two solutions are used up in their stoichiometric proportions is called equivalence point. The equivalence point can be determined using a number of techniques (conductometry, pH determination etc) *The equivalence point determined using an indicator is called the end point.* Ideally end point and equivalence point should be the same. But, the colour change of the indicator generally occurs only *after* the equivalence point and hence end point differs slightly from equivalence point.

(i) **Normality:** It is the number of gram equivalents of the substance dissolved per litre of the solution. It is denoted by N

$$N = \frac{\text{Grams of solute per litre of the solution}}{\text{Gram equivalent weight of the solute}}$$

If one gram equivalent of a substance is dissolved per litre of the solution, it is called Normal solution.

(ii) **Molarity:** It is the number of moles of the solute present per litre of the *solution.* It is denoted by M

$$M = \frac{\text{Grams of solute per litre of the solution}}{\text{Molecular mass of the solute}}$$

(iii) **Molality:** It is the number of moles of the substance dissolved in *1000 gm* or *1 kg of the solvent*. It is denoted by m

$$m = \frac{\text{Grams of solute per 1000g of the solvent}}{\text{Molecular mass of the solute}}$$

(iv) **Formality:** It is the number of gram formula weight of the substance dissolved in one litre of the solution. It is denoted by F

$$F = \frac{\text{Grams of solute per litre of the solution}}{\text{Formula weight}}$$

7. Equivalent Weight: The equivalent weight of a substance is defined as the number of parts by weight of it that combine with or displace one part by weight of hydrogen or eight parts by weight of oxygen or 35.5 parts by weight of chlorine.

Equivalent weight expressed in grams is called *gram equivalent weight*.

(i) Equivalent weight of an acid

$$= \frac{\text{Molecular weight of acid}}{\text{Number of replaceable hydrogen atoms in one molecule of acid}}$$

$$= \frac{\text{Molecular weight of acid}}{\text{Basicity of acid}}$$

(Basicity = Number of replaceable hydrogen atoms in an acid. Example: HCl, H_2SO_4, H_3PO_4 are monobasic, dibasic and tribasic respectively as they contain 1, 2, and 3 replaceable hydrogen atoms respectively).

(ii) Equivalent weight of a base

$$= \frac{\text{Molecular weight of base}}{\text{Number of replaceable } \overline{\text{O}}\text{H groups present in one molecule of a base}}$$

$$= \frac{\text{Molecular weight of base}}{\text{Acidity of base}}$$

(Acidity = Number of replaceable OH groups in one molecule of a base. Example, Na OH, Ba $(OH)_2$ and Al $(OH)_3$ are mono, di and tri-acid bases as they contain 1, 2, and 3 replaceable OH groups respectively)

(iii) Equivalent weight of an oxidising or reducing agent

$$= \frac{\text{Molecular mass of the substance}}{\text{Number of electrons gained / lost by one molecule in the redox reaction}}$$

The equivalent weights of some of the oxidising/reducing agents are given below:

Table 1.1

Substance	*Molecular Mass*	*Balanced ionic equation*	*No. of electrons lost/gained*	*Eq. wt* = $\frac{\text{Mol wt}}{n}$ (*g*)
Ferrous sulphate (Fe $SO_4 \cdot 7H_2O$)	278	$Fe^{2+} \rightarrow Fe^{3+} + e^-$	1	$\frac{278}{1} = 278$
Ferrous ammonium sulphate[(Fe SO_4 $(NH_4)_2$ $SO_4 \cdot 6H_2O$)	392	$Fe^{2+} \rightarrow Fe^{3+} + e^-$	1	$\frac{392}{1} = 392$
Sodium thiosulphate $NaS_2O_3{\cdot}5H_2O$	248	$2S_2O_3^{2-} \rightarrow \frac{1}{2} S_2O_6^{2-} + 2e^-$	1	$\frac{248}{1} = 248$
Iodine (I_2)	254	$I_2 + 2e^- \rightleftharpoons 2I^-$	2	$\frac{254}{2} = 127$
Potassium dichromate $(K_2Cr_2O_7)$	294	$Cr_2O_7^{2-} + 14H^+ + 6e^- \rightarrow 2Cr^{3+} + 7H_2O$	6	$\frac{294}{6} = 49$
Potassium permanganate $(KMnO_4)$	158	$MnO_2^{2-} + 8H^+ + 5e^- \rightarrow Mn^{2+} + 4H_2O$	5	$\frac{158}{5} = 31.6$
Oxalic acid $(H_2C_2O_4{\cdot}2H_2O)$	126	$C_2O_4^{2-} \rightarrow 2CO_2 + 2e^-$	2	$\frac{126}{2} = 63$

8. Standard Solutions: A solution of known concentration is known as standard solution. There are two types of substances which are generally employed for preparing solutions as explained below:

(i) *Primary standard substance:* Any substance whose standard solution can be prepared directly by dissolving an accurately weighed amount of it in water and making up the solution to a known volume by dilution with water. A primary standard substance possesses the following characteristics:

- It should be easily available in highly pure state. The impurities, if present must be inert with respect to the substance with which it is brought to react.
- It should have a high equivalent weight or molar mass so that error due to weighing is minimised.
- It should be readily soluble in water or the desired solvent.
- It should neither be hygroscopic nor efflorescent.
- It should be stable at oven temperature used for drying. The hydrated substances are seldom used as primary standards.
- It should be stable and unaffected by air.
- Its composition should not change on standing during storage.
- Crystalline oxalic acid, anhydrous sodium carbonate, dried KCl, $AgNO_3$, $FeSO_4$ $(NH_4)SO_4 \cdot 6H_2O$ are used as primary standards.

(ii) *Secondary Standard Substance:* The substances which do not fulfil the above requirements and their standard solutions are not prepared directly by weighing are called secondary standard substances.

To prepare the standard solution of secondary standards, an approximate weight of the substance is dissolved in a known volume of the solvent (water). Then the exact strength of the solution is found by titrating it against some primary standard. The process is called *standardization.*

The common secondary standards are:

- Alkali hydroxides like NaOH, KOH etc as they are hygroscopic and their weights keep on changing during weighing.
- $KMnO_4$, as it is not available in pure form and is associated with manganese dioxide which catalyse the auto decomposition of permanganate solution on standing.
- Inorganic acids HCl, H_2SO_4 etc because their concentrations are known only approximately.

EXPERIMENT NO. 1.1

AIM **To prepare standard oxalic acid solution from crystalline oxalic acid.**

THEORY Oxalic acid is a *primary* standard. The balanced ionic equation for the oxidation of oxalic acid is as follows:

$$\begin{array}{c} COO^- \\ | \\ COO^- \end{array} \longrightarrow 2CO_2 + 2e^-$$

Oxalate ion

$$\therefore \text{Equivalent mass of oxalic acid (hydrated)} = \frac{\text{Molar mass of oxalic acid}}{\text{Number of electrons lost}}$$

$$= 126/2 = 63\text{g}$$

Thus to prepare N/10 solution of oxalic acid, 6.3 g of it is dissolved in distilled water and the volume is made to 1000ml. Alternatively, to prepare 250ml of N/10 oxalic acid, the oxalic acid required is $\frac{6.3 \times 250}{1000}$ g = 1.575g

APPARATUS Watch glass or weighing tube, beaker, glass rod, wash bottle etc.

CHEMICALS AR oxalic acid crystals, distilled water.

PROCEDURE

1. Take a watch glass, clean it, wash it with distilled water and dry it.
2. Weigh the watch glass accurately. Let the weight be w_1g.
3. Place oxalic acid in the watch glass and weigh till the total weight of the oxalic acid and the watch glass comes out to be w_1 + 1.575g.
4. Transfer the oxalic acid from the watch glass to a 250ml beaker. Weigh the watch glass again. Let the weight be w_2g.
5. Dissolve the acid in 50 ml of water, stir gently with a glass rod.
6. Transfer the oxalic acid solution from the beaker to a 250ml measuring flask through a funnel carefully.
7. Wash the interior of the beaker with distilled water and transfer these washings into the measuring flask.
8. Now add more of distilled water to the flask carefully from the wash bottle till the level of water is just below the mark on measuring flask.

9. Add last few drops of distilled water with the help of a pipette until the lower meniscus just touches the graduated mark. Shake the flask gently after putting the air tight stopper to make the solution uniform.

OBSERVATIONS

Weight of watch glass = w_1g.

Weight of watch glass + oxalic acid = w_1 + 1.575g = w_2g

Weight of watch glass after transferring the oxalic acid = w_3g

CALCULATIONS

Weight of oxalic acid exactly transferred into the beaker = $w_2 - w_3$g

$$\text{Normality of oxalic acid} = \frac{\text{weight of solute in gram / litre of solution}}{\text{Gram equivalent weight}}$$

$$= \frac{w_2 - w_3}{63} \times \frac{1000}{250} = \frac{4(w_2 - w_3)}{63}$$

RESULT The normality of standard solution of oxalic acid is..

PRECAUTIONS

1. Weighing of oxalic acid crystals should be accurate.
2. Transference of the weighed substance should be done very carefully.
3. Last few milliliters of water should be added using a pipette.
4. The lower meniscus should be on the mark. If by mistake it becomes higher than the mark, reject the solution and start again.

EXERCISE

1. What do you mean by primary and secondary standards ?
2. Why oxalic acid is considered a primary standard?
3. Name a few primary standards used for titrations

Ans. Anhydrous Na_2CO_3, Molar mass = 106.

Potassium dichromate $K_2Cr_2O_7$, Molar mass = 294.2

4. What do you mean by standardization ?

Ans. In the case of secondary standard substances, the standard solution cannot be prepared by direct weighing. Therefore, concentration of these substances is determined by titrating them against primary standard solution. The process is known as standardization.

TYPES OF TITRATIONS

Titrations are classified into various types on the basis of the nature of chemical reaction involved. Some important types of titrations are:

(i) Acid- Base or Neutralization titrations.

(ii) Redox titrations

(iii) Precipitation titrations

(iv) Complexometric titrations.

These are discussed in detail in the succeeding sections.

(I) ACID BASE TITRATIONS (NEUTRALIZATION TITRATIONS)

Acid base titrations are the titrations in which acid is titrated against an alkali or vice versa. These titrations in aqueous solutions are essentially based on the reaction between hydrogen and hydroxyl ions to form unionized water (actually water is ionized to a very small extent) A typical example is the titration of sodium hydroxide against hydrochloric acid. The reaction involved is:

$$NaOH + HCl \longrightarrow NaCl + H_2O$$

or $$Na^+ + \bar{O}H + H^+ + Cl^- \longrightarrow Na^+ + Cl^- + H_2O$$

or $$H^+ + \bar{O}H \longrightarrow H_2O$$

Acid base titrations are also known as acidimetry and alkalimetry.

Acidimetry: To find the strength of an unknown acid with the help of a known base is known as acidimetry. The solution of the given acid is titrated with an alkaline solution and the $\bar{O}H$ ions of the alkali combine with H^+ ions of the acid. The concentration of H^+ decreases gradually and hence the *pH of the solution increases. At a definite pH value, the end point is reached and no more alkali is added.

Alkalimetry: To find the strength of an unknown base with the help of known acid is known as alkalimetry. The solution of an alkali is titrated with an acid solution and the $\bar{O}H$ ions combine with H^+ ions. The concentration of H^+ ions increases gradually and pH of the solution decreases. At a definite pH value, the end point is reached.

---

* $pH = -\log [H^+]$. A solution is acidic if $pH < 7$
A solution is basic if $pH > 7$
A solution is neutral if $pH = 7$

ACID BASE TITRATIONS AND INDICATORS

The process of acid-base titrations is accompanied by a change in pH. The equivalence point in neutralization titrations occurs at a definite pH value. Therefore, substances which change colour in accordance with pH (indicators) are used to indicate the end point.

Acid-Base indicators are complex organic compounds which have one colour in acid solution while an altogether different colour in alkaline solution.

THEORIES OF ACID-BASE INDICATORS

1. Ostwald's Theory. According to this theory the indicators themselves are weak acids or weak bases. They have different colours in ionized and unionized forms. The behaviour of phenolphthalein and methyl orange, the common acid base indicators is described as follows:

Action of phenolphthalein. Phenolphthalein is a weak acid and can be represented as, HPh. Its unionized molecules are colourless while on ionization it gives pink coloured Ph^- ions

$$\underset{\text{(Acid Colour)}}{\underset{\text{Colourless)}}{HPh}} \rightleftharpoons \underset{\text{(Colourless)}}{H^+} + \underset{\text{(Base Colour)}}{\underset{\text{(Pink)}}{Ph^-}} \quad ...(i)$$

In presence of *acid,* due to increase in concentration of common H^+ ions in solution, the dissociation of HPh is suppressed and thus the solution becomes *colourless.* On the other hand, when a base like NaOH is added, the $\bar{O}H$ ions of the base, combine with H^+ of the indicator to form unionized water. Thus, the equilibrium shifts towards the right, producing more of coloured Ph^- ions. The solution therefore appears pink.

Action of methyl orange: Methyl orange is a weak base and can be represented as Me. It changes colour from yellow to red depending upon the pH of the solution. The equilibrium can be represented as:

$$\underset{\text{(Base colour)}}{\underset{\text{(yellow)}}{Me}} + H_2O \rightleftharpoons \underset{\text{(acid colour)}}{\underset{\text{(Red)}}{MeH^+}} + \bar{O}H$$

If the solution is acidic, the hydrogen ions furnished by acid combine with $\bar{O}H$ ions and the equilibrium shifts to right and hence the solution appears red. If on the other hand, a base is added which furnishes $\bar{O}H$ ions then the equilibrium shifts to left (common ion effect) and hence the solution appears yellow.

2. Modern Quinoid Theory. The colour change of acid base indicators have also been explained on the basis of quinoid theory. According to this theory, the acid-base indicators can exist in two tautomeric forms. Ordinarily, one form is *benzenoid* and the other is *quinoid* form. The two forms have different colours and usually quinoid form has deeper colour than benzenoid form. Out of these two forms, one form exist in acidic medium while the other in alkaline medium. Change in pH causes the conversion of benzenoid form to quinoid form or vice versa and hence change in colour.

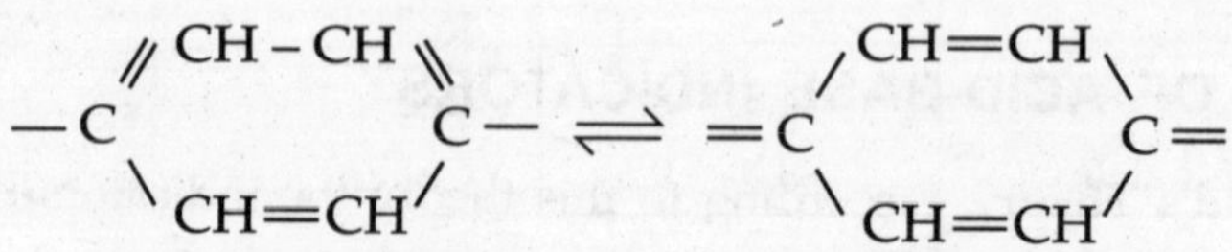

(Benzenoid form) (Quinoid form)

The action of the phenolphthalein and methyl orange on the basis of this theory is explained.

Action of phenolphthalein: Phenolphthalein possesses a benzenoid structure in acidic medium, whereas it changes to pink coloured tautomeric form in alkaline medium having quinoid form.

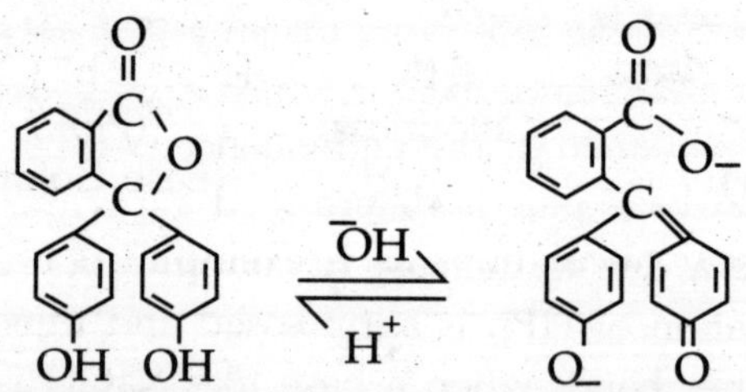

Colourless (Benzenoid form in acid solution) — Red Colour (Quinoid form in base)

Action of Methyl Orange. Methyl orange also exists in two tautomeric forms. In alkaline solution it exists in benzenoid form and is yellow in colour. In acidic solution it exists in red colour having quinoid structure.

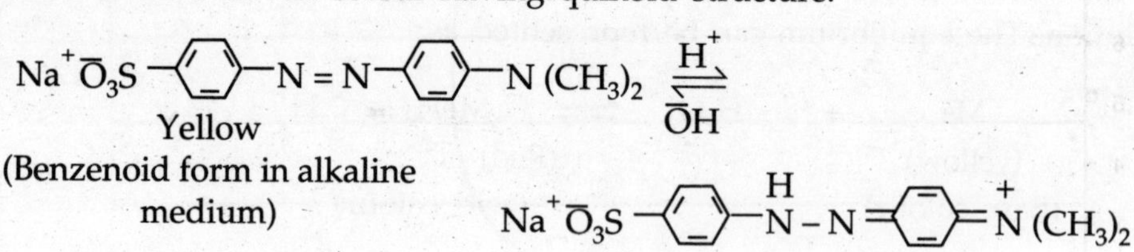

Yellow (Benzenoid form in alkaline medium)

(Quinoid form in acidic medium)

SELECTION OF PROPER INDICATOR

The pH value at the end point in titrations depend upon the relative strengths of acid and base being titrated. Since different indicators have different pH ranges within which they can be used, the selection of a proper indicator for a given titration is very important. As a general rule, one should select an indicator which changes colour at approximately the pH at equivalence point of titration. Some common indicators together with their useful ranges are given in Table 1.2

Table 1.2

Indicator	*pH range*	*Colour in acidic medium*	*Colour in basic medium*
Methyl Orange	3.1 – 4.4	Red	Yellow
Methyl Red	4.2 – 6.3	Red	Yellow
Litmus	4.5 – 8.3	Red	Blue
Phenolphthalein	8.0 – 9.6	Colour less	Pink

ACID BASE TITRATION CURVES

The addition of a base to an acid or vice versa is accompanied by the change in hydrogen ion concentration and hence pH of a solution changes during the course of acid base titration. A plot of pH against the volume of the base or acid added from the burette is called pH titration curve or acid base titration curve.

The pH titration curves are very significant in the detection of equivalence point and selection of suitable indicators for an acid-base pair. For most of acid-base titrations, we can select an indicator which exhibits a distinct colour change at a pH close to that obtained at the equivalence point. The pH at equivalence point for a given acid base pair is determined experimentally by pH titration curves or by theoretical calculations as is exemplified below.

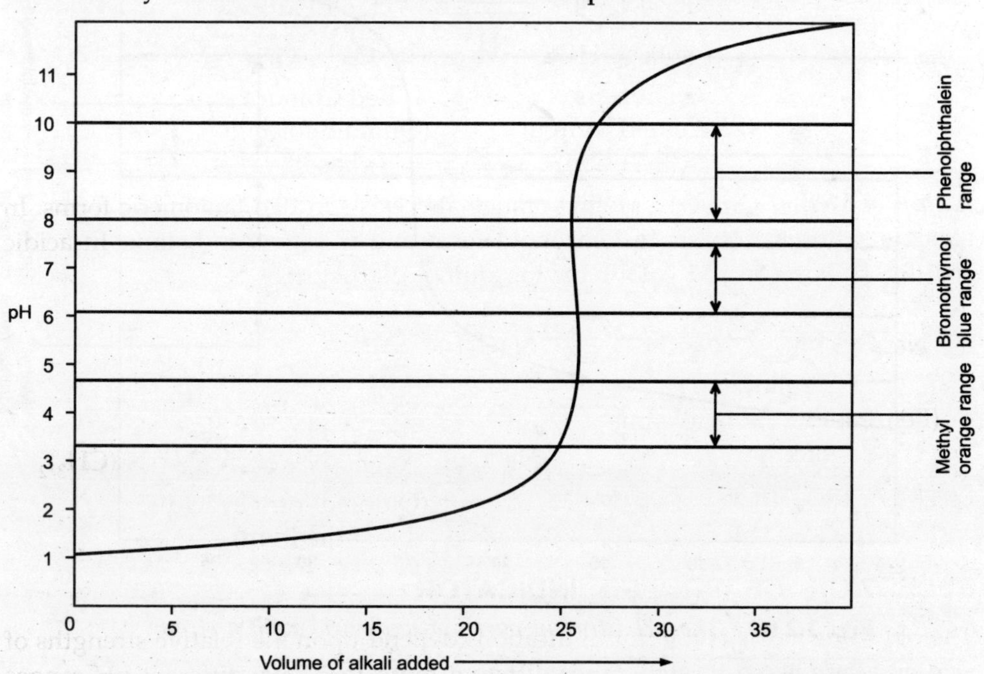

Fig. 1.1 (a) *The pH titration curve of strong acid with a strong base*

(a) A Titration of a strong acid with a strong base The change of pH with the volume of the alkali added in case of titration of strong acid with a strong base is depicted in Fig. 1.1(a). It is seen that the pH changes very slowly at first, increases rapidly near the equivalence point. At the equivalence point, there is a sharp increase in the pH of the solution for even a very small amount of alkali added. This sharp and sudden change in pH from 4 to 10 near the equivalence point is very significant because all the indicators which show colour change in the pH range 4 to 10 can be used to detect end point. Thus, methyl orange, methyl red, phenolphthalein all can be used as indicators for detecting the end point in titrations involving strong acid and strong bases. After the equivalence point, on further addition of alkali, the pH again changes slowly.

It may be emphasized here that if the solutions are dilute (~ 0.01 N), the change in pH and equivalence point is not very sharp. The pH range is limited to 5.5 – 8.5 at equivalence point * and hence the indicators such as methyl red or bromothymol would be a suitable choice.

(b) Titration of a weak acid with a strong base The titration curve of weak acid with a strong base (CH_3COOH Vs NaOH) is depicted in Fig. 1.2. In this case the pH at equivalence point will not be 7, because of the hydrolysis of sodium acetate so formed. As sodium acetate is a salt of a weak acid and a strong base it gives excess of $\overline{O}H$ ions in aqueous solution thereby increasing the pH at equivalence point beyond 7. Thus the indicator used should have pH range slightly on alkaline side. Hence phenolphthalein or thymol blue would be satisfactory indicators.

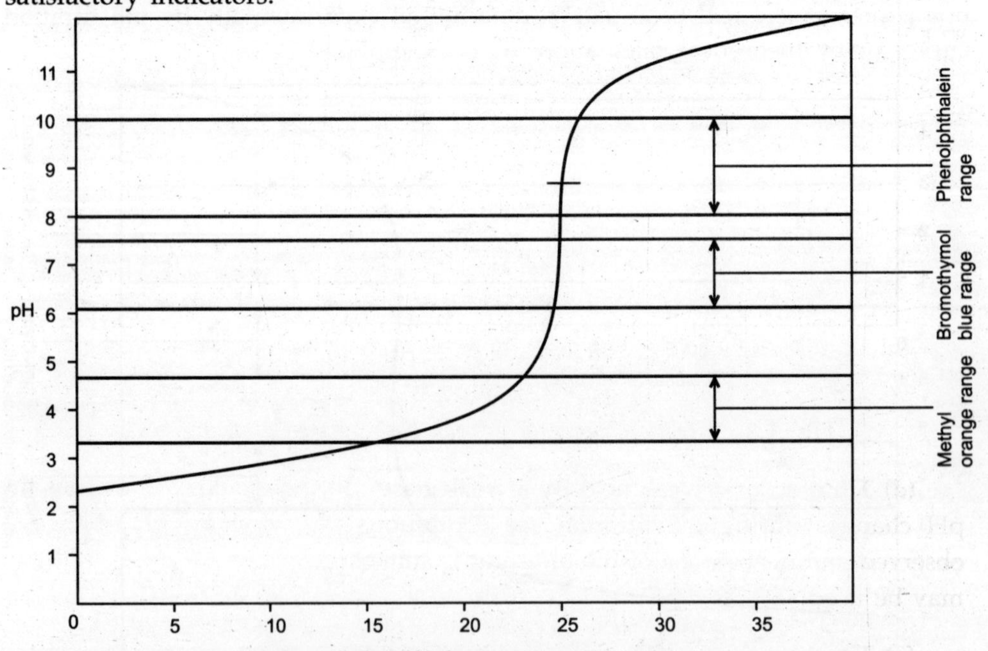

Fig. 1.2 (a) *The pH titration curve of weak acid with a strong base*

* **Note.** The equivalence point in all cases, whether the solution is dilute or concentrated in titration of strong acid with strong base remains the same i.e. 7.

For titration of acids like boric acid against sodium hydroxide the titration curve does not show any sharp rise in pH. This is because the salts so formed (sodium borate) during the titration undergo hydrolysis to a greater extent resulting in alkaline solution. The pH, therefore goes on increasing gradually without showing a sharp increase. Thus very weak acids cannot be successfully titrated.

(c) Titration of a Weak Base with a strong Acid: The titration curve of a weak base against a strong acid (ammonium hydroxide vs hydrochloric acid) is depicted in Fig. 1.3. The pH at equivalence point in this case lies in the acid range i.e. below 7. This is due to the formation of a salt of weak base and a strong acid (NH_4Cl) which undergoes hydrolysis releasing H^+ ions in the solution. Thus the pH change at equivalence lies in the range 6 to 4. Thus the indicator having pH range on acidic side i.e. methyl red, or methyl orange can be used successfully.

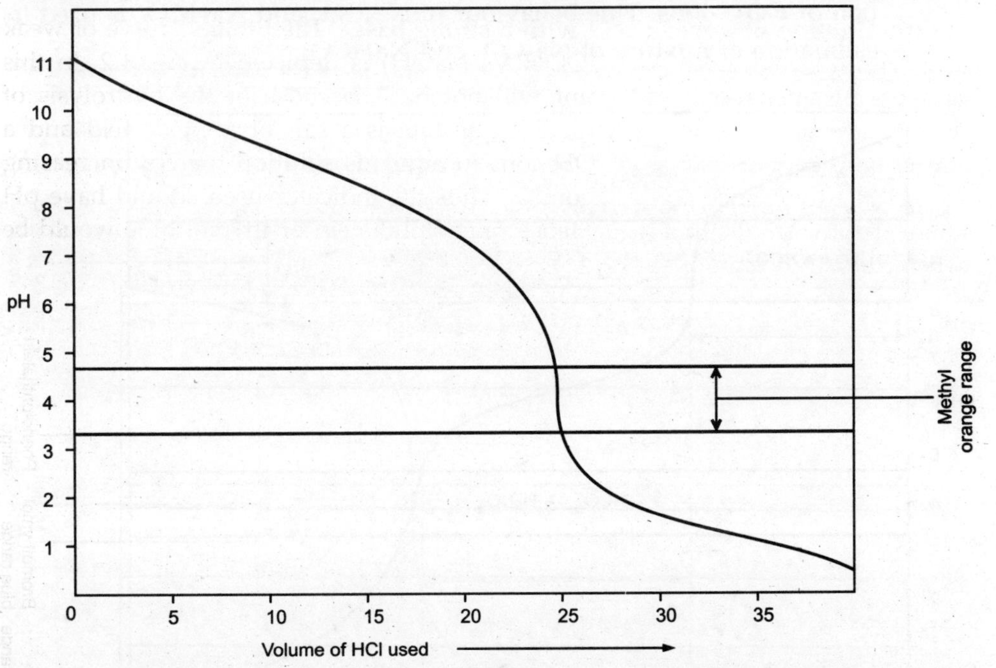

Fig. 1.3 *Titration curve of a weak base against a strong acid*

(d) Titration of a weak acid by a weak base In these types of titrations the pH changes during the titration are continuous i.e. no sharp pH change is observed and hence none of the indicator is suitable. However, a mixed indicator may be used.

(e) Titration of Na_2CO_3 with hydrochloric acid When Na_2CO_3 is titrated with a strong acid, such as hydrochloric acid, the carbonate ions are first converted to bicarbonate ions, and then to carbonic acid.

$$CO_3^{2-} + H^+ \longrightarrow HCO_3^-$$

$$HO_3^- + H^+ \longrightarrow H_2CO_3$$

As the neutralization takes place in two steps, the pH titration curve show two inflection points as shown in Fig. 1.4.

- the first equivalence point corresponds to the neutralization of CO_3^{2-} to HCO_3^- stage. This is completed at about pH 8.5. Hence phenolphthalein can be used to detect the conversion of CO_3^{2-} to HCO_3^- stage.
- the second equivalence point indicates the neutralization of HCO_3^- to H_2CO_3 stage. This reaction gets completed at pH 4.3. Thus, methyl orange can detect the end point in this case i.e. the complete neutralization of CO_3^{2-} ions. This behaviour of Na_2CO_3 and $NaHCO_3$ is used in estimation of mixture of Na_2CO_3 and $NaHCO_3$.

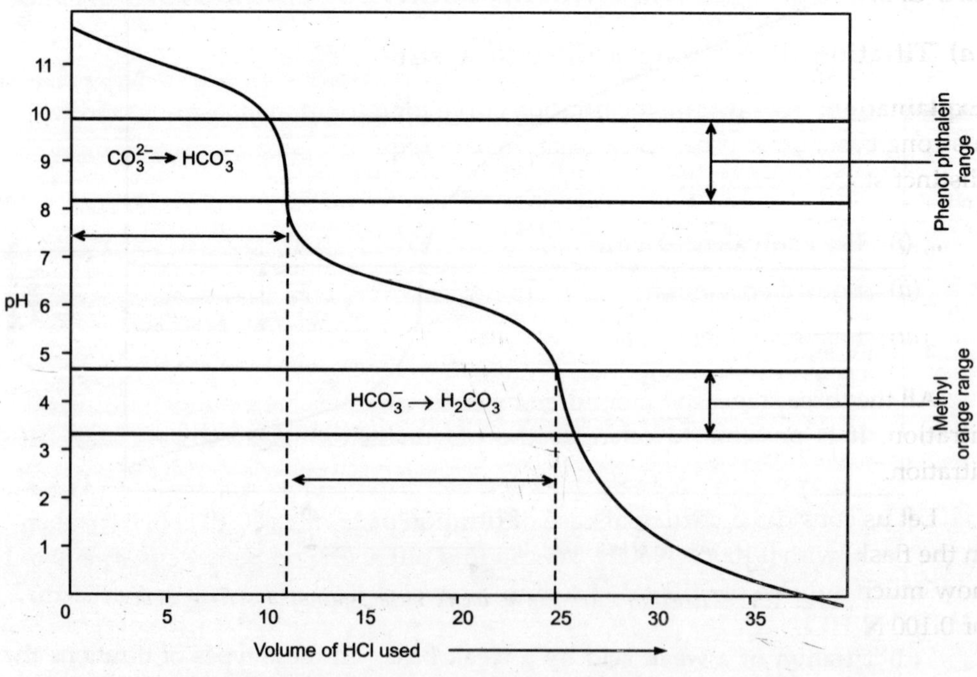

Fig. 1.4 *Titration curve of Sodium carbonate versus HCl*

The choice of indicators in different acid-base titrations can be summarized in Table 1.3.

Table 1.3

Titrations	*Approximate pH range at equivalence point*	*Suitable indicators*	*pH at colour change of indicator*
1. Strong acid and strong base	4 to 10	Methyl Orange Litmus Phenolphthalein	3.1 – 4.4 4.5 – 8.3 8.0 – 9.6
2. Strong acid and weak base	3.5 to 7	Methyl Orange Methyl Red	3.1 – 4.4 4.2 – 6.3
3. Weak acid and strong base	6.5 to 10	Phenolphthalein	8.0 – 9.6
4. Weak acid and weak base –	No sharp inflection at equivalence point	None satisfactory	—

EXPLAINATION FOR THE TITRATION CURVES OF ACIDS AND BASES

(*a*) Titration of a strong acid with a strong base

Explaination: To illustrate the titration curve for a solution of a strong acid with a strong base, three types of calculations are required, each corresponding to a distinct stage in titration:

(*i*) Pre equivalence point

(*ii*) Equivalence point

(*iii*) Post equivalence point

All the three stages are monitored by observing the change in pH during the titration. It is possible to calculate the pH of the solution at every stage of titration.

Let us consider a particular case of titration of 25 mL of 0.100 N HCl (taken in the flask) with 0.100 M NaOH. We can find out, using normality equation that how much volume of 0.100 M is required for complete neutralization of 25 mL of 0.100 N HCl.

$$\underset{\begin{bmatrix}\text{milli equivalents}\\ \text{of base}\end{bmatrix}}{N_1 \times V_2 \text{ mL}} = \underset{\begin{bmatrix}\text{milli equivalents}\\ \text{of acid}\end{bmatrix}}{N_1 \times V_2 \text{ mL}}$$

$$0.100\,N \times V_1 \text{ mL} = 0.100\,N \times 25 \text{ mL}$$

$$V_1 \text{ mL} = 25 \text{ mL}$$

Thus, when 25 mL of the base will be added, the titration/neutralization would be complete.

However, what happens during the course of titration? This can be found out by calculating the pH of the solution at selected points during the course of titration, after adding specific volumes of 0.100 N NaOH, theoretically.

(*i*) **When no NaOH has been added:** Since HCl is a strong acid and is completely dissociated, the solution contains major species as:

$$H^+ \; Cl^- \text{ and } H_2O$$

and the pH is determined by the H^+ from HCl. Since 0.100 N HCl contains 0.100 N H^+ or 0.100 M H^+,

$$\therefore \quad [H^+] = 0.100 \text{ M} = 1 \times 10^{-1} \text{ M}$$

$$\text{pH} = -\text{Log}\,[H^+] = -\text{Log}\,[10^{-1}] = 1.0$$

(*ii*) **Before the equivalence point is reached:** Let us say that only 10 mL of 0.100 N NaOH has added to the titration flask i.e. the reaction is not complete (because we know that completion requires 25 mL of 0.100 N NaOH). As is evident that at this stage H^+ ion concentration will decrease due to the reaction with base and also due to dilution.

$$[H^+] = \frac{\text{Number of milli moles of HCl remaining after addition of NaOH}}{\text{Total volume of solution}}$$

$$= \frac{\begin{bmatrix}\text{Number of millimoles of HCl} \\ \text{present originally}\end{bmatrix} - \begin{bmatrix}\text{Number of millimoles of} \\ \text{NaOH added}\end{bmatrix}}{\text{Total volume of solution}}$$

$$= \frac{(25.0 \text{ mL} \times 0.100\text{M}) - (10.0 \text{ mL} \times 0.100\text{M})}{25\text{mL} + 10\text{mL}}$$

$$= \frac{2.50 \text{ m mol} - 1.00 \text{m mol}}{35\text{mL}} = 0.0429 \text{ M}$$

Thus $\quad [H^+] = 0.0429$ M

$$\therefore \quad \text{pH} = -\log 0.0429 = 1.37$$

Similarly pH values can be calculated for all the volumes of NaOH added before the equivalence point. The results of the calculations are shown in Table 1.4.

(*iii*) **At the equivalence point:** At the equivalence point, neither HCl or NaOH is in excess. The concentration of $[H^+] = [\overline{O}H]$ and the pH of the solution is the pH of water i.e. 7.00.

(*iv*) **After the equivalence point:** After the equivalence point, the solution now contains an excess of NaOH. Let us calculate the pH after adding 35.0 mL of 0.100 N NaOH.

$$[\bar{O}H] = \frac{\begin{bmatrix}\text{Number of millimoles of} \\ \text{NaOH added}\end{bmatrix} - \begin{bmatrix}\text{Number of millimoles} \\ \text{of HCl present}\end{bmatrix}}{\text{Total volume of solution}}$$

$$= \frac{(35.0\,\text{mL} \times 0.100\,\text{M}) - (25.0\,\text{mL} \times 0.100\,\text{M})}{25\,\text{mL} + 35\,\text{mL}}$$

$$= \frac{3.50\,\text{m mol} - 2.50\,\text{m mol}}{60\,\text{mL}} = 0.0167\,\text{M}$$

Thus $[\bar{O}H] = 0.0167\,\text{M}$

$p\bar{O}H = -\log\ 0.0167 = 1.78$

$\therefore$ pH of the solution = $pK_w - pOH = 14.00 - 1.78 = 12.22$

Additional data of pH values beyond equivalence point can be obtained in the same way. The results are shown in Table 1.4:

Table 1.4 Changes in pH during the titration of a strong acid with a strong base

Volume of NaOH added (mL)	*pH of solution (25 mL of 0.100 N HCl titrated with 0.100 N NaOH)*
0.0	1.00
10.0	1.37
20.0	1.95
24.0	2.69
24.5	3.00 ⎤
25.0	7.00 ⎥ inflection point
25.5	11.00 ⎦
26.0	11.29
30.0	11.96
40.0	12.36
50.0	12.52

From the table it is clear that pH changes very gradually until the titration is close to the equivalence point.

Near the vicinity of equivalence point there is a very rapid change in pH. This behaviour is due to the fact that early in titration, there is relatively large amount of H^+ in the solution, and the addition of a given amount of OH^- thus produces a small change in pH. However near equivalence point $[H^+]$ is relatively small, and the addition of $\bar{O}H$ produces a large change.

It should be noted that titration of a strong base with a strong acid requires reasoning very similar to that used above. The only difference will be that $\overline{O}H$ is in excess before the equivalence point and H^+ is in excess after the equivalence point. The pH curve would start at a higher value and proceed to a lower value. Fig. 1.1(*b*)

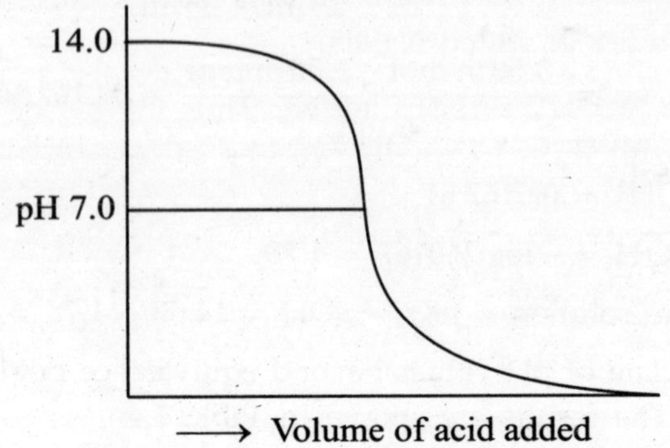

Fig. 1.1(*b*) The pH titration curve of strong base with a strong acid.

(*b*) Titration of a weak acid with a strong base:

Explaination: As the strong acids and strong bases are completely dissociated and hence the calculations to obtain pH curves are quite straight forward.

But this is not the case with titration of weak acid with a strong base. This is because, whenever a weak acid is titrated with a strong base or a weak base is titrated with a strong acid, a *buffer solution** consisting of a conjugate acid/base pair is formed. Hence different types of calculations are required to derive a titration curve for titration of a weak acid with a strong base.

Four types of calculations are involved, which are:

(*i*) At the beginning of the titration, the solution contains only weak acid and the pH is calculated from the concentration of that solute and its dissociation constant.

(*ii*) When the base is added, the solution consists of a series of buffer until immediately before the equivalence point. Hence the pH of each buffer is found out from the concentrations of conjugate base and the residual concentration of the base.

(*iii*) At the equivalence point, all of the weak acid is converted to its conjugate base i.e. a salt and hence pH is calculated from the concentration of the product.

Buffer Solution: A buffer is a mixture of weak acid or its conjugate base or a weak base and its conjugate acid that resists changes in pH of a solution.

(*iv*) Beyond the equivalence point, the excess of base is present. This supresses the basic character of the solution to such an extent that pH of the solution is determined by the concentration of the excess base added.

Let us exemplify this by considering the titration of 50.0 mL of 0.10 M acetic acid with 0.10 M NaOH. As before, we will calculate the pH at various points representing volumes of added NaOH.

(*i*) **When no NaOH has been added:** Thus the pH of the solution having a conc. 0.01 M with $K_a = 1.75 \times 10^{-5}$ can be calculated as follows:

	CH_3COOH	$\rightleftharpoons$	H^+	+	CH_3COO^-
Initial conc. (mol/L)	0.10		0		0
At Equil. the conc.	0.10 - x		x		x

$$K_a = \frac{[H^+][CH_3COO^-]}{[CH_3COOH]} = \frac{x^2}{0.10 - x} \quad \text{or} \quad x^2 + K_a x - 0.10\,K_a = 0$$

The positive solution to this quadric equation is:

$$x = \frac{-K_a + \sqrt{K_a^2 + 4K_a\,(0.10)}}{2}$$

$$x = \frac{-1.75 \times 10^{-5} + \sqrt{(1.75 \times 10^{-5})^2 + 4(1.75 \times 10^{-5})(0.10)}}{2} = 1.32 \times 10^{-3}$$

But $x = [H^+] = 1.32 \times 10^{-3}$ M

$pH = -\log(1.32 \times 10^{-3}) = 2.88$

(*ii*) **Before the equivalence point:** Let us say, that 10 mL of 0.10 NaOH has been added, then this will produce a buffer of weak acid CH_3COOH and its conjugated base CH_3COO^-. The concentrations of the two constituents are

$$[CH_3COOH] = \frac{(50.00\,\text{mL} \times 0.1000\,\text{M}) - (10.00\,\text{mL} \times 0.10\,\text{M})}{60.00\,\text{mL}} = \frac{4.000}{60.00}\,\text{M}$$

$$[CH_3CO\overline{O}] = \frac{10.00\,\text{mL} \times 0.1000\,\text{M}}{60.00\,\text{mL}} = \frac{1.000}{60.00}\,\text{M}$$

Now the dissociation constant expression for acetic acid is given by:

$$CH_3COOH \rightleftharpoons CH_3COO^- + H^+$$

$$K_a = \frac{[CH_3COO^-][H^+]}{[CH_3COOH]} = 1.75 \times 10^{-5}$$

$$\therefore \quad [H^+] = \frac{[CH_3COOH]\,K_a}{[CH_3COO^-]}$$

$$= \frac{(4.000/60.00)\,1.75 \times 10^{-5}}{(1.000/60.00)} = 7.00 \times 10^{-5}$$

$$\therefore \quad pH = -\log 7.00 \times 10^{-5} = \mathbf{4.16}$$

(*iii*) **At the equivalence point:** At the equivalence point, all of acetic acid has been converted to sodium acetate. It is very similar to the solution formed by dissolving that base in water. The base will combine with a proton, the only source for protons at equivalence point, is water.

$$CH_3COO^- + H_2O \rightarrow CH_3COOH + \bar{O}H$$

This is a weak base reaction characterized by K_b

$$K_b = \frac{[CH_3COOH][\bar{O}H]}{[CH_3COO^-]} = \frac{K_w}{K_a} = \frac{1.0 \times 10^{-14}}{1.75 \times 10^{-5}} = 5.6 \times 10^{-10}$$

The concentration of CH_3COO^- before it reacts with H_2O

$$= \frac{50\,mL \times 0.10\,M}{\underbrace{50\,mL + 50\,mL}_{\text{(dilution factor)}}} = 0.050\ M$$

($\because$ the conc. of CH_3COO^- is *not taken* as 0.10 M which is actually the initial conc. of CH_3COOH which gives CH_3COO^-. This is because the solution has been diluted so the dilution factor has to be incorporated).

Substituting the values in the above equation:

$$K_b = 5.6 \times 10^{-10} = \frac{[CH_3COOH][\bar{O}H]}{[CH_3COO^-]}$$

$$\because \quad [CH_3COOH] = [\bar{O}H]$$

$$\therefore \quad 5.6 \times 10^{-10} = \frac{[\bar{O}H]^2}{[CH_3COO^-]}$$

$$\because \quad [\bar{O}H] = \sqrt{0.0500 \times 5.6 \times 10^{-10}} = 5.34 \times 10^{-6}$$

$$pH = 14.00 - (-\log 5.34 \times 10^{-6}) = 8.73.$$

Thus the pH is greater than 7 for the titration of weak acid by a strong base.

(*iv*) **After the equivalence point:** When, say 60 mL of 0.10 M NaOH has been added then, the two bases present in the solution will be $\bar{O}H$ and CH_3COO^-.

However since CH_3COO^- is weak compared to strong $\bar{O}H$, therefore $\bar{O}H$ produced by the reaction

$$CH_3COO^- + H_2O \rightarrow CH_3COOH + \bar{O}H$$

will be very small as compared to $\bar{O}H$ added. Hence the pH is determined by excess of $\bar{O}H$ added.

$$[\overline{O}H] = \frac{\text{m mol of } \overline{O}\text{H in excess}}{\text{Volume in mL}}$$

$$= \frac{1.0 \text{ m mol}}{(50.0 + 60.0) \text{ mL}} = 9.1 \times 10^{-3} \text{ M}$$

and $$[H+] = \frac{1.0 \times 10^{-14}}{9.1 \times 10^{-3}} = 1.1 \times 10^{-12} \text{ M}$$

$$pH = 11.96$$

The other calculated pH values are summarized in Table 1.5:

Table 1.5 Changes in pH during the titration of a weak acid with a strong base.

Volume of NaOH added mL	*pH (50 mL of 0.100 M CH_3COOH titrated with 0.100 M NaOH)*
0.00	2.88
10.00	4.16
25.00	4:76
40.00	5.36
49.00	6.45
49.90	7.46
50.00	8.73] Inflection
50.10	10.00]
60.00	11.96

It is seen from the Table 1.5 that the solution which is a weak acid solution changes into buffer followed by a weak base solution and finally becomes a basic solution with the subsequent addition of strong base. In other words pH continuously increases but the curve shows an inflection near the equivalence point.

The pH curve for this titration is quite similar to the pH titration curve of strong acid and strong base after the equivalence point because excess $\overline{O}H$ controls the pH in both the cases after equivalence point. [Fig. 1.2 (*b*)]

Moreover, the curve for weak acid levels off near the half way point and then increases again. The half way levelling off is due to the buffering effects. The levelling off occurs when volume of the base added = $\frac{1}{2}$ the volume of the acid present or when pH = pKa. This is the region of maximum buffer capacity.

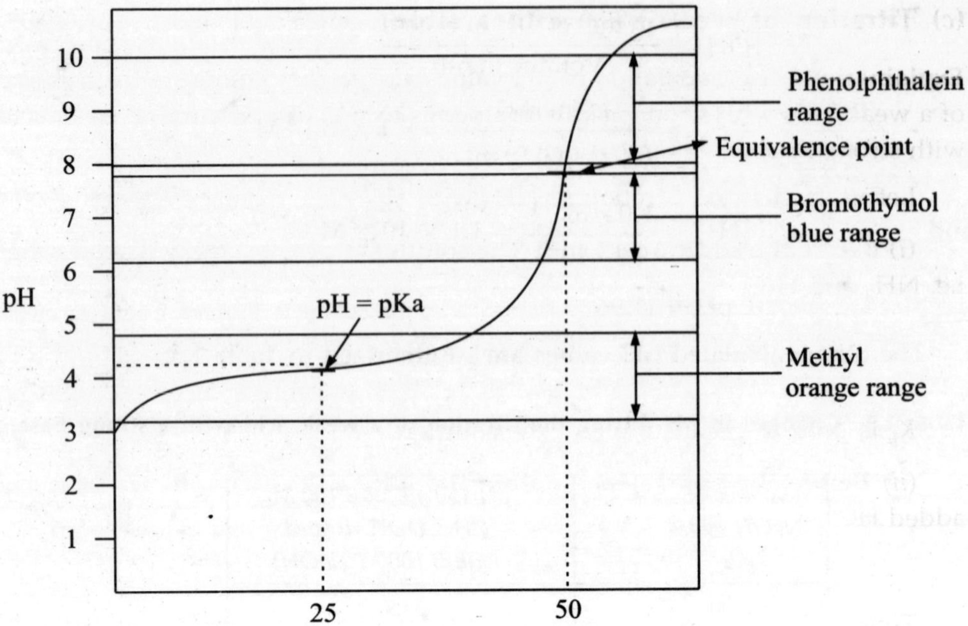

Fig. 1.2 (*b*) *The pH titration curve of weak acid with a strong base.*

Fig. 1.2 (*c*) illustrates the effect of acid strength on titration curves. Each curve is derived for the titration of 50.0 mL of 0.100 M acid with 0.100 M NaOH.

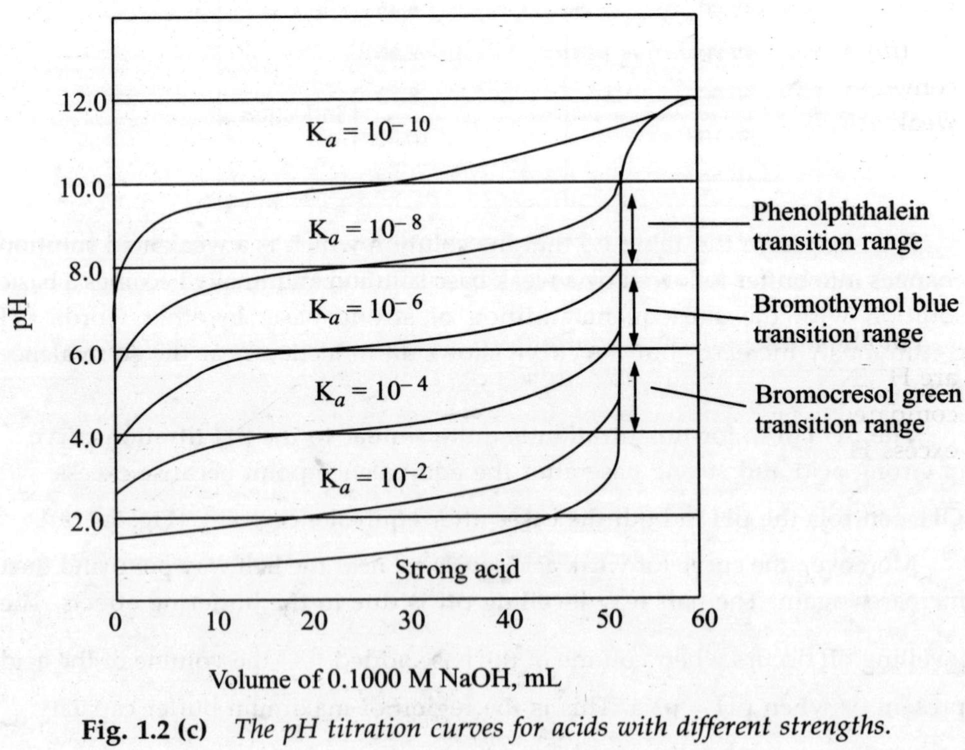

Fig. 1.2 (c) *The pH titration curves for acids with different strengths.*

(*c*) Titration of weak base with a strong acid:

Explaination: The calculation of pH values at different points for the titration of a weak base with strong acid is analogous to that of a titration of weak acid with strong base.

Let us discuss the titration of NH_3 with HCl. The main points are:

(*i*) Before the addition of acid: The solution contains weak base and water i.e. NH_3 and H_2O

$$\underset{(aq)}{NH_3} + \underset{(l)}{H_2O} \overset{K_b}{\rightleftharpoons} \underset{(aq)}{NH_4^+} + \underset{(aq)}{\bar{O}H}$$

K_b is used to calculate pH of the solution.

(*ii*) Before the equivalence point: The NH_3 will react with H^+ from the added HCl

$$NH_{3\,(aq)} + H^+_{(aq)} \rightleftharpoons NH^+_{4\,(aq)}$$

Thus NH_3 reacts with the free proton from acid and the major species in the solution will be: NH_3 (unreacted), $NH_4{}^+$, Cl^- and H_2O. The equilibria of dissociation reaction of NH_4^+ ions is used to determine pH

$$NH^+_{4\,(aq)} \rightleftharpoons NH_{3\,(aq)} + H^+_{(aq)}$$

(*iii*) At the equivalence point: At equivalence point all the original NH_3 is converted into $NH_3{}^+$, Cl^- and H_2O. pH is calculated by the dissociation of the weak acid $NH_4{}^+$

$$NH_4^+ \rightleftharpoons NH_3 + H^+$$

$$K_a = \frac{K_w}{K_b \text{ (for } NH_3)}$$

(*iv*) Beyond equivalence point: After adding excess HCl, the major species are H^+, $NH_4{}^+$, Cl^- and H_2O. Although $NH_4{}^+$ will dissociate but it is weak acid compared to HCl which is being added. Hence the pH will be determined by the excess H^+

$$[H^+] = \frac{\text{m mol of } H^+ \text{ in excess}}{\text{mL of solution}}$$

EXPERIMENT NO. 1.2

AIM **Find the strength in grams per litre of the given sodium hydroxide solution with the help of standard oxalic acid solution.**

REQUIREMENT A standard solution of 0.1 N oxalic acid.

APPARATUS Burette, Pipette, 2 beakers, titration flask, stand.

THEORY A known amount of oxalic acid is titrated with NaOH solution using phenolphthalein as indicator. Using normality equation, the normality and hence the strength of given NaOH is determined.

$$\begin{array}{l}COOH\\ |\\ COOH\end{array}\,2H_2O + 2\,NaOH \rightarrow \begin{array}{l}COONa\\ |\\ COONa\end{array} + 4H_2O$$

INDICATOR Phenolphthalein.

END POINT Pink to colourless.

PROCEDURE

(1) Rinse and fill the buretle with the given oxalic acid solution (N/10).

(2) Pipette out 10 ml of the given NaOH solution in the titration flask.

(3) Add a drop of phenolphthalein to the solution. The solution will turn pink.

(4) Run the oxalic acid solution from the burette into the titration flask till the pink colour of NaOH solution disappears.

(5) Repeat to get three concordant readings.

OSERVATIONS

Normality of oxalic acid solution = 0.1 N

S.No.	*Volume of the solution taken in the titration flask (ml)*	*Burette readings*		*Volume of oxalic acid used (ml) (Final- Initial reading)*
		Initial reading	*Final reading*	
1				
2				
3				x mL

CALCULATIONS

$$\begin{array}{lcl} \text{Acid} & & \text{NaOH Solution} \\ N_1V_1 & = & N_2V_2 \\ \frac{1}{10}\times x & = & N_2 \times 10 \\ N_2 & = & \frac{x}{100} \end{array}$$

$$\therefore \text{Strength} = \text{Normality} \times \text{Eq. wt} = \frac{x}{100}\times 40 \text{ g/L} = y \text{ g/L}$$

$$(\because \text{Eq. wt of NaOH} = 40\text{g})$$

RESULTS The amount of NaOH present in the given mixture = ... g/L.

PRECAUTIONS

1. Wash the apparatus first with water and then with detergent followed by plenty of water.
2. Rinse the burette and pipette with the respective solutions to be taken in.
3. Do not rinse the conical flask with the solution to be taken in it.
4. Use anti parallex card to avoid error in burette reading.

Instructions for Lab Assistant

(1) **0.1 N oxalic acid solution:** Dissolve 6.3 g of oxalic acid $\begin{array}{l}\text{COOH}\\ |\\ \text{COOH}\end{array}\cdot 2H_2O$ in distilled water and make up the volume to 1000 mL.

(2) **NaOH solution:** Dissolve about 4 g of NaOH in distilled water and make up the volume to 1000 mL.

(3) **Phenolphthalein indicator:** Dissolve 1 g of phenolphthalein in 100 mL of distilled ethyl alcohol and add 100 ml of distilled water with constant stirring.

EXERCISE

Based on General Concepts of Volumetric Analysis

1. What do you mean by the term - "volumetric analysis"?

Ans. *Volumetric analysis is the technique of finding the volume of one solution which will exactly react with a definite volume of another solution of known concentration.*

2. What do you mean by the term Titration?

Ans. *The technique known as titration is an analytical method commonly used in chemistry laboratories for determining the quantity or concentration of a substance in a solution. In a titration, an analyte - the substance whose quantity or concentration is to be determined - is reacted with a carefully controlled volume of solution of accurately-known concentration called a standard solution.*

3. Define the terms- titrant, titrand or an analyte.

Ans. ***Titrant:*** *The solution of known concentration which is used to find the strength of an unknown solution is known as a titrant.*

Titrand/Analyte: *The solution whose concentration is to be determined is known as titrand or analyte.*

Example - determination of chlorideions,

$$\underset{\substack{\text{analyte of}\\ \text{unknown}\\ \text{concentration}}}{Cl^-_{aq}} + \underset{\substack{\text{titrant} - AgNO_3\\ \text{- a standard solution}\\ \text{- known concentration}}}{Ag^+_{(aq)}} \xrightarrow{\text{titration}} AgCl_{(s)}$$

4. What do you mean by a standard solution?

Ans. *A solution of known concentration is known as a standard solution Standard solutions are referred to either as primary standards or secondary standards.*

5. What do you mean by a primary standard? Give some examples.

Ans. *A primary standard is a reagent that is extremely pure, stable, has no waters of hydration, and has a high molecular weight.*

Some primary standards for titration of acids:

- *sodium carbonate: Na_2CO_3, mol wt. = 105.99 g/mol*
- *tris-(hydroxymethyl) aminomethane (TRIS or THAM): $(CH_2OH)_3CNH_2$, mol wt. = 121.14 g/mol.*

Some primary standards for titration of bases:

- *potassium hydrogen phthalate (KHP): $KHC_8H_4O_4$, mol wt. = 204.23 g/mol*
- *potassium hydrogen iodate: $KH(IO_3)_2$, mol wt. = 389.92 g/mol*

Some primary standards for redox titrations:

- *potassium dichromate: $K_2Cr_2O_7$, mol wt. = 294.19 g/mol*
- *Mohr's salt*

6. What is a secondary standard?

Ans. *Compounds used as standards that do not meet the criteria of primary substances are called secondary standards. It is prepared in the laboratory for a specific analysis and is usually standardized against a primary standard.*

7. What do you mean by standardization?

Ans. *In case of secondary standard substances, the standard solution cannot be prepared by direct weighing. Therefore the concentration of these substances is determined by titrating them against primary standard solution. The process is called standardization.*

8. Define the term concentration.

Ans. *The* ***concentration*** *of a chemical solution refers to the amount of solute that is dissolved in a solvent. Concentration may be expressed in several different ways, using percent composition by mass, mole fraction, molarity, molality, or normality.*

9. What do you mean by Molarity and Molality.

Ans. *A molality is the number of moles of solute dissolved in one kilogram of solvent. Molarity is the number of moles of a solute dissolved in a liter of solution. Be careful not to confuse molality and molarity. Molality is represented by a small "m", whereas molarity is represented by "M".*

10. How do you prepare 1 molal and 1 molar solution of NaCl?

Ans. *To make a one molal aqueous (water) solution of sodium chloride (NaCl), measure out one kilogram of water and add one mole of the solute, NaCl to it. The atomic weight of sodium is 23 and the atomic weight of chlorine is 35. Therefore the formula weight for NaCl is 58, and 58 grams of NaCl dissolved in 1 kg water would result in a 1 molal solution of NaCl.*

A molar solution of sodium chloride is made by placing 1 mole of a solute into a 1-liter volumetric flask. (Taking data from the example above we will use 58 grams of sodium chloride). Water is then added to the volumetric flask up to the one liter line. The result is a one molar solution of sodium chloride.

11. What is the molarity of a solution made, when water is added to 11 g of $CaCl_2$ to make 100 mL of solution?

Ans. $11\ g\ CaCl_2/(110\ g\ CaCl_2/mol\ CaCl_2) = 0.10\ mol\ CaCl_2$

$100\ mL \times 1\ L/1000\ mL = 0.10\ L$

$Molarity = 0.10\ mol/0.10\ L$

$Molarity = 1.0\ M.$

12. What is the molality of a solution of 10 g NaOH in 500 g water?

Ans. $10\ g\ NaOH/(4\ g\ NaOH/1\ mol\ NaOH) = 0.25\ mol\ NaOH$

$500\ g\ water \times 1\ kg/1000\ g = 0.50\ kg\ water$

$Molality = 0.25\ mol/0.50\ kg$

$Molality = 0.05\ M/kg$

$Molality = 0.50\ m$

13. Define equivalent weight of a substance?

Ans. *The weight of a substance that will combine with or replace one mole of hydrogen or one-half mole of oxygen is known as equivalent weight of a substance. The equivalent weight is equal to the atomic weight divided by the valence.*

For monovalent ions, 1 equivalent = 1 mole

For divalent ions, 1 Eq = 0.5 mol

For trivalent ions, 1 Eq = 0.333 mol

The unit of equivalent weight is the atomic mass unit; the amount of a substance in grams numerically equal to the equivalent weight is called a gram equivalent. Hydrogen has atomic weight 1.008 and always assumes valence 1 in compounds, so its equivalent weight is 1.008.

Some elements exhibit more than one valence in forming compounds and thus have more than one equivalent weight. Iron (atomic weight 55.847) has an equivalent weight of 27.924 in ferrous compounds (valence 2) and 18.616 in ferric compounds (valence 3).

14. What do you mean by the term Normality?

Ans. *Normality is equal to the gram equivalent weight of a solute per liter of solution. A gram equivalent weight or equivalent is a measure of the reactive capacity of a given molecule. Normality is the only concentration unit that is reaction dependent. Example:*

1 M sulphuric acid (H_2SO_4) is 2 N for acid-base reactions because each mole of sulphuric acid provides 2 moles of H^+ ions. On the other hand, 1 M sulphuric acid is 1 N for sulphate precipitation, since 1 mole of sulphuric acid provides 1 mole of sulphate ions.

15. How many milliliters of 5.5 M NaOH are needed to prepare 300 mL of 1.2 M NaOH?

Ans. $5.5\ M \times V_1 = 1.2\ M \times 0.3\ L$

$V_1 = 1.2\ M \times 0.3\ L/5.5\ M$

$V_1 = 0.065\ L$

$V_1 = 65\ mL$

So, to prepare the 1.2 M NaOH solution, we pour 65 mL of 5.5 M NaOH into the container and add water to get 300 mL final volume.

16. How many types of titration are there?

Ans. *There are many types of titrations in common use in the analytical chemistry laboratory.Each type uses a different kind of chemical reaction. Examples of titration types include*

- *acid-base titration,*
- *reduction-oxidation titrations,*
- *precipitation titrations, and*
- *complexometric titrations*

17. What are the common standard reagents used in acid base titrations?

Ans. *The standard reagents used in acid base titrations are always strong acids and strong bases such as HCl, H_2SO_4, NaOH, and KOH.*

18. Why weak acids and bases are not used as standard reagents?

Ans. *Weak Acids and bases are not used as standard reagents because they react incompletely with the analytes.*

19. What is an indicator?

Ans. *An indicator is a substance that undergoes a change in colour when the end-point of a titration is reached.*

20. What is an acid base indicator?

Ans. *In an acid-base titration, addition of titrant near the equivalence point causes the solution pH to change drastically. This pH change is detectable with indicators that change color as a function of pH. An acid-base indicator is a weak acid or a weak base. The undissociated form of the indicator has a different colour than the dissociated form of the indicator.*

Consider methyl orange, if the acid form of the indicator is represented by Hln and its conjugate base form by ln^-, the following equilibrium exists in aqueous solution:

$$\underset{\text{Red}}{Hln_{(aq)}} + H_2O(l) \rightleftharpoons \underset{\text{Yellow}}{ln^-_{(aq)}} + H_3O^+_{(aq)}$$

It must be remembered that an Indicator does not change colour from pure acid to pure alkaline at specific hydrogen ion concentration, but rather, colour change occurs over a range of hydrogen ion concentrations. This range is termed the colour change interval. It is expressed as a pH range.

21. How is an indicator used?

Ans. *Weak acids are titrated in the presence of indicators which change colour under slightly alkaline conditions. Weak bases should be treated in the presence of indicators which change colour under slightly acidic conditions.*

22. What is the chemical law of equivalence?

Ans. *The substance react with each other in a definite ratio by mass.*

23. What is an equivalent point?

Ans. *The point at which the completion of the reaction occurs is called equivalence point.*

24. What is an end point?

Ans. *We call the point at which the indicator changes colour the end point i.e. the end point is the obserable physical change that occurs near the equivalence point. The end point might differ slightly from the equivalence point, and the difference between the equivalence point and the end point is called the titration error. The titration error will usually be negligible if a suitable indicator is available for use.*

25. What is the difference between the end point and the equivalence point?

Ans. *The endpoint of a titration is NOT the same thing as the equivalence point. The equivalence point is a single point defined by the reaction stoichiometry as the point at which the base (or acid) added exactly neutralizes the acid (or base) being titrated. The endpoint is defined by the choice of indicator, as the point at which the indicator changes its colour. Depending on how quickly the colour changes, the endpoint can occur almost instantaneously or be quite wide.*

26. Why is distilled water always used in preparation of standard solution?

Ans. *Because in tap water, Ca^{2+} Mg^{2+} or Cl^- ions are present which may cause turbidity in solution.*

Based on Handling and Use of Glassware

27. Why is it necessary to rinse the burette or pipette with the liquid it is filled with?

Ans. *The water sticking to the sides will alter the strength of the solution taken in it.*

28. The conical flask is only washed with distilled water and not with the solution to be taken in it. Explain

Ans. *Because only a fixed volume of a solution is taken in a conical flask with a pipette and if rinsed with the solution, the solution sticking to it adds to the volume.*

29. Why are hot solutions not poured into a burette?

Ans. *Hot solutions make the glass expand and introduce errors in volume measurements.*

30. *One should not force out the last drop in the pipette by blowing – why?*

Ans. *The pipette is so caliberated that the last drop need not to be forced out.*

31. Pipette should not be held from its bulb. Why?

Ans. *The body temperature may expand the glass and introduce a measuring error.*

32. If some water is added to the measured solution already taken in a conical flask, will it affect the titration?

Ans. *No, because the quantity of the substance dissolved remains the same.*

33. Why is water not added to acid and always acid is added to water?

Ans. *Since the reaction is exothermic and on adding water to acid the heat liberated is enough to change some amount of water into steam. This results in scattering of acid which may cause serious consequences. Hence, always small amounts of acids are added to water.*

34. What is a decinormal and centinormal solution?

Ans. *When one-tenth of the gram equivalent of a substance is dissolved per litre of solution, it is called decinormal solution and when one hundredth of the gram equivalent is dissolved per litre of solution, it is called centinormal solution.*

35. Why is it necessary to remove air bubbles from the jet of burette?

Ans. The air bubbles may escape during the titration and thus introduce error in measuring volume.

36. Sketch a titration curve for a strong acid with a strong base and then sketch on for a weak acid and a strong base. (a) Compare the two shapes (b) How do the end points differ? What is the significance of the point halfway to the end point in a weak acid titration?

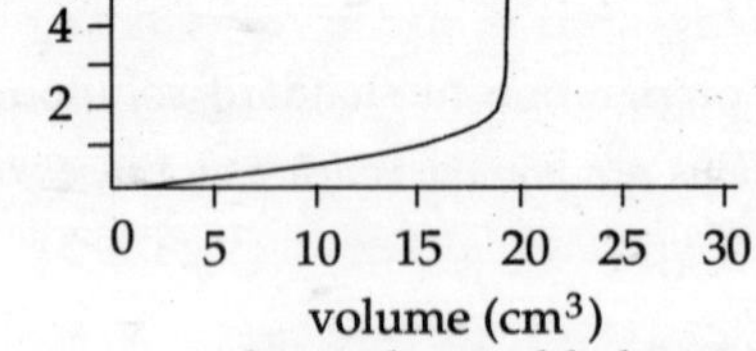

Strong base added to strong acid

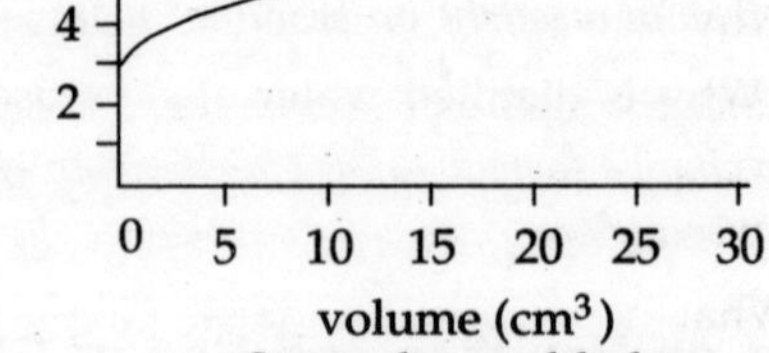

Strong base added to weak acid

A. The pH basically doesn't change as more base is added until the moles of acid and base are equal, in case of strong base added to strong acid. The strong base added to the weak acid shows more gradual changes in pH as more base is added.

B. The end point would be the volume of base added when the indicator used changes colour. That is dependent on the indicator used. The pH of the equivalence point of the strong acid with the strong base is 7 or neutral. The pH of the equivalence point of a weak acid with a strong base is in the basic range.

37. What do you mean by titration?

38. What is a standard solution?

39. What do you understand by the strength and normality of the solution?

40. Define normality, molarity and molality.

41. What is equivalent weight?

42. Define basicity and acidity? How is equivalent weight of an acid calculated.

43. How is equivalent weight of a base and a salt is obtained?

44. Why is distilled water always used in preparation of standard solution?

Ans. *Because in tap water, Ca^{2+} Mg^{2+} or Cl^- ions are present which may cause turbidity in solution.*

45. Why is it necessary to rinse the burette or pipette with the liquid it is filled with?

Ans. *The water sticking to the sides will alter the strength of the solution taken in it.*

46. The conical flask is only washed with distilled water and *not* with the solution to be taken in it. Explain

Ans. *Because only a fixed volume of a solution is taken in a conical flask with a pipette and if rinsed with the solution, the solution sticking to it adds to the volume.*

47. Why are hot solutions not poured into a burette?

Ans. *Hot solutions make the glass expand and introduce errors in volume measurements.*

48. One should not force out the last drop in the pipette by blowing – why?

Ans. *The pipette is so caliberated that the last drop need not to be forced out.*

49. Pipette should not be held from its bulb. Why?

Ans. *The body temperature may expand the glass and introduce a measuring error.*

50. If some water is added to the measured solution already taken in a conical flask, will it affect the titration?

Ans. *No, because the quantity of the substance dissolved remains the same.*

51. Why is water not added to acid and always acid is added to water?

Ans. *Since the reaction is exothermic and on adding water to acid the heat liberated is enough to change some amount of water into steam. This results in scattering of acid which may cause serious consequences. Hence, always small amounts of acids are added to water.*

52. What is primary standard? Give example.

53. What is a decinormal and centinormal solution?

Ans: *When one-tenth of the gram equivalent of a substance is dissolved per litre of solution, it is called decinormal solution and when one hundredth of the gram equivalent is dissolved per litre of solution, it is called centinormal solution.*

EXPERIMENT NO. 1.3

AIM **Find the strength in g/L of given NaOH solution with the help of standard sodium-carbonate solution (N/10) and intermediate solution of an acid.**

REQUIREMENT A standard solution of 0.1 N sodium carbonate and HCl solution having approx. normality N/10.

APPARATUS Burette, pipette, 2 beakers, titration flask and stand.

THEORY The experiment involves two titrations.

I Titration: In the first titration, HCl (which is a secondary standard), is titrated against Na_2CO_3 to determine its exact normality. The titration is carried out using methyl orange as indicator.

II Titration: Once the normality of HCl is known, the normality of given NaOH can be determined by titrating it with HCl using phenolphthalein as indicator.

The reactions involved are:

I Titration:

$$2HCl + Na_2CO_3 \rightarrow 2NaCl + H_2O + CO_2$$

II Titration:

$$HCl + NaOH \rightarrow NaCl + H_2O$$

INDICATOR Methyl orange (I titration)

Phenolphthalein (II titration)

END POINT Orange to red (I titration)

Disappearance of pink colour (II titration)

PROCEDURE

Titration I

(1) Rinse and fill the burette with the given HCl.

(2) Pipette out 10 ml of Na_2CO_3 in a titration flask and add a few drops of methyl orange.

(3) Titrate it with the acid till a permanent red colour is obtained.

(4) Repeat to get 3 concordant readings.

Titration II

(5) Pipette out 10 ml of the given NaOH in a clean conical flask.

(6) Add a few drops of phenolphthalein. The solution will turn pink.

(7) Titrate the above solution with the HCl taken in the burette till the pink colour disappears.

(8) Repeat to get three concordant readings.

OBSERVATIONS

Titration I

S. No.	*Volume of the solution taken in the titration flask*	*Burette readings*		*Volume of the HCl used (Final-Initial reading) (ml)*
		Initial reading	*Final reading*	

Titration II

S. No.	*Volume of the solution taken in the titration flask*	*Burette readings*		*Volume of the HCl used (Final-Initial reading) (ml)*
		Initial reading	*Final reading*	

CALCULATION

I Titration: Let the volume of the acid used = x mL

$$\begin{aligned} &\text{Acid} && Na_2CO_3 \\ N_1V_1 &= N_2V_2 \\ N_1 \times x &= \frac{1}{10} \times 10 \\ N_1 &= \frac{1}{10} \times \frac{10}{x} = \frac{1}{x} \end{aligned}$$

II Titration: Let the volume of the acid used = y ml

$$
\begin{aligned}
&\text{Acid} \quad \text{NaOH} \\
N_1' V_1' &= N_2' V_2' \\
\frac{1}{x} \times y &= N_2' \times 10 \\
N_2' &= \frac{1}{x} \times \frac{y}{10}
\end{aligned}
$$

Strength of NaOH solution = N_2' × Eq. wt = N_2' × 40 g/L

RESULT The strength of the given NaOH solution = ... g/L.

PRECAUTIONS

Same as in previous experiment.

Instructions for Lab Assistant

(1) **0.1 N oxalic acid solution:** Dissolve 6.3 g of oxalic acid $\begin{matrix} \text{COOH} \\ | \\ \text{COOH} \end{matrix} \cdot 2H_2O$ in distilled water and make up the volume to 1000 mL.

(2) **NaOH solution:** Dissolve about 4 g of NaOH in distilled water and make up the volume to 1000 mL.

(3) **Phenolphthalein indicator:** Dissolve 1 g of phenolphthalein in 100 mL of distilled ethyl alcohol and add 100 ml of distilled water with constant stirring.

EXERCISE

1. HCl is a primary standard or a secondary standard. Explain.

Ans. *HCl is a secondary standard because its concentration is known only approximately. It has to be standardized before it is used.*

2. Why anhydrous Na_2CO_3 is considered as primary standard?

Ans. *(i) It is easily available in highly pure state.*

(ii) Readily soluble in water.

(iii) Neither hydroscopic nor efflorescent.

(iv) Stable and unaffected by air.

3. NaOH is a secondary standard. How an exact N/10 solution of NaOH can be prepared.

Ans. *Firstly, a solution of NaOH having approx. strength of N/10 is prepared by dissolving about 4g (a little more) of NaOH in 1000 ml of water. Then this is standardized by N/10 oxalic acid using phenolphthalein as indicator to know the exact normality of NaOH solution. The solution of required normality (N/10) is prepared by diluting it with appropriate amount of water.*

4. **Give the structures of phenolphthalein and methyl orange.**

Ans.

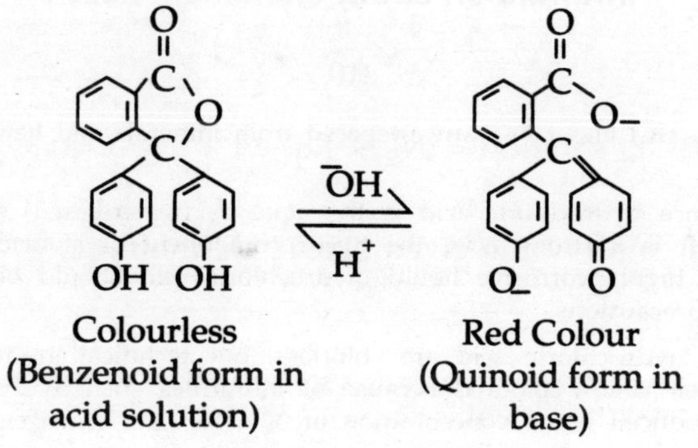

Phenolphthalein

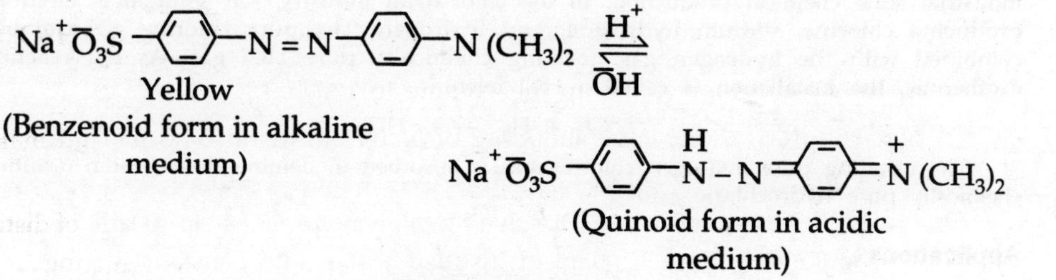

Methyl Orange

5. **Why phenolphthalein cannot be used as an indicator for titrating a weak base against a strong acid.**

Ans. *The approximate pH range at equivalence point in the case of a titration of weak base against a strong acid lies within 3.5 to 7. Phenolphthalein do not change colour in this range and hence cannot be used as an indicator for titrating a weak base against a strong acid. Methyl orange would be a suitable choice because it changes colour in the pH range 3.1 – 4.4 which very well lies within the range of equivalence point in the above case.*

6. **How will you prepare 100ml of N/30 solution from N/10 solution of NaOH?**

Ans. *Apply* $N_1V_1 = N_2V_2$; $\frac{N}{30} \times 100 = \frac{N}{10} \times V_2$ *i.e.* $V_2 = 33.3$ *ml.*

Therefore, 33.3 ml of N/10 NaOH is taken in the measuring flask and then its volume is made 100ml by adding distilled water to it.

7. What is a neutralization reaction?

Ans. *A reaction in which hydronium ions from an acid and a hydroxide ions from a base react to produce water molecules.*

8. What is the molarity of an HCl solution if 30.0 mL of the solution is just neutralized by 48.0 mL of 0.100 M NaOH?

Ans. $M_1 \times V_1 = M_2 \times V_2$, $M_{HCl} \times 30\ mL = 0.1\ M \times 48\ mL$, $M_{HCl} = 0.16\ M$.

Information about chemicals used

Hydrochloric acid

The acids like H_2SO_4, HCl and HNO_3 are prepared from minerals and hence are called mineral acids.

The chemical substance hydrochloric acid is the aqueous (water-based) solution of hydrogen chloride (HCl) gas. It is a strong acid, the major component of stomach acid and of wide industrial use. As a highly corrosive liquid, hydrochloric acid should be handled only with appropriate safety precautions.

Pure grades of hydrochloric acid are colorless, but technical grades, commonly called *muriatic acid,* are often yellow-coloured because of impurities such as dissolved metals. Most hydrochloric acid produced has a concentration of 30% to 35% hydrogen chloride by weight.

Direct synthesis

The large scale production of hydrochloric acid is almost always integrated with other industrial scale chemical production. In the chlor-alkali industry, salt solution is electrolyzed producing chlorine, sodium hydroxide, and hydrogen. The pure chlorine gas can be re-combined with the hydrogen gas, forming chemically pure HCl gas. As the reaction is exothermic, the installation is called a HCl oven.

$$Cl_2 + H_2 \longrightarrow HCl$$

The resulting pure hydrogen chloride gas is absorbed in demineralized water, resulting in chemically pure hydrochloric acid.

Applications

Hydrochloric acid is a strong inorganic acid that is used in many industrial processes. The application often determines the required product quality.

1. Regeneration of ion exchangers.

An important application of high-quality hydrochloric acid is the regeneration of ion exchange resins. Cation exchange is widely used to remove ions such as Na^+ and Ca^{2+} from aqueous solution, producing demineralized water.

Ion exchangers and demineralized water are used in all chemical industries, drinking water production, and many food industries.

2. pH control and neutralization.

A very common application of hydrochloric acid is to regulate the basicity (pH) of solutions.

$$OH^- + HCl \longrightarrow H_2O + Cl^-$$

In industry demanding purity (food, pharmaceutical, drinking water), high-quality hydrochloric acid is used to control the pH of process water streams. In less demanding industry, technical quality hydrochloric acid suffices for *neutralizing* waste streams and *swimming pool* treatment.

3. Pickling of steel

Pickling is an essential step in *metal* surface treatment, to remove *rust* or *iron oxide* scale from *iron* or *steel* before subsequent processing, such as *extrusion, rolling, galvanizing,* and other techniques. Technical quality HCl at typically 18% concentration is the most commonly used pickling agent for the pickling of carbon steel grades.

$$Fe_2O_3 + Fe + 6HCl \longrightarrow 3FeCl_2 + 3H_2O$$

The *spent acid* has long been re-used as *ferrouschloride* solutions, but high *heavy metal* levels in the pickling liquor has decreased this practice.

4. *Production of inorganic compounds*

Numerous products can be produced with hydrochloric acid in normal *acid-base reactions*, resulting in *inorganic* compounds. These include water treatment chemicals such as *iron(III) chloride* and *poly aluminium chloride (PAC)*

$$Fe_2O_3 + 6HCl \longrightarrow 2FeCl_3 + 3H_2O$$

5. *Production of organic compounds*

The largest hydrochloric acid consumption is in the production of *organic compounds* such as *vinyl chloride* for *PVC*, and *MDI* and *TDI* for *polyurethane*. Other *organic* compounds produced with hydrochloric acid include *bisphenol A* for *polycarbonate*, *activated carbon* and *ascorbic acid*, as well as numerous *pharmaceutical* products.

6. *Other applications*

Hydrochloric acid is a fundamental chemical, and as such it is used for a large number of small scale applications, such as *leather* processing, household *cleaning*, and *building construction*.

7. *Hydrochloric acid and living organisms*

Gastric acid is one of the main *secretions* of the *stomach*. Chemically it is an acid solution consisting mainly of *hydrochloric acid*, but also containing small quantities of *potassium chloride* (KCl) and *sodium chloride* (NaCl).

Safety Concern

Hydrogen chloride forms corrosive hydrochloric acid on contact with body tissue. Inhalation of HCL can cause coughing, choking, inflammation of the nose,throat, and upper respiratory tract, and in severe cases, pulmonary edema, circulatory system failure, and death. Skin contact can cause redness, pain, and severe skin burns. Hydrogen chloride may cause severe burns of the eye and permanent eye damage.

Ingestion can result in corrosion of gastrointestinal tract, vomiting, diarrhoea, and abdominal pain.

In case of skin contact, immediately wash contaminated area with plenty of water. Remove contaminated clothing and wash before re-use. If swelling, redness, blistering or irritation occurs seek medical advice.

In case of ingestion, thoroughly rinse mouth with water. Give plenty of water to drink. DO NOT induce vomiting. If vomiting occurs give further water. Seek immediate medical assistance.

SODIUM HYDROXIDE

Sodium hydroxide (NaOH), also known as caustic soda or lye, is a *caustic* metallic *base* used in industry (mostly as a strong *chemical base*) in the manufacture of *paper*, *textiles*, and *detergents*.

It is a white crystalline substance that readily absorbs carbon dioxide and moisture from the air. It is very soluble in water, alcohol, and glycerin. It is a *caustic* and a strong base. It is available commercially in various solid forms, e.g. pellets, sticks, or chips, and in water solutions of various concentrations; both solid and liquids forms vary in purity.

Synthesis

It is prepared by the electrolytic method by passing the direct current through an aqueous solution of sodium chloride. Hydrogen and chlorine gases are released at cathode and anode respectively./ The electrolytic solution remaining at the end is NaOH.

Applications

1. The major use of sodium hydroxide is as a chemical and in the manufacture of other chemicals; because it is inexpensive, it is widely used wherever a strong base is needed.
2. ***Food preparation:*** Food uses of NaOH include washing or chemical peeling of *fruits* and *vegetables, chocolate* and *cocoa* processing, *caramel* color production, *soft drink* processing, and thickening *ice cream.*
3. ***Biodiesel:*** For *biodiesel*, sodium hydroxide is used as a *catalyst*. This only works with *anhydroous* sodium hydroxide, because water and NaOH would turn biodiesel into *soap (saponification).*
4. NaOH it is also used in producing rayon and other textiles, in making paper, in etching aluminium, in making soaps and detergents, and in a wide variety of other uses.
5. Sodium hydroxide is occasionally used in the home as an agent for unblocking drains.

Safety Concern

It is highly caustic and has a high danger of causing *chemical burns,* permanent injury or scarring, and *blindness,* due to its high *reactivity.* Therfore, it should be stored separtely.

When sodium hydroxide reacts with water and fluids, it can become hot enough to cause *fires.* For this reason, it is important to have the proper type of chemical fire extinguisher on hand before working with sodium hydroxide. Store NaOH in an airtight container to prevent it from absorbing water and CO_2 from the air.

EXPERIMENT NO. 1.4

AIM **To determine the percentage composition of a given mixture of NaCl and NaOH, 8 gms. of which is dissolved per litre of the solution.**

REQUIREMENT A standard solution of 0.1 N HCl.

APPARATUS Burette, pipette, 2 beakers, titration flask, stand.

THEORY For the titration of a solution of NaOH and NaCl the other solution needed is HCl solution. NaOH will react with acid solution whereas NaCl will remain as such.

$$NaOH + HCl \rightarrow NaCl + H_2O$$

The mixture solution will be titrated with the acid solution and in this way we can determine the normality of NaOH and hence its strength. By subtracting the strength of NaOH from the total strength of solution we can find out the strength of NaCl and hence find percentage composition.

INDICATOR Phenolphthalein (as strong acid is reacting with a strong base)

END POINT Pink to colourless

PROCEDURE

1. Rinse the burette with the given HCl solution (N/10 HCl)
2. Take HCl in the burette and note initial reading.
3. Pipette out 10ml. of solution in the titration flask.
4. Add a drop of phenolphthalein to the solution.
5. Run the acid solution into the mixture from the burette till the solution changes colour from pink to colourless
6. Repeat to get three concordant readings.

OBSERVATIONS

Normality of the HCl solution = 0.1 N

Volume of mixture taken for each titration = 10 ml.

S. No.	*Volume of the solution taken in the titration flask (ml)*	*Burette Readings*		*Volume of the titrant used (Final-Initial reading) (ml)*
		Initial Reading	*Final Reading*	
1.				
2.				
3.				

CALCULATIONS

$$\begin{array}{ll} \text{Acid} & \text{Mixture (NaOH)} \\ N_1V_1 & = N_2V_2 \\ \frac{1}{10} \times x & = N_2 \times 10 \\ \therefore \quad N_2 & = \frac{x}{100} \end{array}$$

$$\text{Strength} = \text{Normality} \times \text{Eq. Wt.} = \frac{x}{100} \times 40 = y \text{ gms/litre.}$$

(Equivalent weight of NaOH = 40g)

∴ The amount of NaOH in the mixture is y gm/litre

Amount of NaCl of in the mixture = $(8 - y)$ g/L = zg/L

Now percentage of NaOH = $\frac{y}{8} \times 100\%$ and percentage of NaCl = $\frac{z}{8} \times 100\%$.

RESULT The percentage of NaOH present in the given mixture = ... and the percentage of NaCl present in the given mixture = ...

PRECAUTIONS

1. Wash the apparatus first with water and then with detergent followed by plenty of water.
2. Rinse the burette and pipette with the respective solutions to be taken in.
3. Do not rinse the conical flask with the solution to be taken in it.
4. Use anti parallex card to avoid error in burette reading.

Instructions for Lab Assistant

Provide the following.

1. 0.1 N HCl: Dissolve 8.4 ml of conc. HCl (12N) in 1000ml of distilled water. Standardize it with N/10 Na_2CO_3 solution using methyl orange as indicator.
2. A mixture of NaOH and NaCl: Dissolve 4g of NaOH and 4g of NaCl in 1000 ml of water.
3. Phenolphthalein indicator: Dissolve 1gm of phenolphthalein in 100ml of distilled methyl alcohol and add 100ml of distilled water with constant stirring Filter if required.

EXPERIMENT NO. 1.5

AIM **To determine the amount of Na_2CO_3 and $NaHCO_3$ in the given mixture of Na_2CO_3 and $NaHCO_3$.**

APPARATUS Burette, pipette, 2 beakers, titration flask, stand.

THEORY The determination of the strength of Na_2CO_3 and $NaHCO_3$ present in a solution is based on the selective use of indicators.

- When methyl orange is used as an indicator, the volume of 0.1 N HCl in the titration will neutralize both, Na_2CO_3 and $NaHCO_3$ completely.

$$Na_2CO_3 + 2HCl \rightarrow 2NaCl + H_2O$$

$$NaHCO_3 + HCl \rightarrow NaCl + H_2O + CO_2$$

Let the volume of acid used is x ml when methyl orange is used as indicator.

- When phenolphthalein is used as indicator the volume of 0.1 N HCl used in the titration will correspond to half neutralization of Na_2CO_3 or neutralization upto half stage

$$Na_2CO_3 + HCl \rightarrow NaHCO_3 + NaCl$$

Let the volume of the acid used for neutralization be y *ml* when phenolphthalein is used as an indicator. In fact, it does not give colour in $NaHCO_3$ solution.

- Therefore, the volume of the acid used for complete neutralization of Na_2CO_3 will be $2y$ ml. And volume of the acid used for neutralization of $NaHCO_3$ will be $x - 2y$ ml.

Therefore, two different titrations are carried out with the given mixture using methyl orange and phenolphthalein indicator respectively.

INDICATORS Phenolphthalein, Methyl orange.

END POINT Pink to colourless (Using phenolphthalein as indicator), Orange to red (Using methyl orange as indicator)

PROCEDURE

Using methyl orange as indicator:

1. Rinse and fill the burette with 0.1 N HCl.
2. Pipette out 10 ml. of alkali mixture solution into a conical flask. To it add 2 – 3 ml. of methyl orange.
3. Run HCl into the flask till the yellow colour of the solution changes to red. Let the volume of the acid used be x ml.

Using phenolphthalein as indicator

1. Pipette out 10ml of alkali mixture solution into a conical flask.
2. To it add 2 – 3 drops of phenolphthalein. Run 0.1 N HCl from the burette into the flask till the pink colour of the solution changes to colourless. Let the volume of the acid used be y ml.

Observation Table – I

S. No.	*Volume of the solution taken in the titration flask (ml)*	*Burette Readings*		*Volume of the titrant used (Final-Initial reading) (ml)*
		Initial Reading	*Final Reading*	
1.				
2.				
3.				

Observation Table– II

S. No.	*Volume of the solution taken in the titration flask (ml)*	*Burette Readings*		*Volume of the titrant used (Final-Initial reading) (ml)*
		Initial Reading	*Final Reading*	
1.				
2.				
3.				

CALCULATIONS

(i) Volume of acid required for neutralizing Na_2CO_3 upto half stage i.e. upto $NaHCO_3$ stage = y ml

∴ Volume of acid required for neutralizing Na_2CO_3 completely = $2y$ ml

$$\begin{aligned} \text{Acid} \quad & \quad Na_2CO_3. \\ N_1V_1 &= N_2V_2 \\ 0.1 \times 2y &= N_2 \times 10 \\ N_2 &= \frac{0.1 \times 2y}{10} = \frac{2y}{100} \end{aligned}$$

Hence strength of Na_2CO_3 in the solution = Normality × Eq. wt

$$= \left(\frac{2y}{100}\right) \times 53 \text{ g/L}$$

(ii) Volume of acid required for complete neutralization of Na_2CO_3 and $NaHCO_3$ present in the mixture = x ml

∴ Volume of acid required for neutralization of $NaHCO_3$ only = $(x - 2y)$ ml

$$\begin{aligned} \text{(Acid)} \quad & \quad (NaHCO_3) \\ N_1' V_1' &= N_2' V_2' \\ 0.1 \times (x - 2y) &= N_2' \times 10 \\ N_2' &= \frac{0.1 \times (x - 2y)}{10} = \frac{x - 2y}{100} \end{aligned}$$

Hence strength of $NaHCO_3$ in the solution = $\left(\frac{x - 2y}{100}\right) \times 84$ g/L.

RESULT The amount of Na_2CO_3 present in the given mixture = ... g/L

The amount of $NaHCO_3$ present in the given mixture = ... g/L

PRECAUTIONS Same as previous experiment number 1.4.

Instructions For Lab Assistant

Provide the following

1. 0.1 N HCl. As prepared in experiment..1.4
2. Solution of mixture of Na_2CO_3 and $NaHCO_3$ (8.4g of $NaHCO_3$ + 5.3g of Na_2CO_3 per litre of water)
3. Phenolphthalein. As prepared in experiment..1.4
4. Mehtyl orange. Dissolve 50mg of yellow powder in 100ml of water and filter.

EXERCISE

1. What air the commercial names of $NaHCO_3$ and Na_2CO_3?

Ans. *$NaHCO_3$ — Baking Soda*
Na_2CO_3 — Washing Soda

2. Find equivalent weights of $NaHCO_3$ and Na_2CO_3.

Ans. *Equivalent weight of $NaHCO_3$ = Molecular weight = 84 g.*

$$\text{Equivalent weight of } Na_2CO_3 = \frac{\text{Molecular weight}}{2} = \frac{106}{2} = 53\,g.$$

3. What is the significance of estimating Na_2CO_3 and $NaHCO_3$ in a water sample.

Ans. *The presence of Na_2CO_3 and $NaHCO_3$ impart alkalinity to water. Hence their estimation is used to find out the alkalinity of water. Highly alkaline water may lead to the formation of scales and sludges in boiler tubes and also lead to caustic embrittlement.*

4. What is the use of using 2 indicators in the above titration?

Ans. *The pH value at the end point in titrations depend upon the relative strengths of acid and base being titrated. Different indicators have different pH ranges within which they can be used. In this particular titration, phenolphthalein having a pH range of 8.0 – 9.6 will give the end point when CO_3^{2-} ions are neutralized to HCO_3^- stage, (the solution is at about pH 8.5.)*

$$CO_3^{2-} + H^+ \longrightarrow HCO_3^-$$

The methyl orange having a pH range 3.1 – 4.4 will give the end point when the CO_3^{2-} and HCO_3^-, both are completely neutralized (the solution is at about pH 4.3)

$$CO_3^{2-} + H^+ \longrightarrow HCO_3^-$$

Thus, this behaviour of the indicators is used to estimate CO_3^{2-} and HCO_3^- ions in a mixture.

Information about chemicals used

Sodium bicarbonate

Sodium bicarbonate ($NaHCO_3$), or **sodium hydrogen carbonate**, also known as **baking soda** and **bicarbonate soda,** is a soluble white *anhydrous* or *crystalline* compound, with a slight alkaline taste resembling that of *sodium carbonate.* It is found in many mineral springs and also produced artificially.

Synthesis

One important industrial process is the Solvay process, in which sodium hydrogen carbonate (sodium bicarbonate, $NaHCO_3$) and sodium carbonate (soda ash, Na_2CO_3) are produced from ammonia, carbon dioxide, water, and concentrated brine solution.

Applications:

1. It is used as an *antacid* to treat *acid indigestion* and *heartburn.* However, excess consumption can raise blood pH. Sodium bicarbonate should only be consumed in solution with water.
2. The anhydrous form is also used to absorb moisture and odours; a box can be left in the refrigerator for this purpose. Additionally, a paste from baking soda can be very effective when used in cleaning and scrubbing.
3. Used in toothpaste, baking soda helps to gently remove stains, whiten teeth, and freshen breath.
4. It is used as a leavening agent in baking (it is "baking soda"), and in the manufacture of sodium carbonate, Na_2CO_3. "Baking powder" is a mixture composed mainly of $NaHCO_3$. In addition, it contains anti-caking agents such as starch and weak acids such as alum or tartaric acid. These weak acids react with sodium bicarbonate, releasing CO_2 gas which causes cake batter and bread dough to rise and produces the tiny holes in cakes and breads.

Safety Concern:

Eye: Causes eye irritation. Causes redness and pain.

Skin: May cause skin irritation. Repeated or prolonged exposure may cause drying and cracking of the skin.

Ingestion: May cause irritation of the digestive tract.

Inhalation: May cause respiratory tract irritation.

First Aid:

Eyes: Flush eyes with plenty of water for at least 15 minutes, occasionally lifting the upper and lower eyelids. Get medical aid. Do NOT allow victim to rub or keep eyes closed.

Skin: Flush skin with plenty of soad and water for at least 15 minutes while removing contaminated clothing and shoes. Get medical aid if irritation develops or persists. Wash clothing before reuse.

Ingestion: Never give anything by mouth to an unconscious person. Do NOT induce vomiting. If conscious and alert, rinse mouth and drink 2 - 4 cupfuls of milk or water. Get medical aid if irritation or symptoms occur.

Inhalation: Remove from exposure to fresh air immediately. If not breathing, give artificlial respiration. If breathing is difficult, give oxygen. Get medical aid if cough or other symptoms appear.

Sodium Carbonate

Soda ash is a common name for sodium carbonate and as the name suggests, it was originally prepared from ashes obtained by burning wood. Sodium carbonate is used in detergents (washing soda), glass and other household products.

Soda ash production is the 11th largest of all domestic inorganic and organic chemicals, excluding petroleum products. Reagent grade anhydrous sodium carbonate is suitable for use as a primary standard for titrations of strong acids.

Synthesis

"Soda" or "soda ash" is prepared industrially by dissolving ammonia in a brine solution followed by treatment with carbon dioxide. Most of the "natural" soda ash comes from the mineral, trona.

Applications

1. This compound is one of the most common inorganic industrial chemicals. Almost everywhere you go, soda ash surrounds you since it is such an essential ingredient: From dry laundry detergent to baking soda to cat litter, soda ash will be there
2. It helps control pH levels and it softens water. The carbonate ions from dissolved sodium carbonate precipitate magnesium and calcium ions from hard water. These metal ions would otherwise combine with soap or detergent and form an insoluble scum that would stain the laundered clothes.
3. It helps clean and boil compounds and cleans electrical components.
4. It is part of the food process and is used in many different types of manufactures, particularly in chrome tannages.
5. It is widely used in glass, textile, and chemical industries, as well as construction material industies, detergent industries, and other types of industries.
6. It is also used in processing wood pulp to make paper.

Safety Concern

Sodium cabronate is neither toxic nor corrosive. However, sodium cabronate dust can cause temporary irritation of the skin, throat or eyes and a running nose.

First Aid:

- wear lightweight face masks and goggles;
- if appropriate,use protective crams.
- in the event of irritation, dab the affected areas with a 2% solution of acetic acid or apply soothing creams based on zinc oxide;
- in the event of sodium carbonate getting into the eyes, wash with large quantites of water and consult an eyespecialist.

II OXIDATION REDUCTION TITRATIONS: (REDOX TITRATIONS)

Titrations in which a reducing agent is titrated against an oxidizing agent or vice versa are called *redox titrations.*

Oxidation. Oxidation may be defined as the process in which

- oxygen or any electronegative element is added up

or

- hydrogen or any electropositive element is removed

or

- loss of electrons from the element takes place.

Example

$Fe^{2+} \rightarrow Fe^{3+} + e^-$

Ferrous ions is said to be oxidised to Fe^{3+} ions.

Reduction. Reduction may be defined as the process in which

- hydrogen or any electropositive element is added up

or

- oxygen or any electronegative element is removed

or

- electrons are accepted by the element or the ion.

Example.

$Cu^{2+} + 2e^- \rightarrow Cu$

$Cl + e^- \rightarrow Cl^-$

Cu^{2+} and Cl are said to be reduced to Cu and Cl^- respectively.

Further, the reagent undergoing oxidation is called reducing agent and the reagent undergoing reduction is called oxidizing agent.

Oxidizing agent. It is a substance that

- gives up oxygen or any other electronegative element
- accepts hydrogen or any other electropositive element
- accepts electrons.

Note. An oxidizing agent itself gets reduced.

Reducing agent. It is a substance that

- gives up hydrogen or any other electropositive element
- accepts oxygen or any electropositive element
- gives up electrons.

Note. A reducing agent itself gets oxidized.

A number of substances in solution can be estimated by oxidation-reduction reactions. As is well known that oxidation and reduction are complementary to one another and take place simultaneously, while one substance is being oxidized, the other is being reduced e.g.

$$\underset{\substack{\text{Reduced}\\\text{(Oxidizing agent)}}}{MnO_4^-} + \underset{\substack{\text{Oxidized}\\\text{(Reducing agent)}}}{5Fe^{2+}} + 8H^+ \rightarrow Mn^{2+} + 5Fe^{3+} + 4H_2O$$

Hence these reactions are termed as **redox reactions.**

REDOX TITRATIONS AND INDICATORS

A redox indicator is defined as a substance whose oxidized form is of different colour from that of its reduced form. The oxidation and reduction of the indicator is readily reversible. As the change in colour in acid-base indicator is dependent upon the pH of the solution, similarly change in colour of redox indicator is dependent upon the electrode potential of the system. Therefore, while choosing a redox indicator for a particular titration, it is necessary to ensure that its potential lies within that of the system. Moreover the colour change should occur as close to the equivalence point as possible.

With respect to the use of indicators, redox indicators can be classified as:

(i) Self indicator
(ii) Internal indicator
(iii) External indicator

Self Indicator.

If the reagent (reacting solution) is strongly coloured and is decolourized or changes into a different colour by the substance that is being determined then it is called self indicator. For example, potassium permanganate is purple in its oxidized form, but its reduced product i.e. manganese (II) ion is almost colourless. In acidic solution, the reaction is,

$$\underset{\text{purple}}{MnO_4^-} + 8H^+ + 5e^- \rightarrow \underset{\text{Colourless}}{Mn^{2+}} + 4H_2O$$

Thus, during titration of $KMnO_4$ with a reducing agent such as oxalic acid, manganese (II) is produced which is colourless. As the titration proceeds and when whole of reducing agent is consumed up, then the addition of an excess drop of $KMnO_4$ solution gives its own colour (pink) to the solution. This marks the completion of the reaction (end point). Hence $KMnO_4$ acts as self indicator since no other indicator is added to mark the end point.

Internal Indicator.

Broadly speaking, internal indicators are the substances which are added to the volumetric flask in which titration is carried out (The acid base indicators like phenolphthalein, methyl orange etc are all examples of internal indicators). In redox titrations, diphenylamine is one of the first internal indicator discovered by Knop. This is used for the titration of iron (II) with potassium dichromate. This indicator has different colours in oxidized and reduced form. When all the iron (II) ions are oxidized to iron (III) ions by $K_2Cr_2O_7$ then an extra drop of $K_2Cr_2O_7$ added in the solution will oxidize the indicator (diphenylamine) which changes to blue-violet colour and hence marks the completion of reaction.

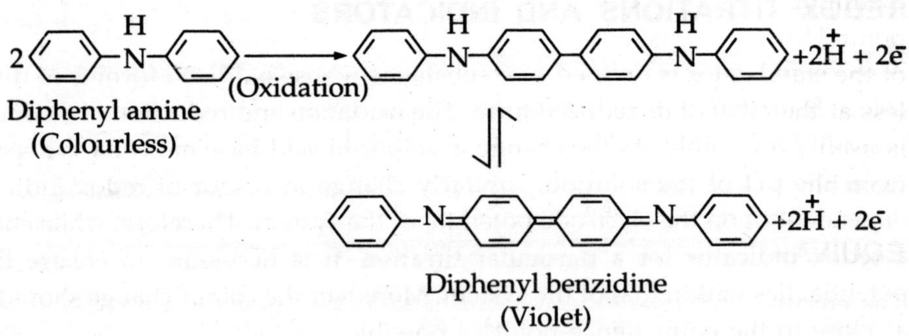

External Indicator.

The indicators which cannot be added to the volumetric flask in which the titration is carried out, but are used externally are called external indicators. For example, potassium ferricyanide is used as an external indicator in the titration of $K_2Cr_2O_7$ with ferrous ammonium sulphate. Potassium ferricyanide ($K_3[Fe(CN)_6]$) gives a blue colour with Fe^{2+} ions due to the formation of the complex ferrous ferricyanide, according to the following equation:

$$2\,K_3Fe(CN)_6 + 3\,FeSO_4 \longrightarrow \underset{\text{Ferrous ferri cyanide (Dark blue complex)}}{Fe_3[Fe(CN)_6]_2} + 3\,K_2SO_4$$

Thus, as long as Fe^{2+} ions are present, $K_3[Fe(CN)_6]$ will react with it to give blue colour. When all Fe^{2+} ions are converted to Fe^{3+} ions, blue colour will no longer be obtained.

The drops of external indicator $[K_3[Fe(CN)_6]$ are placed on a groove tyle by means of a dropper. During the titration, the drops of the reaction mixture are taken out with the help of glass rod at intervals and tested by mixing with the drops of indicator solution in the groove tile. Appearance of blue colour indicates that the end point has not reached and the titration is continued further, till the reaction mixture gives no blue colour. This marks the end point. This is a rough titration. An almost exact end point is obtained in a subsequent titration in which number of drops taken out from the titration solution are minimum.

Adsorption Indicator

In some cases, though the reagent is coloured but colour change at the end point is not distinct to ensure accuracy. For example, the iodine is brown in solution whereas iodide ion is colourless,

$$\underset{\text{Brown in solution}}{I_2} + 2e^- \longrightarrow \underset{\text{Colourless}}{2I^-}$$

As the iodine is reduced, the colour fades to a pale yellow. It is difficult to distinguish the pale yellow colour from the colourless iodide ion at the end point. However if starch is used as an indicator, iodine is *adsorbed* at the surface of the starch molecules and produce blue colour. The solution becomes colourless at the end point as the iodine is removed from the solution. Hence starch is used as an indicator in the iodine titrations as it gives distinct end point i.e. from blue to colourless.

EQUIVALENT WEIGHT OF A REAGENT IN REDOX TITRATION

The equivalent weight of a reagent in an redox reaction is

$$= \frac{\text{Molecular Weight}}{\text{Change in oxidation number}} = \frac{\text{Molecular Weight}}{\text{Number of electrons lost or gained}}$$

Same substance can have different equivalent weight under different conditions of the reaction. For example, $KMnO_4$ has three different values of equivalent weights depending upon the reaction,

(i) $MnO_4^- + e^- \rightarrow MnO_4^{2-}$ (In neutral medium)

$$\text{Eq. wt of } KMnO_4 = \frac{\text{Molar mass}}{1}$$

(ii) $MnO_4^- + 2H_2O + 3e^- \rightarrow MnO_2 + 4\overline{O}H$

$$\text{Eq. wt of } KMnO_4 = \frac{\text{Molar mass}}{3}$$ (In alkaline medium)

(iii) $MnO_4^- + 8H^+ + 5e^- \rightarrow Mn^{2+} + 4H_2O$

$$\text{Eq. wt of } KMnO_4 = \frac{\text{Molar mass}}{5} \qquad \text{(In acidic medium)}$$

The equivalent weight of a few oxidizing and reducing agents is given in Table-1.1 Page 4

Common oxidizing agents used in volumetric analysis are:

(a) Potassium permanganate

(b) Potassium dichromate

(c) Iodine

II. A POTASSIUM PERMANGANATE TITRATIONS

Potassium permanganate is a very strong oxidizing agent and is employed in the estimation of reducing agent like oxalic acid, Mohr's salt etc. It acts as an oxidizing agent in acidic and basic medium, but the most common application of $KMnO_4$ is in acidic medium according to the equation,

$$MnO_4^- + 8H^+ + 5e^- \rightarrow Mn^{2+} + 4H_2O$$

Some of the important points in $KMnO_4$ titrations are

1. *For acidification of $KMnO_4$ solution, only H_2SO_4 is used whereas other mineral acids like HCl and HNO_3 are not used.*
   - *HCl cannot be used because some of $KMnO_4$ will oxidize HCl to Cl_2 gas and thus interferes in the quantitative estimations.*

     $$2MnO_4^- + 16H^+ + 10Cl^- \rightarrow 2Mn^{2+} + 8H_2O + 5Cl_2$$

   - *HNO_3 cannot be used because it is itself a strong oxidizing agent and may oxidize the reducing agent, thereby introducing error.*
2. *$KMnO_4$ is not a primary standard.*
   - *It cannot be obtained in perfectly pure state as it always contains traces of manganese dioxide.*
   - *$KMnO_4$ may react with organic matter present in water in which it is dissolved.*
   - *$KMnO_4$ solution slowly decomposes when exposed to bright sun light, reducing gases etc. Therefore, it is stored in dark coloured bottles.*
   - *$KMnO_4$ solution can be standardized using a suitable primary standard such as oxalic acid or Mohr's salt, which can be directly weighed.*
3. ***$KMnO_4$ is a self indicator.*** *This point is already discussed under the heading "redox indicators".*

EXPERIMENT NO. 1.6

AIM **To determine the amount of oxalic acid and sulphuric acid in 1 litre of solution given .1 N NaOH solution and 0.1 N $KMnO_4$ solution.**

APPARATUS Burette, pipette, 2 beakers, titration flask, stand

THEORY This involves double titration.

1st Titration: NaOH reacts with oxalic acid as well as H_2SO_4 according to the following equations:

$$\begin{matrix} COOH \\ | \\ COOH \end{matrix} 2H_2O + 2NaOH \longrightarrow \begin{matrix} COONa \\ | \\ COONa \end{matrix} + 4H_2O$$

$$H_2SO_4 + 2NaOH \longrightarrow Na_2SO_4 + 2H_2O$$

By titrating NaOH against the given mixture of oxalic acid and sulphuric acid, the total normality of oxalic acid and H_2SO_4 can be found out.

2nd Titration. The mixture solution is titrated with N/10 $KMnO_4$ solution which will react with oxalic acid (redox titration) in the presence of H_2SO_4.

$$2KMnO_4 + 3H_2SO_4 \longrightarrow K_2SO_4 + 2MnSO_4 + 3H_2O + 5[O]$$

$$\left[\begin{matrix} COOH \\ | \\ COOH \end{matrix} + [O] \xrightarrow[\Delta]{60 - 70^{o}\,C} 2CO_2 + H_2O\right] \times 5$$

$$2KMnO_4 + 3H_2SO_4 + 5H_2C_2O_4 \longrightarrow K_2SO_4 + 2MnSO_4 + 8H_2O + 10CO_2$$

Thus from second titration, normality of oxalic acid alone can be found out and hence its strength is determined.

From the normality of the solution obtained in the first titration, the normality of oxalic acid is subtracted and hence the normality of H_2SO_4 and its strength can be found out.

The reaction of oxalic acid with $KMnO_4$ is very slow therefore the oxalic acid solution is heated to 60–70° C initially. Once the reaction has started, its rate automatically increases due to the formation of Mn^{2+} ions which catalyze the reaction (auto catalysis)

INDICATOR USED Phenolphthalein (1st titration)

$KMnO_4$ is self-indicator (2nd titration)

END POINT Colourless to Pink (1st Titration)

Colourless to Pink (2nd Titration)

PROCEDURE

1st Titration

1. Rinse the apparatus thoroughly. Take NaOH in the burette and pipette out 10ml. of mixture in the titration flask.
2. Add a drop of phenolphthalein and run NaOH from the burette.
3. Titrate till end point is achieved and note down the volume of NaOH used.
4. Repeat to get 3 concordant readings.

2nd Titration

1. Rinse the apparatus thoroughly. Take $KMnO_4$ in the burette. Since it is a coloured solution we note the upper meniscus for taking the initial reading.
2. Pipette out 10ml. of the mixture in the titration flask. Add half a test tube (almost 10ml.) of H_2SO_4 and then heat the conical flask slowly until the moisture appears on the neck of the flask or it is unbearable to touch (60–70°)
3. Titrate to achieve the end point and note volume of $KMnO_4$ used.
4. Repeat to get 3 concordant readings.

Observation Table I

NaOH Versus Mixture

S. No.	*Volume of the solution taken in the titration flask (ml)*	*Burette Readings*		*Volume of the titrant used (Final-Initial reading) (ml)*
		Initial Reading	*Final Reading*	
1.				
2.				
3.				

Observation Table II

$KMnO_4$ versus Mixture (only oxalic acid reacts)

S. No.	*Volume of the solution taken in the titration flask (ml)*	*Burette Readings*		*Volume of the titrant used (Final-Initial reading) (ml)*
		Initial Reading	*Final Reading*	
1.				
2.				
3.				

CALCULATIONS

Titration – 1st

$$\underset{(H_2SO_4 + \text{oxalic acid})}{N_1V_1} = \underset{(NaOH)}{N_2V_2}$$

$$N_1 = \frac{0.1 \times V_2}{10}$$

Titration– 2nd

$$\underset{(\text{oxalic acid})}{N_1' V_1'} = \underset{(KMnO_4)}{N_2' V_2'}$$

$$N_1' = \frac{0.1 \times V_2'}{10}$$

Normality of oxalic acid +

$$H_2SO_4 = N_1$$

$$\text{Normality of oxalic acid} = N_1'$$

$$\therefore \quad \text{Strength of oxalic acid} = N_1' \times \text{Eq. wt}$$

$$= N_1' \times 63\text{g/L}$$

$$\text{Normality of } H_2SO_4 = (N_1 - N_1')$$

$$\therefore \quad \text{Strength of } H_2SO_4 = (N_1 - N_1') \times \text{Eq. wt.}$$

$$= (N_1 - N_1') \times 49$$

RESULT Amount of oxalic acid present in the given mixture = ... g/L

Amount of H_2SO_4 present in the given mixture = ... g/L

PRECAUTIONS

1. All the apparatus should be washed before use.
2. Rinse the burette with $KMnO_4$ and pipette with the given solution.
3. Titration flask should not be rinsed with the given solution but only with distilled water.
4. Never use a pinch cock burette as $KMnO_4$ attacks rubber.
5. Read upper meniscus because $KMnO_4$ is a coloured solution.
6. Sulphuric acid should be added in excess, otherwise a brown precipitate of MnO_2 may appear.
7. Always use freshly prepared $KMnO_4$ solution.

Instructions For Lab. Assistant

Provide the following

1. N/10 NaOH solution: Dissolve a little above 4g of NaOH pellets in distilled water and make up the volume to 1000ml. Standardize it with N/10 oxalic acid and prepare N/10 NaOH by diluting with requisite amount of water.
2. N/10 $KMnO_4$ solution: Dissolve a little above 3.16g of $KMnO_4$ crystals per litre of solution and standardize it with N/10 oxalic acid to find out the exact normality. Prepare N/10 $KMnO_4$ by diluting with water.
3. Mixture of oxalic acid and H_2SO_4: Add 6.3g of oxalic acid and 2.8ml of conc. H_2SO_4 (36N) in distilled water and make up the volume to 1000ml.
4. Phenolphthalein.

EXERCISE

1. For acidification of $KMnO_4$ solution, only H_2SO_4 is used whereas other mineral acids like HCl and HNO_3 are not used.

Ans. • *HCl cannot be used because some of $KMnO_4$ will oxidize HCl to Cl_2 gas and thus interferes in the quantitative estimations.*

$$2MnO_4^- + 16H^+ + 10Cl^- \rightarrow 2Mn^{2+} + 8H_2O + 5Cl_2$$

• *HNO_3 cannot be used because it is itself a strong oxidizing agent and may oxidize the reducing agent, thereby introducing error.*

2. $KMnO_4$ is not a primary standard. Explain

Ans. • *It cannot be obtained in perfectly pure state as it always contains traces of manganese dioxide.*

• *$KMnO_4$ may react with organic matter present in water in which it is dissolved.*

- *$KMnO_4$ solution slowly decomposes when exposed to bright sun light, reducing gases etc. Therefore, it is stored in dark coloured bottles.*
- *$KMnO_4$ solution can be standardized using a suitable primary standard such as oxalic acid or Mohr's salt, which can be directly weighed.*

3. The titration involving $KMnO_4$ are carried out in acidic medium. Explain

Ans. *This is due to higher oxidizing power of permanaganate ion in acidic medium than in neutral or alkaline medium. Moreover, the formation of brown coloured MnO_2 in alkaline medium interferes with the detection of the end point.*

4. Why the oxalic acid solution is heated before titration with $KMnO_4$ solution?

Ans. *The reaction between oxalic acid and $KMnO_4$ is very slow. Thus, to increase the rate of reaction the solution is heated.*

5. What are the type of indicators according to their method of usage?

Ans. *Three types of indicators are there:*

(i) ***Internal indcators:*** *There are added to the solution taken in the titration flask. Examples include phenolphthalein, methyl orange and starch solution.*

(ii) ***External indcators:*** *These indicators are not added in the titration flask but are kept outside on a porcelain plate. The solution in the titration flask is withdrawn with a dropper after each addition of the titrant from the burette and placed on indicator to see the colour change. Example of external indicator $K_3Fe(CN)_6$.*

(iii) ***Self indicator:*** *When one of the reactant (titrant or titrand) acts as an indicator.*

6. Why is brown turbidity sometimes seen while titrating $KMnO_4$ solution with oxalic acid.

Ans. *The brown turbidity is due to MnO_2 produced as a result of insufficient amount of H_2SO_4.*

$$2KMnO_4 + 3MnSO_4 + 7H_2O \rightarrow K_2SO_4 + 2H_2SO_4 + \underset{(Brown)}{5MnO_2 \times H_2O}$$

7. Why does colour of $KMnO_4$ disappears slowly in the beginning and quite fast afterwards when titrated against a hot solution of oxalic acid containing dil. H_2SO_4.

Ans. *The reaction is slow in the beginning but quite fast afterwards due to the formation of $MnSO_4$ as one of the products. The reaction is accelerated because $MnSO_4$ acts as a catalyst.*

8. A brown layer is developed on the glass containing $KMnO_4$ solution.

Ans. *This is due to the formation of MnO_2.*

9. Why it is not advisable to filter $KMnO_4$ through a filter paper?

Ans. *The organic matter of filter paper is oxidized by $KMnO_4$.*

10. Water is also oxidized by $KMnO_4$ then why do we prepare aqueous solutions of $KMnO_4$.

Ans. *Though $KMnO_4$ can oxidize water but the rate is very slow and hence freshly prepared $KMnO_4$ solutions can be used.*

Information about chemicals used

Potassium Permanganate

Potassium permanganate ($KMnO_4$) is an *inorganic chemical compound* composed of *potassium* (K^+) and permanganate (KnO_4^-) *ions*. It is also known as "permanganate of potash" and "Condy's crystals". The permanganate acts as a strong *oxidizing agent*. As a *solid*, it forms deep purple prisms. It is purple when dissolved in an *aqueous solution*. It has a sweet taste and is odorless.

Applcations

1. Potassium permanganate is used as an oxidizing agent in many different kinds of *chemical reactions* in a laboratory and in industry.
2. It is also used a *disinfectant* and in deodorizers and dyes. It is used to treat some parasitic diseases of fish, and used in treatment of drinking water, as well as an antidote in *phosphorus* poisoning.
3. It can be used as a *reagent* for the *synthesis* of many different kinds of chemical compounds. For example, a dilute solution of $KMnO_4$ can convert an organic compound with a carbon-carbon *double bond* into a *diol* (glycol) compound.
4. It has also been used to make *cocaine* 100% pure.
5. In a diluted solution, it can be used as a *mouthwash* (1/4%), or to disinfect the hands (1%).

Safety Concern

1. This material is harmful if swallowed on inhaled. It is also harmful if absorbed through the skin.
2. Potassium permanganate is a strong oxidizing agent and may react very exothermically with organic materials. Potassium permanganate stains the hand and clothing and should be handled with care.
3. Clothing stains may be washed away using *acetic acid* or hydrochloric acid. Skin stains go off within 48 hours.
4. It causes corrosive burns on the skin, while swallowing it may lead to *gastroenteritis*. In addition, mixing solid $KMnO_4$ with concentrated *hydrochloric acid* genrates lethal *chlorine gas*.

Wear safety glasses and keep the solid or solution away from contact with the skin. Take care not to allow the solid to come into contact with flammable materials.

Oxalic Acid

Oxalic acid (*IUPAC name*; ethanedioic acid, formula $H_2C_2O_4$) is a bi-*carboxylic acid* with structure (*HOOC*)-(COOH). Because of the joining of two *carboxyl groups*, this is one of the strongest organic *acids*. The *anions* of oxalic acid as well as its *salts* and *esters* are known as *oxalates*.

It is a colourless, crystalline organic carboxylic acid that melts at 189ºC with sublimation. Oxalic acid and oxalate salts are poisonous.

Oxalic acid is found in many plants, e.g., sorrel and rhubarb, usually as its calcium or potassium salts.

Synthesis

oxalic acid is usually prepared by heating sodium formate with sodium hydroxide to form sodium oxalate, which is converted to calcium oxalate and treated with sulphuric acid to obtain free oxalic acid.

Uses

Unlike other carboxylic acids (except *formic acid*), it is readily oxidized; this makes it useful as a reducing agent for photography, bleaching and ink removal.

Vaporized oxalic acid is used by some *beekeepers* as an insecticide against the parasitic *varroa mite*.

Safety Concern

Oxalic acid is corrosive to tissue. When ingested, oxalic acid removes calcium from the blood. Kidney damage can be expected as the calcium is removed from the blood in the form of calcium oxalate. The calcium oxalate then obstructs the kidney tubules and may cause kidney damage.

Because it binds vital *nurients* such as calcium, long-term consumption of foods high in oxalic acid can lead to *nurient* deficiencies. Healthy individuals can safely consume such foods in moderation, but those with *kidney* disorders, *gout*, or *heumatoid arthritis* are typically advised to avoid foods high in oxalic acid or oxalates.

Wear impervious protective clothing, including boots, gloves, lab coat, apron or coveralls, as appropriate, to prevent skin contact.

Sulphuric acid

Sulphuric acid, H_2SO_4, is a strong *mineral acid*. It is soluble in *water* at all concentrations. The old name for sulphuric acid was oil of vitriol.

When high concentrations of $SO_3(g)$ are added to sulphuric acid, $H_2S_2O_7$ forms. This is called fuming sulphuric acid or *oleum* or, less commonly, Nordhausen acid.

Applications

- Sulphuric acid has many applications, including in many *chemical reactions* and production processes. It is the most widely used industrial chemical. Principal uses include fertilizer manufacturing, ore processing, chemial synthesis, wastewater processing and oil refining.
- The *hydration* reaction of sulphuric acid is highly *exothermic*. If water is added to concentrated sulphuric acid, it can boil. Always add the acid to the water rather than the water to the acid. This problem is due to the relative densities of the two liquids. Water is less *dense* than sulphuric acid and will tend to float above the acid.
- Because the *hydration* of sulphuric acid is *thermodynamically* favorable, sulphuric acid is an excellent dehydration agent, and is used to prepare many dried fruits.
- The affinity of sulphuric acid for *water* is sufficiently strong that it will take hydrogen and oxygen molecules out of other compounds; for example, mixing *glucose* ($C_6H_{12}O_6$) and concentrated sulphuric acid will give elemental *carbon* and *water* which is absorbed by the sulphuric acid (which becomes slightly diluted): $C_6H_{12}O_6 \rightarrow 6C + 6H_2O$.
- When in the atmosphere it is part of many chemicals that make up *acid rain*.

II. B POTASSIUM DICHROMATE TITRATIONS

Potassium dichromate acts as an oxidizing agent in the presence of the acid. The chemical equation in presence of an acid being,

$$K_2Cr_2O_7 + 4H_2SO_4 \rightarrow K_2SO_4 + Cr_2(SO_4)_3 + 4H_2O + 3[O],$$

or Ionic equation $Cr_2O_7^{2-} + 14H^+ + 6e^- \rightarrow 2Cr^{3+} + 7H_2O$

As the above equation involves the transfer of 6 electrons during reduction of $Cr_2O_7^{2-}$ to Cr^{3+}, hence the equivalent weight of $K_2Cr_2O_7$ will be,

$$= \frac{\text{Molar mass of } K_2Cr_2O_7}{6} = \frac{294}{6} = 49$$

Some of the important points involved in $K_2Cr_2O_7$ titrations are

1. *It is a less powerful oxidizing agent than $KMnO_4$.*
2. *It is a excellent primary standard.*
   - *It is available in pure state.*
   - *It is very stable. Its solutions do not decompose even on long standing.*
   - *Its solutions are stable towards light and are little affected by dust, organic matter or by autocatalytic reactions.*

Thus, $K_2Cr_2O_7$ has above advantages over $KMnO_4$.

3. *For acidification of $K_2Cr_2O_7$, apart from H_2SO_4, HCl can also be used. $K_2Cr_2O_7$ do not oxidize chlorides at room temperature.*
4. *$K_2Cr_2O_7$ acts as on oxidizing agent only in acidic medium. In presence of an alkali, $K_2Cr_2O_7$ reacts to give chromate salt and cannot act as oxidizing agent.*
5. ***Indicators.** A dilute solution of $K_2Cr_2O_7$ has a faint orange colour and its reduction product chromium (III) is green in colour. Thus, a drop of $K_2Cr_2O_7$ in excess at the end point is not sufficient to give a distinct colour to the solution (unlike $KMnO_4$) and hence it cannot act as self indicator. Therefore a redox indicator should be used in $K_2Cr_2O_7$ titrations. There are two types of indicators used in $K_2Cr_2O_7$ titrations:-*

   ***(a) External Indicator.** Potassium ferricyanide is used as an external indicator.*

   ***(b) Internal Indicators.** Commonly used internal indicators are:*

Indicator	*Colour in reduced state*	*Colour in oxidized state*
• *N Phenylanthranilic acid*	*green*	*violet-red*
• *Diphenylamine in conc. H_2SO_4*	*colourless*	*violet*
• *Diphenyl benzidine in conc. H_2SO_4*	*colourless*	*violet*
• *Ferroin in conc. H_2SO_4*	*Red*	*faint blue*

EXPERIMENT NO. 1.7

AIM **To determine the number of water molecules of crystallization in Mohr's salt, provided standard potassium dichromate solution (0.1 N) using diphenylamine as internal indicator.**

APPARATUS Burette, pipette, 2 beakers, titration flask, stand.

THEORY $K_2Cr_2O_7$ oxidizes $FeSO_4$ present in Mohr's salt to $Fe_2(SO_4)_3$ in the presence of dil. H_2SO_4.

$$Cr_2O_7^{2-} + 14H^+ + 6e^- \longrightarrow 2Cr^{3+} + 7H_2O$$

$$[Fe^{2+} \longrightarrow Fe^{3+} + e^-] \times 6$$

$$Cr_2O_7^{2-} + 14H^+ + 6Fe^{2+} \longrightarrow 2Cr^{3+} + 6Fe^{3+} + 7H_2O$$

In this titration diphenylamine is used as an internal indicator. At the end point all the ferrous ions present in the solution get completely oxidized to ferric ions by chromate ions and as soon as a slight excess of dichromate is added, it will bring about the oxidation of diphenylamine and results in the formation of a blue coloured complex. This indicates the end point of the titration.

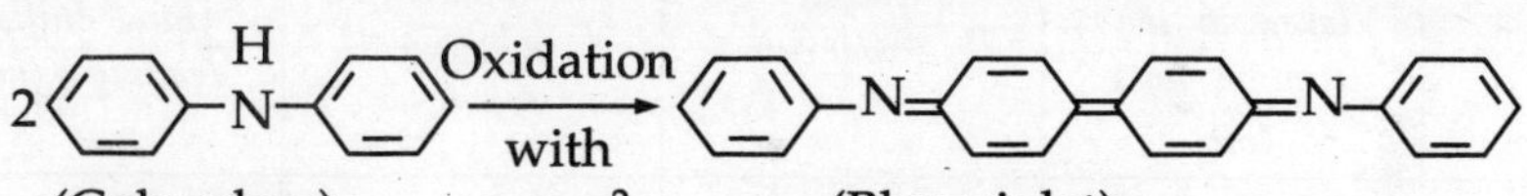

(Colourless) Diphenylamine — $Cr_2O_7^{2-}$ — (Blue violet) Diphenylbenzidine

From this titration, strength of anhydrous Mohr's salt can be calculated. Then applying the following relation, the number of water molecules of crystallization may be found out.

$$\frac{\text{*Strength of hydrated salt}}{\text{Strength of anhydrated salt}} = \frac{284 + 18x}{284}$$

Given strength of hydrated salt = 20g/litre.

***Note.** Molecular mass of anhydrated Mohr's salt = 284

INDICATOR USED Diphenylamine

END POINT Colourless to deep blue

PROCEDURE

1. Rinse the burette with $K_2Cr_2O_7$. Take $K_2Cr_2O_7$ in the burette and note initial reading.
2. Pipette out 10ml. of Mohr's salt solution in the titration flask. Add 5ml of dil H_2SO_4 and 2–3ml of 1 : 1 H_3PO_4 to the flask. Also add a drop of indicator.
3. Start titrating and stir at regular intervals.
4. When end point is reached, note down the final reading of the burette.
5. Repeat to get 3 concordant readings.

OBSERVATIONS

Volume of Mohr's solution taken for each titration = 10ml

Normality of $K_2Cr_2O_7$ solution = 0.1 N.

Observation Table

S. No.	*Volume of the solution taken in the titration flask (ml)*	*Burette Readings*		*Volume of the titrant used (Final-Initial reading) (ml)*
		Initial Reading	*Final Reading*	
1.				
2.				
3.				

CALCULATIONS

$$\text{(Mohr's solution)} \quad (K_2Cr_2O_7)$$

$$N_1V_1 = N_2V_2$$

$$N_1 \times 10 = \frac{1}{10} \times V_2$$

$$N_1 \text{ (Mohr's solution)} = \frac{V_2}{100}$$

Strength of anhydrated Mohr's Salt

$$= \text{Normality x Eq. wt}$$
$$= N_1 \times 284 = yg/L$$

$$\text{Strength of hydrated salt} = 20g/L \text{ (Given)}$$

$$\frac{\text{Strength of hydrated salt}}{\text{Strength of anhydrated salt}} = \frac{284+18x}{284}$$

Calculate the value of x**.** x is the number of water molecules of crystallization

RESULT The numbers of water molecules of crystallization is.....

Instructions for the Lab assistant

Provide the following

1. **0.1 N $K_2Cr_2O_7$ Solution.** Dissolve 4.9g of dry solid $K_2Cr_2O_7$ in distilled water and make up the volume to 1000ml.
2. **Mohr's salt.** Weigh 20g of hydrated Mohr's salt (A. R). Transfer it to a measuring flask and add a few drops of conc. H_2SO_4 or 20ml of dil. H_2SO_4 (to prevent the hydrolysis of Mohr's salt). Then add distilled water and dissolve the salt. Make up the volume to 1000ml.
3. **H_3PO_4.**
4. **Dil H_2SO_4 (5N).** Add slowly 139 ml of conc. H_2SO_4 (36N, 98%) to distilled water and make up the volume to 1000ml with distilled water.
5. **Diphenylamine indicator.** Dissolve 1g of diphenylamine in 100ml of conc. H_2SO_4.

EXERCISE

1. $K_2Cr_2O_7$ is used as an oxidizing agent only in presence of acid. Explain.

Ans. *In presence of alkali, $K_2Cr_2O_7$ reacts to give chromate salt and hence cannot act as oxidising agent.*

2. What do you mean by water of crystallisation?

Ans. *In a hydrated crystal definite amount of water molecules are combined with each molecule. This is called water of crystallisation.*

Notes. 1. In the titration of Fe^{2+} ions with $K_2Cr_2O_7$, using diphenylamine as indicator, H_3PO_4 is always added. This is because during the titration when about only 50% of Fe^{2+} ions are oxidized to Fe^{3+} ions, the high conc. of Fe^{3+} ions may also cause the oxidation of diphenylamine. Thus blue colour may result even before the end point is reached. In order to prevent this H_3PO_4 is added which forms a stable complex with ferric ions $[Fe(PO_4)_2]^{3-}$ and prevent the oxidation of the indicator before the end point.

2. Mohr's salt is the name given to Ferrous ammonium sulphate

Formula = $[FeSO_4 (NH_4)_2 SO_4 \cdot 6H_2O]$
Molar mass = 392.10 (hydrated) 284 (anhydrated)
Eq. wt = 392.10 (hydrated) 284 (anhydrated)

3. Dil H_2SO_4 is always added while preparing the standard solution of Mohr's salt?

Ans. *To prevent the hydrolysis of Mohr's salt dil H_2SO_4 is added. $FeSO_4$ in Mohr's salt undergoes hydrolysis $FeSO_4 + 2H_2O \rightleftharpoons Fe(OH)_2 + H_2SO_4$.*

4. Why Mohr's salt is considered a primary standard?

Ans. *It is readily soluble in water, has high molecular mass, stable and is not readily oxidized by air.*

5. Why $K_2Cr_2O_7$ do not act as self indicator?

Ans. *A dilute solution of $K_2Cr_2O_7$ has a faint organic colour and its reduction product chromium(III) is green in colour. Thus a drop of $K_2Cr_2O_7$ in excess at the end point is not sufficient to give a distinct colour to the solution.*

6. What are the indicators used in $K_2Cr_2O_7$ titrations?

Ans. *Two types of indicators are used in $K_2Cr_2O_7$ titrations:*

(a) **External Indicator.** *Potassium ferricyanide is used as an external indicator.*

(b) **Internal Indicators.** *Commonly used internal indicators are:*

Indicator	*Colour in reduced state*	*Colour in oxidized state*
• *N Phenylanthranilic acid*	*green*	*violet-red*
• *Diphenylamine in conc. H_2SO_4*	*colourless*	*violet*
• *Diphenyl benzidine in conc. H_2SO_4*	*colourless*	*violet*
• *Ferroin in conc. H_2SO_4*	*Red*	*faint blue*

7. Why phosphoric acid is added when diphenylamine is used as an indicator?

Ans. *During titration when about only 50% of Fe^{2+} ions are oxidized to Fe^{3+} ions, the high conc. of Fe^{3+} may also cause oxidation of diphenylamine, which may indicate a wrong end point. In order to prevent this H_3PO_4 is added which forms a complex with ferric ions $[Fe(PO_4)_2]^{3-}$ and prevent the oxidation of indicator before the end point.*

8. What is the structure of diphenylamine.

Ans.

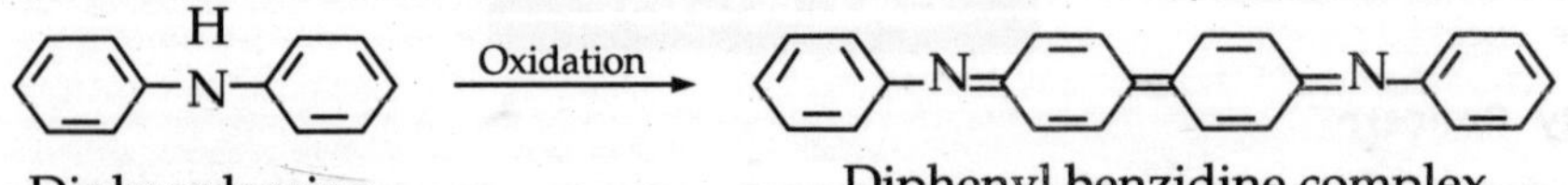

Diphenylamine → Diphenyl benzidine complex (Blue)

Information about chemicals used

Potassium dichromate

A bright yellowish-red crystalline compound, $K_2Cr_2O_7$, is used as an oxidizing agent, and in pyrotechnics, explosives, and safety matches.

Uses

1. As a poweful oxidizing agent, it is the preferred compound for cleaning laboratory glassware of any possible organics.
2. Potassium dichromate also has important uses in photography and in photographic screen printing. In both uses it is used an oxidising agent together with a strong mineral acid. Cr^6 ions have the property of tanning animal proteins when exposed to strong light. This quality is used in photographic screen printing. In screen printing a fine screen of bolting silk of similar material has ink squeezed through it onto paper or cloth

underneath. If a mask or design is placed on the screen so that ink cannot pass, the desing is transferred onto the material underneath. Many artistic prints, T-shirts etc. are printed in this manner.

3. *Chromium intensification* uses potassium dichromate together with equal parts of concentrated *hydrochloric acid* diluted down to approximately 10% vol to treat weak and thin negatives of black and white film. This solution reconverts the elemental silver particles in the film to silver chloride. After thorough washing and exposure to light, the film can be redeveloped to its end-point yielding a stronger negative which is able to produce a more satisfactory print.

Safety concern

- Potassium dichromate is toxic if swallowed, inhaled or absorbed through the skin. It is corrosive and may produce severe eye damage.
- Potassium dichromate is one of the most common culprits in causing chromum dermatitis.
- More significantly chromium is highly likely to induce sensistisation leading to chronic and dificult to treat dermatitis, especially of the hand and fore-arms.
- Hexavalent chromium compounds are generally more toxic than trivalent chromium compounds. Contain chromium(VI), a known cancer hazard.
- All Cr^6 products should be handled with gloves and appropriate health and safety protection.
- Wear Safety glasses, gloves, good ventilation. Treat as a carcinogen.
- Work in a well ventilated area, preferably using a fume cupboard. Ensure that any spills are cleaned up without delay.

Chromic acid is, in any case, a very dangerous material and should not be used unless it is absolutely necessary.

MOHR'S SALT

A ferrous-ammonium sulphate, $FeSO_4(NH_4)_2SO_4 . 6H_2O$ is a light green crystalline salt. It is a double salt and is used as a standard substance for volumetric analysis.

Uses

Used as a route to iron(II) compounds for use in analytical chemistry for quantitative analysis of oxidants. The method is to treat the sample with an excess of iron(II) followed by titration of the excess using a standard solution of potassium dichromate or cerium(IV).

Safety Concern

May cause gastrointestinal irritation with nausea, vomiting and diarrehea. May cause systemic toxic efects on the heart, liver, and kidneys.The toxicological properties of this substance have not been fully investigated. Prolonged or repeated skin contact may cause dermatitis.

If the solution is ingested, give 2 - 4 cupfuls of milk or water to the victim. Get medical aid immediately. Flush skin with plenty of soap and water in case of skin contact.

PHOSPHORIC ACID

Phosphoric acid, also known as **orthophosphoric acid** or **phosphoric(V) acid**, is a weak *mineral acid* with the chemical formula H_3PO_4.

It is usually available as a greenish liquid, suitable for dipping (acid bath), but is more generally used as a component in a gel, commonly called *Naal jelly*.

Phosphoric acid is prepared by adding *sulphuric acid* to *calcium phosphate* rock.

Uses

- Phosphoric acid may be used by direct application to rusted iron or steel tools or surfaces to convert *iron(III) oxide (rust)* to a water soluble *phosphate* compound.

- Another specialized use is the provision of local anesthesia. Phosphoric acid is used in *dentistry* and *orthodontics* as an *etching* solution, to clean and roughen the surfaces of teeth where dental appliances or fillings will be placed.
- It is also used to acidify foods and beverages such as avarious *colas*, but not without controversy as to its health effects. It provides a tangy taste, and being an agro-industrial chemical, is available cheaply and in large quantities.

Safety Concern

Corrosive, causes severe irritation and burns to every area of contact. Harmful if swallowed or inhaled.

Phosphoric acid, used in many soft drinks, has become suspect in the increase of the *bone* deficiency disease *osteoporosis* in young women.

Keep in a tightly closed container. Since it is corrosive to mild steel, store in rubber lined or stainless steel designed for phosphoric acid. Do not wash out container and use it for other purposes. When diluting, the acid should always be added slowly to water and in small amounts. Never use hot water and never add water to the acid. Conainers of this material may be hazardous when empty since they retain product residues (vapors, liquid).

IIC IODINE TITRATIONS (IODOMETRY AND IODIMETRY)

Iodine is a mild oxidizing agent and is used in volumetric estimations. Iodine acts as an oxidizing agent in neutral aqueous medium according to the reaction,

$$I_2 + 2e^- \rightarrow 2I^-$$

On the other hand an oxidizing agent can oxidize I_2^- ions to I_2. Both these reactions are used in iodine titrations.

Some of the important points regarding iodine titrations are:

1. *Iodine is a much weaker oxidizing agent than potassium permanganate and dichromate. It is used to estimate the amount of various reducing agents like sodium thiosulphate, sodium sulphite, sodium arsenite and antimony salts in their solutions.*
2. *Iodine can also be employed to estimate the strength of oxidizing agents as well.*
3. *Hence, Iodine Titrations are of two types:*

   (a) ***Iodimetry.*** *All such titrations which involve the direct titration of iodine with a reducing agent are grouped under iodimetry. Iodimetry is employed*

* **Note.** The starch indicator is not added in the beginning of the titration but towards the end of the titration. This is because the concentration of iodine is very high in the beginning and it is strongly adsorbed by colloidal starch particles. It is not released even at the end point i.e. at the end point, though the solution becomes colourless but the blue coloured starch flocks make the detection of end point difficult.

The sodium thiosulphate (or the reducing agent) is added to the reaction mixture (which is brown in colour due to free iodine). When this I_2 is used up, the brown colour fades to straw (pale yellow) colour, indicating that only a small amount of iodine is left. Then starch solution is added near the end. The solution acquires the blue colour and the titration is continued till the solution just becomes colourless.

to determine the strength of reducing agents such as sodium thiosulphate, sulphite, arsenite, stannous and ferrous ions in neutral or moderately acidic solutions.

(b) ***Iodometry:*** *In these titrations, iodine is liberated as a result of chemical reaction involving oxidation of an iodide ion by a strong oxidizing agent in neutral or acidic medium. The liberated iodine is then titrated with a reducing agent. Iodometry is used to estimate the strength of oxidizing agents. For example, the estimation of* Cu^{2+} *with thiosulphate–*

$$2Cu^{2+} + 4I^- \rightarrow Cu_2I_2 + 2I_2$$

$$I_2 + 2S_2O_3^{2-} \rightarrow S_4O_6^{2-} + 2I^-$$

4. ***Indicator:*** *The indicator used in these types of titrations is starch solution* which is used as an internal indicator. Starch forms a dark blue complex with triiodate ions. The blue colour disappears when there is no more of free iodine.*

IODOMETRIC TITRATIONS

As is discussed earlier the estimation of oxidizing agents can be done by iodometric titrations.

Following points must be kept in mind during iodometry.

1. *Since iodine is volatile, titrations should be performed in cold. Moreover, the indicator starch also loses its sensitivity at high temperatures.*
2. *Iodometry titrations cannot be performed in strongly alkaline solutions. This is because iodine reacts with alkalies to give hypoiodite which is a strong oxidizing agent and may oxidize thiosulphate ions.*

$$I_2 + 2NaOH \rightarrow NaIO + NaI + H_2O$$

or $$I_2 + 2\overline{O}H \rightarrow I\overline{O} + I^- + H_2O$$

$$S_2O_3^{2-} + 4I\overline{O} + 2\overline{O}H \rightarrow 4I^- + 2SO_4^{2-} + H_2O$$

3. *Always use excess of KI: This is because the solubility of* I_2 *in water is very low. But in the presence of KI it forms an unstable complex salt,* KI_3*. This soluble complex will release iodine slowly during titration due to reversible reaction,*

$$KI + I_2 \rightleftharpoons KI_3$$

$$I^- + I_2 \rightleftharpoons I_3^-$$

When free iodine is used up in the reaction with thiosulphate, the equilibrium shifts to the left releasing more I_2.

4. *Since the rate of reaction between I^- ion and the oxidizing agent is low, therefore the solution is allowed to stand for 2 – 3 minutes before titrating.*
5. *The reaction mixture is kept in the dark before titrating. This is due to the reason that light accelerates the oxidation of iodide ions to iodine by atmospheric oxygen*

$$4I^- + 4H^+ + O_2 \rightarrow 2I_2 + 2H_2O$$

EXPERIMENT NO. 1.8

AIM **To determine the strength in g/L of a given $K_2Cr_2O_7$ (potassium dichromate) solution, provided approximately N/10 sodium thiosulphate and N/10 $K_2Cr_2O_7$.**

APPARATUS Burette, pipette, 2 beakers, titration flask, stand

THEORY When $K_2Cr_2O_7$ reacts with KI, in the presence of dil. acids, the dichromate is reduced to a green chromic salt and liberates an equivalent amount of I_2, which is then titrated against hypo solution using starch as indicator. At end point the blue colour of iodo-starch complex suddenly disappears and pure green solution due to chromic salt is left.

$$K_2Cr_2O_7 + 7H_2SO_4^- + 6KI \rightarrow 4K_2SO_4 + Cr_2\ (SO_4)_3 + 7H_2O + 3I_2$$

$$2NaS_2O_3 + I_2 \rightarrow Na_2S_4O_6 + 2NaI$$

It involves double titration. In the first titration the exact normality of $Na_2S_2O_3$ is calculated by titrating against given standard solution of $K_2Cr_2O_7$. Then this $Na_2S_2O_3$ solution of known normality is used in the second titration for calculating the normality of the given sample of $K_2Cr_2O_7$ of unknown strength.

Small amount of sodium bicarbonate (solid) is added to the acidified solution before the titration. This is to prevent aerial oxidation of excess of iodide ions which is prone in presence of Cr^{3+} ions. Moreover CO_2 evolved from $NaHCO_3$ on reacting with acid keeps the atmospheric oxygen off the solution.

INDICATOR Starch

END POINT Blue to Green

PROCEDURE

1. Rinse the burette with $Na_2S_2O_3$ solution. Take $Na_2S_2O_3$ in the burette and pipette out 10ml of standard solution of $K_2Cr_2O_7$ in the titration flask.
2. Add 5ml. of dil. H_2SO_4 to it. Then add 1gm. of KI and leave in dark for 2–5 minutes.

3. The solution in the titration flask will now become dark brown . Add a small amount of solid $NaHCO_3$ and titrate the solution against $Na_2S_2O_3$ till a pale brown colour is obtained.
4. Add 2ml. of starch solution. The solution in the conical flask turns blue. Now titrate it further to get end point.
5. Repeat and get 3 concordant readings.
6. Repeat the above procedure with the given sample of $K_2Cr_2O_7$. Since now the normality of the hypo solution is exactly known and hence the strength of given $K_2Cr_2O_7$ can be found.

OBSERVATIONS AND CALCULATIONS

(For finding Normality of $Na_2S_2O_3$)

S. No.	*Volume of the solution taken in the titration flask (ml)*	*Burette Readings*		*Volume of the $Na_2S_2O_3$ used (Final-Initial reading) (ml)*
		Initial Reading	*Final Reading*	
1.				
2.				
3.				

$$\text{Standard } K_2Cr_2O_7 \qquad Na_2S_2O_3$$

$$N_1V_1 = N_2V_2$$

$$\frac{N}{10} \times 10 = N_2 \times V_2$$

$$N_2 = \frac{N}{10} \times \frac{10}{V_2}$$

$\therefore$ Normality of $Na_2S_2O_3 = N_2$

OBSERVATIONS AND CALCULATIONS

(*For finding Normality and Strength of given $K_2Cr_2O_7$*)

S. No.	*Volume of the solution taken in the titration flask (ml)*	*Burette Readings*		*Volume of the titrant used (Final-Initial reading) (ml)*
		Initial Reading	*Final Reading*	
1.				
2.				
3.				

Given Solution of $K_2Cr_2O_7$ $Na_2S_2O_3$

$$N_1V_1 = N_2V_2$$

$$N_1 \times 10 = N_2 \times V_2$$

$$N_1 = \frac{N_2V_2}{10}$$

Strength of the given $K_2Cr_2O_7$

$$= \frac{N_2V_2}{10} \times \text{Eq. Wt. of } K_2Cr_2O_7$$

$$= \frac{N_2V_2}{10} \times 49 \text{ g/L}$$

RESULT The strength of given $K_2Cr_2O_7$ solution is ... g/L

PRECAUTIONS Same as in previous experiment.

Instructions to the Lab Assistant

Provide the following.

1. 0.1 $Na_2S_2O_3$ solution: 0.1 N hypo solution ($Na_2S_2O_3$ solution) Since equivalent weight of $Na_2S_2O_3 \cdot 5H_2O$ is 248.18 g therefore dissolve 24.818 g of $Na_2S_2O_3 \cdot 5H_2O$ in distilled water and make up the volume to 1000ml.
2. 0.1 N $K_2Cr_2O_7$ solution (known): Dissolve 4.9 g of $K_2Cr_2O_7$ in distilled water and make up the volume to one litre.
3. Unknown $K_2Cr_2O_7$ solution: Dissolve about 5g of $K_2Cr_2O_7$ in distilled water and make up the volume to one litre.
4. Starch solution
5. $NaHCO_3$ (solid)
6. KI (solid).

EXPERIMENT NO. 1.9

AIM **Determine the strength of Cu in the copper ore solution provided hypo solution (0.1N)**

APPARATUS Burette, pipette, 2 beakers, titration flask, stand

THEORY This experiment comes under the category of iodometry which is used widely in the analysis of ores, alloys etc.

When an excess of KI is added to the solution containing Cu^{2+} in neutral or slightly acidic medium, quantitative liberation of iodine takes place.

$$2CuSO_4 \cdot 5H_2O + 4KI \rightarrow 2CuI_2 + 2K_2SO_4 + 5H_2O$$

$$2CuI_2 \rightarrow Cu_2I_2 + I_2$$

This liberated iodine is then titrated against standard $Na_2S_2O_3$ solution using starch solution as indicator near the end point.

$$2Na_2S_2O_3 + I_2 \rightarrow \underset{\text{Sodium tetrathionate}}{Na_2S_4O_6} + 2NaI$$

Ionic equation will be:

$$2Cu^{2+} + 2I^- \rightarrow Cu_2^{2+} + I_2 + 2e^-$$

$$I_2 + 2S_2O_3^{2-} + 2e^- \rightarrow S_4O_6^{2-} + 2I^-$$

From the above equation it is evident,

$$2Cu^{2+} \equiv I_2 \equiv 2S_2O_3^{2-} \equiv 2e^-$$

The equivalent weight of Cu^{2+} will be one half of twice the molecular weight since the reaction involves two electrons per two moles of Cu^{2+}

$$\text{Eq. weight of } Cu^{2+} = \frac{2 \times 63.5}{2} = 63.5.$$

From the normality equation, normality of Cu^{2+} and hence the strength of Cu^{2+} in hydrated $CuSO_4$ crystals is determined.

The titration fails when any mineral acid is present in the solution and therefore before commencing the titration the acid should be neutralized. This is done by dropwise addition of a solution of Na_2CO_3 until a slight precipitate or turbidity appears

$$2H^+ + CO_3^{2-} \rightarrow CO_2 + H_2O$$

$$Cu^{2+} + CO_3^{2-} \rightarrow \underset{\text{(turbidity)}}{CuCO_3}$$

The turbidity is removed by dropwise addition of dilute CH_3COOH.

$$CuCO_3 + 2CH_3COOH \rightarrow Cu^{2+} + CO_2 + 2CH_3COO^- + H_2O$$

Moreover, the precipitate of Cu_2I_2 absorbs I_2 from the solution and releases it slowly making the detection of sharp end point difficult so a small amount

---

Note. The equivalent weight of $K_2Cr_2O_7 = \frac{\text{Molecular weight}}{6}$ since the reduction of dichromate to the chromium (III) ions involves a change of six electrons.

of NH_4SCN is added near the end point to displace the adsorbed iodine from Cu_2I_2 precipitate.

INDICATOR Starch Solution

END POINT Blue to colourless

PROCEDURE

1. Rinse and fill the burette with 0.1 N $Na_2S_2O_3$ solution.
2. Measure 20ml. of the $CuSO_4$ solution into a conical flask.
3. Neutralize any free acid present by adding Na_2CO_3 solution drop by drop till a faint permanent precipitate remains even on shaking.
4. Add dil. acetic acid drop wise until the precipitate just dissolves.
5. Add 1 gm. of solid KI, cover the mouth of conical flask by filter paper and allow the mixture to stand for 2–5 minutes in the dark. The solution becomes brown due to liberated iodine.
6. Now titrate the liberated iodine with hypo solution added from the burette, the brown colour of iodine becomes fainter as the addition by hypo solution proceeds. When only a faint yellow colour remains add 2ml. of starch solution and 1g of NH_4SCN. This immediately forms a deep blue iodo-starch complex.
7. Now add hypo solution further, drop by drop, when only milky white colour remains.

OBSERVATIONS

S. No.	*Volume of the solution taken in the titration flask (ml)*	*Burette Readings*		*Volume of the titrant used (Final-Initial reading) (ml)*
		Initial Reading	*Final Reading*	
1.				
2.				
3.				

CALCULATIONS

$$(CuSO_4) \quad (Na_2S_2O_3)$$

$$N_1V_1 = N_2V_2$$

$$N_1 \times 20 = 0.1 \times V_2$$

$$N_1 = \frac{0.1 \times V_2}{20}$$

$$N_1(Cu^{2+}) = \frac{V_2}{200}$$

$$\therefore \text{ Strength of } Cu^{2+} = \frac{V_2}{200} \times 63.5 g/L$$

$$\textit{Percentage of } Cu^{2+} \; CuSO_4 \textit{ Crystals} = \frac{V_2}{200} \times \frac{63.5}{\text{St. of } CuSO_4 \text{ crystals}} \times 100$$

RESULT The amount of Cu present in copper sulphate solution is...

PRECAUTIONS

1. The solution of $Na_2S_2O_3$ is always taken in the burette in iodometric titrations.
2. Sufficient amount of KI solution is to be added for the reason explained earlier.
3. The indicator should be added just before the end point.
4. NH_4SCN should be added near the end point to displace the adsorbed iodine and to get a sharp end point.

Instructions For Lab Assistant

Provide the following

1. 0.1 N hypo solution ($Na_2S_2O_3$ solution) Since equivalent weight of $Na_2S_2O_3 \cdot 5H_2O$ is 248.18 g therefore dissolve 24.818 g of $Na_2S_2O_3 \cdot 5H_2O$ in distilled water and make up the volume to 1000ml.
2. $CuSO_4 \cdot 5H_2O$ solution: Weigh accurately x g of $CuSO_4 \cdot 5H_2O$ crystals and dissolve it in water, add 1–2 ml of Conc. H_2SO_4 or glacial acetic acid, shake well and make up the volume to 1000ml.
3. Starch indicator: Prepare a paste of 1g of starch with 10ml of distilled water and add it gradually to 100 ml of boiling water with constant stirring. Boil and cool it.
4. Dilute Acetic acid (2N)

Notes 1. Sodium thiosulphate crystals are efflorescent, i.e. lose their water of crystallisation on keeping in air and decompose due to bacterial action in solution. Thus it is not a primary standard. The equivalent weight of $Na_2S_2O_3$ in iodometry will be $\frac{2 \times \text{Molecular weight}}{2}$ since the reaction involves 2 electrons per two moles of $Na_2S_2O_3$.

2. $CuSO_4 \cdot 5H_2O$ is a primary standard. Its equivalent weight in above titration $= \frac{\text{Molecular weight} \times 2}{2}$.

5. Na_2CO_3 solution: Dissolve 5.3g of Na_2CO_3 in boiled out distilled water and make up the volume to one litre.
6. KI (solid)
7. NH_4CNS (solid)

EXERCISE

1. What are iodine titrations?

Ans. *The titrations involving the use of iodine are termed as iodine titrations. Iodine is a mild oxidising agent and is used in volumetric estimations.*

2. What type of substances are estimated using iodine titrations?

Ans. *Iodine is a much weaker oxidizing agent than potassium permanganate and dichromate. It is used to estimate the amount of various reducing agents like sodium thiosulphate, sodium sulphite, sodium arsenite and antimony salts in their solutions.*

Iodine can also be employed to estimate the strength of oxidizing agents as well.

3. What is the difference between iodimetery and iodometery?

Ans. *Iodine Titrations are of two types:*

(a) ***Iodimetry.*** *All such titrations which involve the direct titration of iodine with a reducing agent are grouped under iodimetry. Iodimetry is employed to determine the strength of reducing agents such as sodium thiosulphate, sulphite, arsenite, stannous and ferrous ions in neutral or moderately acidic solutions.*

(b) ***Iodometry:*** *In these titrations, iodine is liberated as a result of chemical reaction involving oxidation of an iodide ion by a strong oxidizing agent in neutral or acidic medium. The liberated iodine is then titrated with a reducing agent. Iodometry is used to estimate the strength of oxidizing agents. For example, the estimation of* Cu^{2+} *with thiosulphate–*

$$2Cu^{2+} + 4I^- \rightarrow Cu_2I_2 + 2I_2$$

$$I_2 + 2S_2O_3^{2-} \rightarrow S_4O_6^{2-} + 2I^-$$

4. What type of indicator is used in these type of titrations?

Ans. ***Indicator:*** *The indicator used in these types of titrations is starch solution which is used as an internal indicator. Starch forms a dark blue complex with triiodate ions. The blue colour disappears when there is no more of free iodine.*

5. Why iodine titrations are performed in cold?

Ans. *Since iodine is volatile, titrations should be performed in cold. Moreover, the indicator starch also loses its sensitivity at high temperatures.*

6. Iodometers should not be performed in strongly alkaline solutions?

Ans. *Iodometry titrations cannot be performed in strongly alkaline solutions. This is because iodine reacts with alkalies to give hypoiodite which is a strong oxidizing agent and may oxidize thiosulphate ions.*

$$I_2 + 2NaOH \rightarrow NaIO + NaI + H_2O$$

or $$I_2 + 2\bar{O}H \rightarrow I\bar{O} + I^- + H_2O$$

$$S_2O_3^{2-} + 4I\bar{O} + 2\bar{O}H \rightarrow 4I^- + 2SO_4^{2-} + H_2O$$

7. Why always excess of KI is used?

Ans. *This is because the solubility of I_2 in water is very low. But in the presence of KI it forms an unstable complex salt, KI_3. This soluble complex will release iodine slowly during titration due to reversible reaction,*

$$KI + I_2 \rightleftharpoons KI_3$$

$$I^- + I_2 \rightleftharpoons I_3^-$$

When free iodine is used up in the reaction with thiosulphate, the equilibrium shifts to the left releasing more I_2.

8. Why the solution is allowed to stand for 2 - 3 minutes before the titration?

Ans. *Since the rate of reaction between I^- ion and the oxidizing agent is low, therefore the solution is allowed to stand for 2 – 3 minutes before titrating.*

9. Why the reaction mixture is kept in dark?

Ans. *The reaction mixture is kept in the dark before titrating. This is due to the reason that light accelerates the oxidation of iodide ions to iodine by atmospheric oxygen*

$$4I^- + 4H^+ + O_2 \rightarrow 2I_2 + 2H_2O$$

10. Why iodine is not used as self indicator, even though it is strongly coloured?

Ans. *It is possiblke to use iodine solution as indicator but its brown colour becomes very faint towards end point and hence is difficult to detect. It is better to use a little starch, which gives intense bluecolour with iodine. The colour disappears when the last traces of iodine has been reduced.*

11. What is the advantage of using starch as indicator in iodine titrations?

Ans. *Starch reacts with iodine in the presence of iodide ion to form an intensely colored blue complex, which is visible at very low concentrations of iodine, making it a very good indicator in both direct and indirect iodometric titrations.*

12. The iodine titrations should be performed as pH below 8?

Ans. *The blue color produced is also destroyed by alkali and heat, so titrations involving Starch indicator should be conducted at a pH below 8 and at temperatures near room temperature (this will also minimize loss of volatile iodine from the titration vessel).*

13. Why starch should also not be used in a strongly acid medium.

Ans. *Starch should also not be used in a strongly acid medium because hydrolysis of the starch occurs.*

14. Why starch gives bluecolour with iodine in presence of iodide?

Ans. *This is due to the formation of Starch iodine complex.*

Starch can be separated into two fractions - amylose and amylopectin.Amylose in starch is responsible for the formation of a deep blue color in the presence of iodine. The iodine molecule slips inside of the amylose coil to form a blue coloured complex.

Only iodine element in the presence of iodide ion will give the characteristic blue black colour. Neither iodine element alone nor iodide ions alone will give the colour result.

15. Why starch is added near the endpoint of an iodine titration and not in the beginning?

Ans. *This is because the concentration of iodine is very high in the beginning and it is strongly adsorbed by colloidal starch particles. It is not released even at the end point thus making the detection difficult. The additioin of the starch solution is thus delayed till the reaction is nearly complete (near the end point) as shown by a colour change* *of solution from deep red-brown to pale yellow. Moreover, starch tends to decompose at high concentrations of iodine.*

16. Why only freshly prepared starch solution should be used?

Ans. *Aqueous starch suspensions decompose within a few days, primarily because of bacterial action. The decomposition product is glucose, a reducing agent. This may interfere with a redox titration.*

17. What is the substitute for starch as an indicator in iodometric titrations.

Ans. *Sodium starch glycollate.*

18. Why sodium thiosulphate solution is not considered a primary standard. What is its equivalent weight in iodometric titration?

Ans. *Sodium thiosulphate crystals are efflorescent, i.e. lose their water of crystallisation on keeping in air and decompose due to bacterial action in solution. This it is not a primary standard. The equivalent weight of $Na_2S_2O_3$ in iodometry will be* $\frac{2 \times \text{Molecular weight}}{2}$ *since the reaction involves 2 electrons per two moles of $Na_2S_2O_3$.*

19. Why is it essential to use a neutral solution of copper sulphate in iodometric titration?

Ans. *Iodometic estimation of $CuSO_4$ involves the precipitation of Cu_2I_2 and liberation of I_2 on adding KI. Mineral acids dissolve Cu_2I_2 and therefore estimation of I_2 is not possible.*

20. Why a small amount of NH_4SCN is added near the end point?

Ans. *The Cu_2I_2 absorb I_2 and releases it slowly. The NH_4SCN is added near the end point to displace the absorbed iodine from Cu_2I_2 precipitate.*

21. What is the equivalent weight of $CuSO_4 . 5H_2O$ in above titration?

Ans. *Equivalent weight in above titration*

$$= \frac{\text{Molecular weight}}{2} \times 2 = \text{Molecular wt.} = 63.5$$

Information about chemicals used

Sodium Thiosulphate

Sodium thiosulphate ($Na_2S_2O_3$) is a colour less *crystalline* compound that is more familiar as the pentahydrate, $Na_2S_2O_3 . 5H_2O$, an *efflorescent, monoclinic* crystalline substance, also called sodium hyposulphite or "hypo".

Sodium thiosulphate has a tendency to cake on storage due to the presence of surface moisture.

Synthesis

Sodium thiosulphate is produced chiefly from liquid waste products of *sodium sulphide* or suphur dye manufacture. It is also produced from *sodium carbonate* and *sulphur dioxide.*

Uses

1. Because sodium thiosulphate dissolves *silver* salts, its major use is in *photography* for developing *film.* (As a fixer)
2. It is also used in the *tanning* of *leather,* in chemical manufacture in *bacteriological* water assessment, and in treating *cyanide* poisoning.
3. As an aid in the bending of tubes.

Safety Concern

1. This material may be harmful if swallowed. It is moderately toxic when ingested.
2. It may also irritate the eyes or lungs. Flush the eye with plenty of water.

III. PRECIPITATION TITRATIONS (ARGENTOMETRIC TITRATIONS)

The titrations accompanied by the formation of sparingly soluble salts, when the solutions of two reacting substances are brought into contact with each other, are called precipitation titrations. For example, when a solution of silver nitrate is added to a solution of sodium chloride, a white precipitate of silver chloride is formed.

$$Ag\ NO_3 + NaCl \rightarrow AgCl \downarrow + NaNO_3$$

Thus the complete precipitation of AgCl marks the end point of the reaction.

Some of the important points regarding the precipitation titrations are

1. *These reactions are limited to those involving $AgNO_3$ solution and hence are called Argentometric Titrations.*
2. *In these type of titrations, solubility product pays a key role. The precipitation occurs only when the actual ionic product exceeds the solubility product.*
3. *The application of volumetric analysis by precipitation requires that*
   - *The precipitate must be practically insoluble*
   - *Formation of precipitate should be rapid.*
   - *Equivalence point should be easily detectable.*
4. ***Indicators in precipitation titrations:*** *The use of indicators in precipitation is based on the formation of coloured precipitates near the end point.* ***There are two types of Argentometric titrations depending upon the indicator used:***

   (a) ***Mohr's Method.*** *This method was developed by Mohr to estimate Chloride. This method involves the titration of $AgNO_3$ against halides in neutral solution using potassium chromate, K_2CrO_4 as indicator. The end point is marked by the appearance of a brick red precipitate or colouration, due to the formation of silver chromate with excess of $AgNO_3$*

   $$Ag^+ + Cl^- \rightarrow \underset{\text{(white ppt)}}{AgCl \downarrow}$$

   $$2Ag+ + CrO_4^{2-} \rightarrow \underset{\text{(Brick red)}}{Ag_2CrO_4}$$

   When AgCl is titrated with $AgNO_3$ using K_2CrO_4 as indicator, white precipitate of AgCl is formed. When all the Cl ions are precipitated, the extra drop of $AgNO_3$ reacts with K_2CrO_4 to form brick red precipitate of Ag_2CrO_4, which marks the end point. The precipitate of Ag_2CrO_4 is produced only after the complete precipitation of Cl^- as AgCl. This is due to *the difference in solubility product of AgCl and Ag_2CrO_4. The solubility product of AgCl is reached earlier at a lower concentration of Ag^+ ions than that of Ag_2CrO_4. Consequently AgCl is precipitated first. At equivalence point Ag_2CrO_4 begins to precipitate and the solution acquires a brick red colour.*

Limitations of Mohr's method

(i) Ag_2CrO_4 is soluble in acid and hence the solution to be titrated should be neutral . If it is acidic, it is first neutralized by adding $CaCO_3$ and then indicator is added.

(ii) If the solution is alkaline, $AgNO_3$ is precipitated as AgOH. In that case, alkali is neutralized by dil. HNO_3. The excess acid is then destroyed by $CaCO_3$.

(ii) This method is not suitable for the determination of iodides because colour of the indicator is masked by that of precipitate and end point is not easily detectable.

(iv) The titration must be carried out at a pH between 7 to 10.

(b) **Volhard's Method.** *This method was developed by Volhard for the estimation of silver in dilute nitric acid solution by titrating against a standard thiocyanate solution in presence of ferric salt as indicator.*

$$Ag^+ + SCN^- \rightarrow Ag\,SCN$$

$$Fe^{3+} + SCN^- \rightarrow \underset{Red}{Fe\,(SCN)_3}$$

When a standard solution of ammonium or potassium thiocyanate is titrated with solution of $AgNO_3$ in presence of nitric acid and indicator a precipitate, of AgSCN is formed. When silver has been completely precipitated, the addition of further drop of thiocyanate reacts with the ferric ion to form red-coloured ferric thiocyanate complex.

The concentration of nitric acid should be from 0.5 to 1.5N. This method can also be used for the estimation of bromides in which case the colour at the end point is stable.

Limitations of Volhard's Method

1. *This method cannot be used if the solution itself is coloured e.g. cobalt, nickel or copper halides.*
2. *The presence of peptizing agents also interferes in determination of end point because of the increase in surface area of the precipitate.*
3. *The solution must be free from oxidizing agents which might oxidize CNS^- ions.*
4. *In case of estimations of chloride by Volhard's method, the end point disappears on stirring and hence this method is not suitable.*

EXPERIMENT NO. 1.10

AIM **Determine the strength in grams per litre of a given $AgNO_3$ solution being provided N/30 NaCl solution by Mohr's Method**

APPARATUS Burette, pipette, 2 beakers, titration flask, stand.

THEORY When silver nitrate solution is added to a solution of sodium chloride containing a few drops of potassium chromate, white silver chloride is precipitated initially. As soon as all the chloride ions have been precipitated out, even a drop of silver nitrate added in excess gives a red precipitate of silver chromate. This indicates the end point.

The solubility product of silver chloride is lower than that of silver chromate. Hence as long as the chloride ions are available the less soluble silver chloride is precipitated. As soon as the chloride ions have been precipitated out, even a slight excess of Ag^+ produces insoluble silver chromate which is red in colour.

$$AgNO_3 + NaCl \rightarrow AgCl \downarrow + NaNO_3$$

$$2Ag^+ + CrO_4^{2-} \rightarrow \underset{\text{Brick Red Precipitate}}{Ag_2CrO_4}$$

PROCEDURE

1. Rinse and fill the burette with given $AgNO_3$ solution
2. Pipette out 10ml. of NaCl solution in the titration flask.
3. Add 0.5 ml of K_2CrO_4 indicator.
4. Add $AgNO_3$ from the burette, shaking the flask constantly. A white precipitate of AgCl is obtained. After the addition of a few ml of $AgNO_3$ a red colour appears in the flask but disappears quickly upon shaking.
5. Continue the addition drop by drop till a permanent reddish brown colour is obtained. Take three concordant readings.

OBSERVATIONS

S. No.	*Volume of the solution taken in the titration flask (ml)*	*Burette Readings*		*Volume of the $AgNO_3$ used (Final-Initial reading) (ml)*
		Initial Reading	*Final Reading*	
1.				
2.				
3.				

CALCULATIONS

$$AgNO_3 \quad aCl$$

$$N_1V_1 = N_2V_2$$

$$N_2 = \frac{1}{30} \times \frac{10}{V_1}$$

Strength of the given $AgNO_3$ solution

$$= N_2 \times \text{Eq. Wt.}$$

$$= N_2 \times 169.9\text{g/L}$$

RESULT The strength of a given $AgNO_3$ solution... g/L

PRECAUTIONS

1. Store $AgNO_3$ in a coloured bottle.
2. The whole apparatus must be washed with distilled water.
3. Same amount of indicator must be added each time.
4. The reaction mixture should be briskly shaken during the titration.

Instructions to the Lab Assistant

1. N/30 NaCl solution: A. R. NaCl is pure and very little hygroscopic. (Equivalent weight = 58.45) 1.95 gm of NaCl is dissolved in distilled water and the solution is made to 1L.
2. Approx. N/30 $AgNO_3$ solution: (Equivalent weight = 169.9). Dissolve 5.663 gm of $AgNO_3$ in distilled water and make up the volume to 1L.
3. K_2CrO_4 solution (indicator): Dissolve 5gm A.R K_2CrO_4 in 100ml of water.

EXPERIMENT NO. 1.11

AIM **Determine volumetrically the strength of a given ammonium thiocyanate solution by Volhard's method, provided N/30 silver nitrate solution.**

APPARATUS Burette, pipette, 2 beakers, titration flask, stand

THEORY In Volhard's method the indicator employed is a solution of *Ferric alum*. $[Fe_2(SO_4)_3 \cdot (NH_4)_2SO_4 \cdot 12H_2O]$. When thiocyanate solution is added to the silver nitrate solution in the presence of nitric acid containing some ferric alum, a white precipitate of silver thiocyanate is produced initially. But when all the Ag^+ ions have been precipitated out, even a slight excess of thiocyanate added further, gives a reddish-brown colouration due to the formation of ferric thiocyanate.

$$Ag^+ + CNS \rightarrow \underset{\text{(White ppt.)}}{AgCNS}$$

$$Fe^{3+} + 3CNS \rightarrow \underset{\substack{\text{(Ferric Thiocyanate)} \\ \text{(Reddish Brown)}}}{Fe(CNS)_3}$$

PROCEDURE

1. Rinse and fill the burette with the given thiocyanate solution.
2. Pipette out 10ml. of N/30 $AgNO_3$ solution.
3. To it add 1 ml. of ferric alum indicator solution.
4. Titrate it against $AgNO_3$ solution while shaking the flask vigorously. A white precipitate of AgCNS will be formed.
5. Appearance of brownish red colour indicates the end point. Take 3 concordant readings.

OBSERVATIONS

S. No.	*Volume of the solution taken in the titration flask (ml)*	*Burette Readings*		*Volume of the NH_4CNS used (Final-Initial reading) (ml)*
		Initial Reading	*Final Reading*	
1.				
2.				
3.				

CALCULATIONS

$$\begin{aligned} NH_4CNS \quad & AgNO_3 \\ N_1V_1 &= N_2V_2 \\ N_1 \times V_1 &= \frac{1}{30} \times 10 \\ N_1 &= \frac{1}{30} \times \frac{10}{V_1} \end{aligned}$$

Strength of the given ammonium thiocyanate solution

$$= N_1 \times \text{Eq. Wt.} = \left(\frac{1}{30} \times \frac{10}{V_1}\right) \times 76.12 \text{ g/L}$$

RESULT The strength of the given ammonium thiocyanate solution = ...g/L

PRECAUTIONS Same as experiment number 8.

Instructions to the Lab Assistant

Provide the following

1. N/30 $AgNO_3$ solution: It is obtained in high degree of purity. It may be dried in oven at 250–350° C for one hour or so and cooled in a dessicator Dissolve 5.663 gm of $AgNO_3$ in distilled water and make the volume to 1L.
2. **Approx.** N/30 NH_4CNS solution: This is a deliquescent solid and hence is secondary standard. Dissolve 2.537g (Eq. wt = 76.124) of NH_4CNS in distilled water and make up the volume to 1L.
3. **Ferric alum** (indicator). Its 40% solution is prepared by dissolving 40g in 100ml of water containing a few drops of pure conc. HNO_3 (prepared by boiling conc. HNO_3 and water in the ratio of 4 : 1 until the solution becomes colourless)

EXERCISE

1. What are precipitation titrations?

Ans. *The titrations accompanied by the formation of sparingly soluble salts, when the solutions of two reacting substances are brought into contact with each other, are called precipitation titrations.*

2. Why precipitation titrations are called Argentometric titrations?

Ans. *These titrations are limited to those involving $AgNO_3$ and hence these are called Argentometric titrations.*

3. When will the precipitation take place?

Ans. When ionic product exceeds the solubility product, precipitation will take place.

4. What are the requisites for precipiation titrations?

Ans. *The requisites for precipitation titraitons are:*

- *The precipitate formed must be insoluble.*
- *Formation of precipitate should be rapid.*
- *Equivalence point should be easily detectable.*

5. What is the basis for an indicator for a precipitation titration?

Ans. *The basis for an indicator for a precipitation titration is the formation of coloured precipitate near the end point.*

6. Name few indicators used in precipitation titrations?

Ans. *The indicators used in precipitation titrations are: Potassium chromate, Ferric Alum and Fluorescien.*

7. Name the two methods used in Argentometery titrations?

Ans. *Mohr's method and Volhard's method.*

8. What is the difference between the two methods?

Ans. *Mohr's method is used to estimate chloride ions using chromate ion as indicator in neutral medium. In Volhard's method silver ions are titrated with a standard solution of thiocyanate ion in presence of ferric alum as indicator in an acidic medium.*

9. Why the titration is carried out in neutral medium in Mohr's method?

Ans. *The precipitates of Ag_2CrO_4 are soluble in acidic medium and if the solution is alkaline, $AgNO_3$ is precipitated as AgOH. Therefore, the titrations are carried out in neutral medium.*

10. In Volhard's method, why the titration is carried out in acidic medium?

Ans. *To prevent the precipitation of Fe^{3+} present in Ferric alum as hydrated oxide.*

11. Why Mohr's method is not suitable for the determination of iodides?

Ans. *Because colour of the end point is masked by that of precipitate and end point is not easily detectable.*

12. Why chlorides cannot be estimated using Volhard's method?

Ans. *In this case the end point disappears on stirring and hence this method is not suitable.*

13. What are the important applications of $AgNO_3$ titrations?

Ans. *These are used to estimate a number of ions such as Br^-, Cl^-, BH_4^-, etc.*

14. Sodium chloride is hygroscopic then how is it used as primary standard?

Ans. *Pure NaCl of A.R. quality is not hygroscopic. The ordinary common salt is hygroscopic due to the presence of $MgCl_2$.*

15. Why a standard solution of thiocyanate cannot be prepared by direct weighing.

Ans. *Because it is diliquescent in nature.*

Information about chemicals used

Silver nitrate

Silver nitrate, chemical compound, $AgNO_3$, a colourless crystalline material that is very soluble in water.

A poisonous, corrosive compound, silver nitrate crystals can be produced by dissolving silver in nitric acid and evaporating the solution.

Uses

1. The most important compound of silver, it is used in the preparation of silver salts for photography, this nitrate of silver is a light-sensitive ingredient in photographic film.
2. It is also used in chemical analysis, in silver plating, in inks and hair dyes, and to silver mirrors.
3. It is used in medicine in the treatment of eye infections and gonorrhea. Silver nitrate has been used as an antiseptic, dropped into newborn babies' eyes at birth.
4. Fused silver nitrate is also called lunar caustic. Taken internally silver nitrate is a poison.

Safely Concern

1. Long-term exposure can cause permanent blue-grey staining of eyes, mouth, throat and skin, (argyria) and may cause eye damage.
2. Skin contact with silver nitrate solid or solutions is likely to leave silver stains on the skin. These develop slowly over a period of hours and are initially brown, but darken gradually to black. Once the stains become apparent, they cannot usually be removed with soap and water, but gradually disappear as new skin grows.
3. Very destructive of mucous membranes.
4. Poisonous if swallowed or inhaled

   Wear safety glasses. Do not breathe dust. Do not allow solution or solid to come into contact with the skin. In case of contact with eyes, rinse immediately with plenty of water. In case of accident or if you feel unwell, seek medical advice immediately.

(IV) COMPLEXOMETRIC TITRATIONS

Complexometric titration is a generic term involving the titrations which are based on the formation of a soluble complex between a metal ion and a complexing agent. The metal ion is called the central atom and the complexing agent is called the ligand. Titrimetric methods based on complex formation have been used for more than a century.

Some important points regarding the complexometric titrations are

1. The ligands or the complexing agents can be unidentate, bidentate or polydentate having one, two and more than two donor atoms respectively which can form coordinate bond with the metal atom. The polydentate ligands are effective in forming stable complexes with a wide variety of metals because they form ring structures called chelates.

2. One of the remarkable ligand used widely in complexometric titrations is Ethylene diaminetetraacetic acid (EDTA) because it forms stable complexes with number of metal atoms. EDTA has the structure.

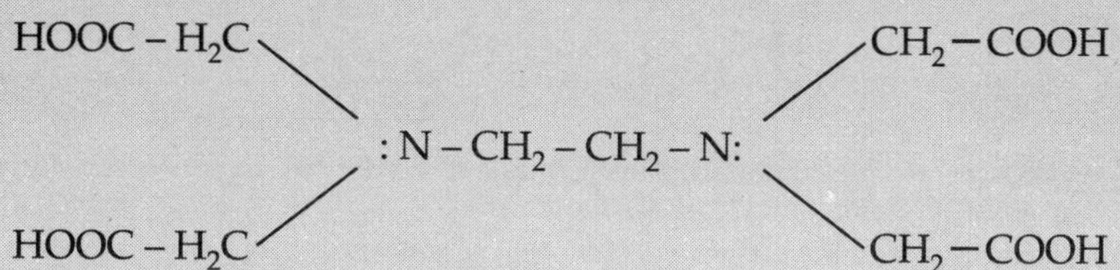

It has six potential sites (hexadentate) for bonding a metal ion, the four carboxy groups and two amino groups.

When EDTA is dissolved in water, EDTA forms a zwitterion which has the structure

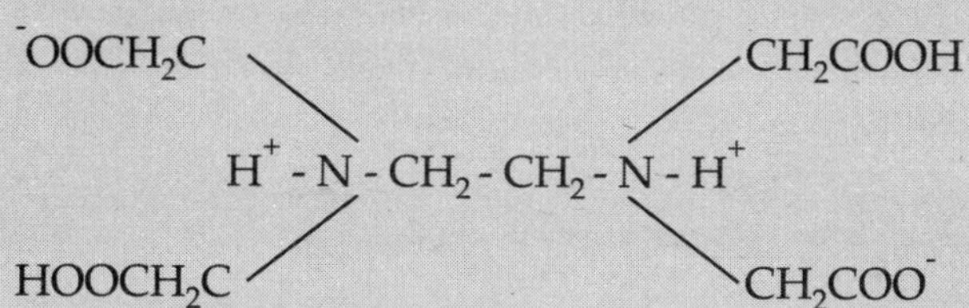

The various EDTA species are abbreviated as H_4Y, H_3Y^-, H_2Y^{2-}, HY^{3-} and Y^{4-}. The relative amounts of these species is a function of pH.

At pH > 12, EDTA exists as Y^{4-}

pH ≥ 8, EDTA exists as HY^{3-}

pH ≃ 5, EDTA exists as H_2Y^{2-}

3. The free acid H_4Y is not employed for preparation of standard solution of EDTA because of its limited solubility. The disodium salt Na_2H_2Y is most useful for analytical work because it can be obtained in high state of purity as the dihydrate (Mol. wt 372.24).

4. ***Reactions of EDTA with metal ions:*** *Solutions of EDTA are used as titrants because it combines with metal ion in a 1 : 1 ratio regardless of the charge on the cation, to form a stable cage like structure. The reactions of EDTA with cations may be written as:*

$$M^{2+} + H_2Y^{2-} \rightleftharpoons MY^{2-} + 2H^+$$

$$M^{2+} + H_2Y^{2-} \rightleftharpoons MY^{-} + 2H^+$$

$$M^{2+} + H_2Y^{2-} \rightleftharpoons MY + 2H^+$$

The structure of one such complex can be depicted as:

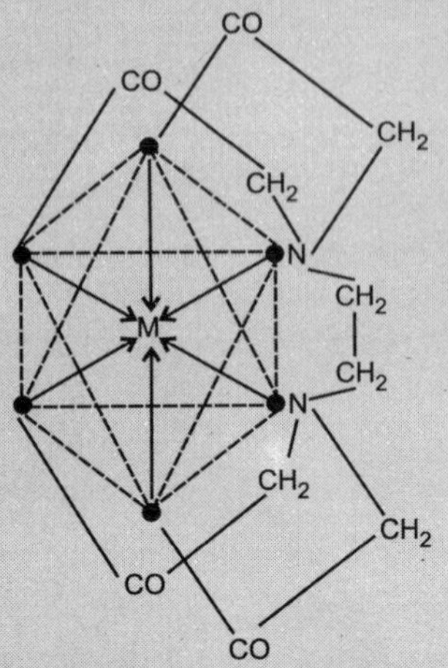

EDTA Complex

It is quite obvious from the above reactions that the equilibrium is markedly affected by the pH of the solution. The complexes of divalent metals are stable in ammoniacal solution.

5. Indicators used in EDTA titrations*. The indicators used in these titrations are the organic colouring substances that undergo change of colour when they form metal complexes. The indicator forms a less stable complex with the metal atom than EDTA-metal complex. The indicator releases metal ion at equivalence point and this results into change in colour.*

A large number of compounds have been mentioned in literature for use as metal ion indicators in complex formation titrations. The most widely used is Eriochrome Black T

Eriochrome Black T. Eriochrome Black T (EBT) is a complex organic chelating agent possessing the formula:

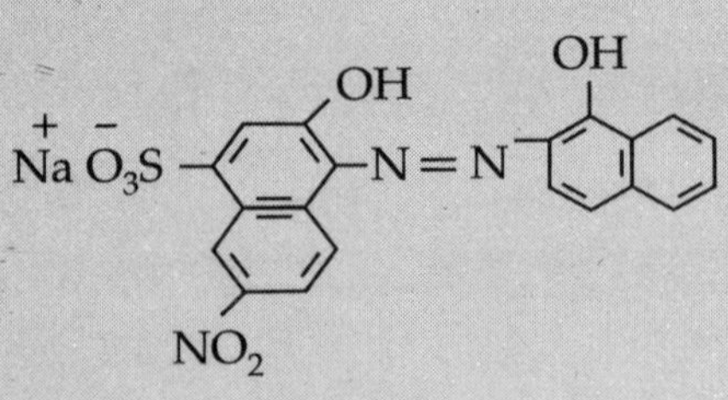

1-(hydroxy 2-naphthylazo) 6 nitro – 2 – napthol – 4 – sulphonate

The ionized form of Eriochrome Black T is red in colour. The phenolic group on EBT may also get ionized at higher pH values. The indicator is usually abbreviated as H_3In. Its dissociation is represented as:

$$\underset{\text{(Black)}}{H_3In} + H_2O \rightarrow \underset{\text{(Red)}}{H_2In^-} + H_3O^+$$

$$\underset{\text{(Red)}}{H_2In^-} + H_2O \rightarrow \underset{\text{(Blue)}}{HIn^{2-}} + H_3O^+$$

$$\underset{\text{(Blue)}}{HIn^{2-}} + H_2O \rightarrow \underset{\text{(Orange)}}{In^{3-}} + H_3O^+$$

Thus, it follows that indicator would exhibit different colours, depending upon its dissociation, which in turn depends upon the pH of the solution i.e. it is a pH sensitive indicator.

At $pH < 6$ the indicator exists as H_2In^- (red)

$pH > 7$ the indicator exists as HIn^{2-} (blue)

$pH > 10$ the indicator exists as In^{3-} (orange)

Thus, the pH of the titrating solution has to be maintained by adding a suitable buffer solution.

The EDTA titrations are performed in buffers of pH 8–10, the range in which the predominant form of the indicator is blue (HIn^{2-}). When the metallic salts are added, the colour of the indicator changes from blue to red as indicator complexes with metal ion. At the end point in EDTA titration, the indicator will be set free because of the formation of a stable metal EDTA complex. The colour change will be red to blue.

$$\underset{\text{(Indicator) Blue}}{HIn^{2-}} + M^{2+} \rightarrow \underset{\text{Red Complex}}{MIn^-} + H^+$$

$$MIn^- + H_2Y^{2-} \rightarrow \underset{\text{(Colour less)}}{MY^{2-}} + \underset{\text{(Blue)}}{HIn^{2-}} + H^+$$

6. Buffer solution in EDTA titration. As is evident from the foregoing discussion that the stability of the metal-EDTA complex and the colour change of the indicator are sensitive to pH changes, hence to maintain the pH to appropriate range (8 to 10) ammoniacal buffer solution is added.

EXPERIMENT NO. 1.12

AIM (a) **To determine the strength of calcium ions in given $CaCO_3$ solution by Complexometric Titrations.**

(b) **To determine the strength of Magnesium ions in given $MgSO_4$ Solution by Complexometric Titrations.**

THEORY Many metals form complexes with such reagent which contain appropriate ligands. EDTA (Ethylenediamine tetracetic acid) is one such reagent which forms complexes with metal. In form of its disodium salt it is used to estimate Ca^{++} and Mg^{++} ions.

Estimation of Ca^{2+} or Mg^{2+} by EDTA method is based on the following principle

- First, the indicator Eriochrome black T, forms an unstable complex with Ca^{2+} or Mg^{2+} solution taken in the flask at a pH of 9 to 10. The complex is wine red in colour.

  Mg^{2+}/Ca^{2+} + EBT $\rightarrow$ (Mg/Ca) – EBT
  [Unstable Complex (wine red)]

- As this solution is titrated against EDTA, the free Ca^{2+} or Mg^{2+} ions in water form stable metal-ion-EDTA complex.

  Mg^{2+}/Ca^{2+} + EDTA $\rightarrow$ [Mg/Ca–EDTA]
  stable complex (colourless)

- Once the free metal ions are complexed, the EDTA replaces Ca or Mg ions from the unstable indicator and the indicator is set free. Since the indicator is blue at above pH, the end point is appearance of blue colour

  [Mg/Ca–EBT] + EDTA $\rightarrow$ [Mg/Ca – EDTA] + EBT
  wine red Colourless Blue.

ESTIMATION OF CALCIUM

APPARATUS Burette, pipette, conical flask, measuring cylinder beakers etc.

CHEMICALS Calcium carbonate solution (unknown), 0.1 M EDTA solution, ammonia buffer solution, indicator solution.

PROCEDURE

1. In a 250ml. conical flask pipette out 5ml. of prepared calcium solution.
2. Add 10ml. distilled water, 10ml. buffer solution and 2-3 drops of the indicator solution.
3. Add EDTA solution from the burette dropwise till the red colour of the solution changes to permanent blue.
4. Repeat the titration to get concordant values.

OBSERVATIONS

Observation Table.

S. No.	*Volume of the solution taken in the titration flask (ml)*	*Burette Readings*		*Volume of the titrant used (Final-Initial reading) (ml)*
		Initial Reading	*Final Reading*	
1.				
2.				
3.				

CALCULATION

Calculate as follows:

EDTA forms 1 : 1 complex,

$\therefore$ 1 Molar solution of EDTA reacts with = 1 mole of Ca^{2+} ions

= 40.08 g of Ca^{2+} ions

or 1000 ml of 1 M EDTA react with = 40.0 g of Ca^{2+} ions

1 ml of 0.1 M EDTA

= 4.008mg of calcium ions

Hence V ml of 0.1 M EDTA

= V × 4.008mg. of calcium ions

or 5ml $CaCO_3$ solution has

= V × 4.008mg of calcium ions

$\therefore$ 1000ml $CaCO_3$ solution has

$$= \frac{V \times 4.008}{5} \times 1000 \text{ mg/L of calcium ions}$$

$$= \frac{V \times 4.008}{5} \times 1000 \text{ ppm of calcium ions}$$

ESTIMATION OF MAGNESIUM

CHEMICALS Magnesium Sulphate (unknown),
0.01 M EDTA Solution, Buffer Solution, Indicator,

PROCEDURE

1. Take 5ml. of magnesium solution in a conical flask.
2. To this add 5ml. buffer solution and dilute it with 10ml of distilled water and 2–3 drops of the indicator.
3. Now run in EDTA solution from the burette gradually till the colour of the solution changes from red to blue and there remains not even a red tinge.
4. Repeat the titration to get concordant values.

OBSERVATION

Observation Table.

S. No.	*Volume of the solution taken in the titration flask (ml)*	*Burette Readings*		*Volume of the titrant used (Final-Initial reading) (ml)*
		Initial Reading	*Final Reading*	
1.				
2.				
3.				

CALCULATIONS

$$1 \text{ ml of } 0.01 \text{ M. EDTA} = 0.243 \text{ mg of magnesium}$$

$$\therefore \quad \text{V ml of } 0.01 \text{ M EDTA} = \text{V} \times 0.243 \text{ mg of magnesium}$$

$$\text{or} \quad 5 \text{ ml } MgSO_4 \text{ solution} = \text{V} \times 0.243 \text{ mg of magnesium}$$

$$\therefore \ 1000 \text{ ml } MgSO_4 \text{ Solution} = \frac{\text{V} \times 0.243}{5} \times 1000 \text{ mg/L of magnesium}$$

$$= \frac{\text{V} \times 0.243}{5} \times 1000 \text{ ppm of magnesium.}$$

Note. For taking known volume of Buffer solution use measuring cylinder.

RESULT (i) Strength of Ca^{2+} in $CaCO_3$ solution = ... ppm

(ii) Strength of Mg^{2+} in $MgSO_4$ solution = ... ppm.

PRECAUTIONS

1. The burette, pipette and conical flask should be rinsed with distilled water.
2. Redistilled water should be employed for preparing the EDTA solution.
3. The colour change near the end point is very slow and thus should be observed carefully.

Instructions to Lab Assistant

Provide the following:

1. 0.1 M EDTA solution. Dissolve 37.225 g of EDTA (disodium salt) into 1000 ml volumetric flask in redistilled water and make up the volume to 1 Liter.
2. EBT Indicator solution: Dissolve 0.5g of the dye in 50ml of methanol.
3. Buffer solution: Dissolve 70g of NH_4Cl in 570ml of conc. NH_3 and the solution is diluted to one litre with distilled water.
4. 0.1 M calcium solution: Weigh 10gm of $CaCO_3$ and transfer it to a 1000 ml measuring flask. Add dil. HCl in drops till there is effervescence and the salt completely dissolves. Now add distilled water to make up the volume to 1L.
5. 0.01 M $MgSO_4$ solution: Weigh 2.46gm of $MgSO_4$ and dissolve it in distilled water and make up the volume to 1L.

EXERCISE

1. What is Complexometric titration?

Ans. *Complexometric titration is a type of titration based on complex formation between the analyte and titrant*

2. *Give the structure of EDTA and its dissociated form.*

Ans. *EDTA*
Ethylenediaminetetraacetic Acid

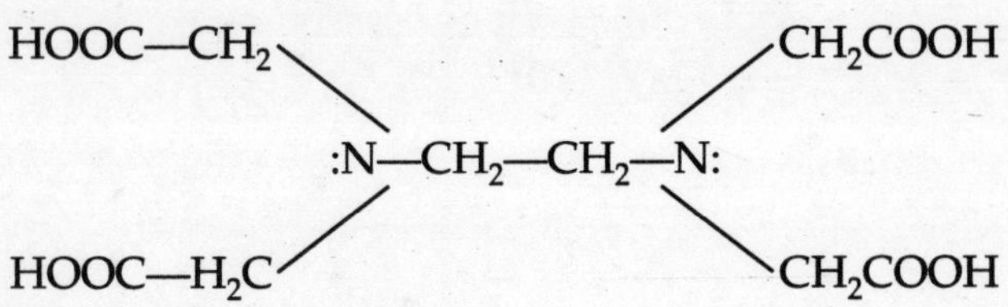

Note. Determination of hardness of water is given in the chapter 'Water Analysis'.

EDTA (H_4Y)

- *Completely dissociated EDTA (Y^{4-}):*

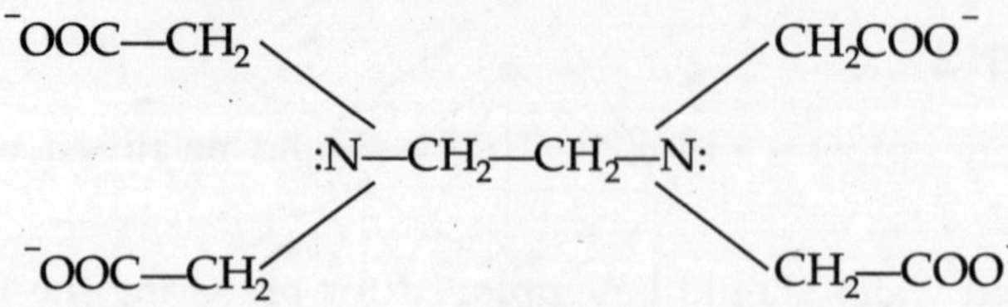

Four acidic protons:

$K_{a1} = 1.02 \times 10^{-3}$ $K_{a2} = 2.14 \times 10^{-3}$

$K_{a3} = 6.92 \times 10^{-3}$ $K_{a4} = 5.50 \times 10^{-11}$

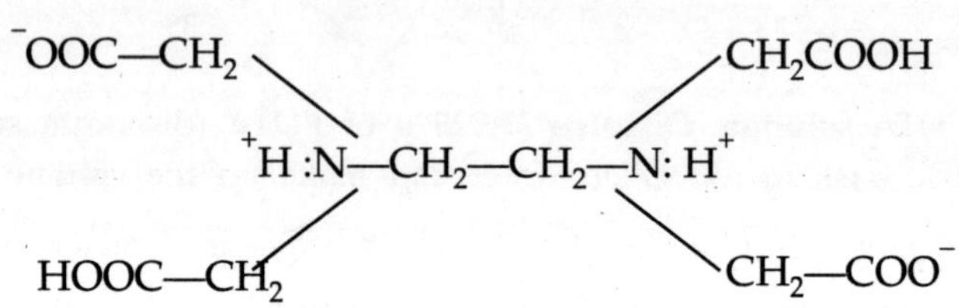

Zwitterion

3. Discuss the dissociation of EDTA as a function of pH.

Ans. *The various EDTA species are abbreviated as H_4Y, H_3Y^-, H_2Y^{2-}, HY^{3-} and Y^{4-}. The relative amounts of these species is a function of pH.*

At *pH > 12, EDTA exists as* Y^{4-}

pH ≥ 8, EDTA exists as HY^{3-}

pH ≃ 5, EDTA exists as H_2Y^{2-}

4. Why always disodium salt of EDTA is used?

Ans. *The free acid H_4Y is not employed for preparation of standard solution of EDTA because of its limited solubility. The disodium salt Na_2H_2Y is most useful for analytical work because it can be obtained in high state of purity as the dihydrate (Mol. wt 372.24).*

5. Why EDTA is widely used for detection of metal ions?

Ans. *Solutions of EDTA are used as titrants because it combines with metal ion in a 1 : 1 ratio regardless of the charge on the cation, to form a stable cage like structure. The reactions of EDTA with cations may be written as:*

$$M^{2+} + H_2Y^{2-} \rightleftharpoons MY^{2-} + 2H^+$$

$$M^{2+} + H_2Y^{2-} \rightleftharpoons MY^- + 2H^+$$

$$M^{2+} + H_2Y^{2-} \rightleftharpoons MY + 2H^+$$

The structure of one such complex can be depicted as:

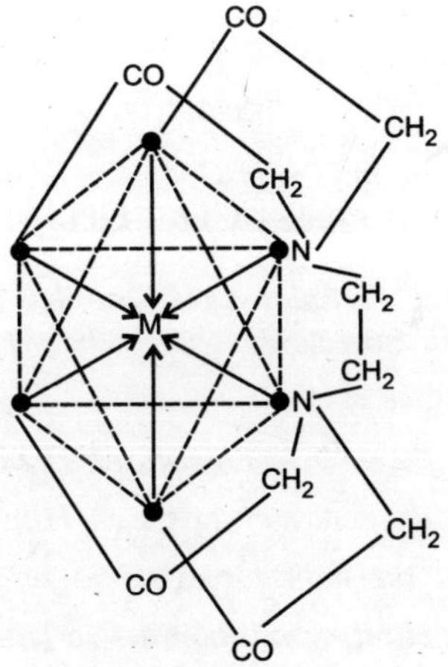

EDTA Complex

It is quite obvious from the above reactions that the equilibrium is markedly affected by the pH of the solution. The complexes of divalent metals are stable in ammoniacal solution.

6. What are the indicators used in EDTA titrations?

Ans. *The indicators used in these titrations are the organic colouring substances that undergo change of colour when they form metal complexes. The indicator forms a less stable complex with the metal atom than EDTA-metal complex. The indicator releases metal ion at equivalence point and this results into change in colour.*

7. What are indicators used in EDTA titrations? Give structures.

Metal Ion Indicators

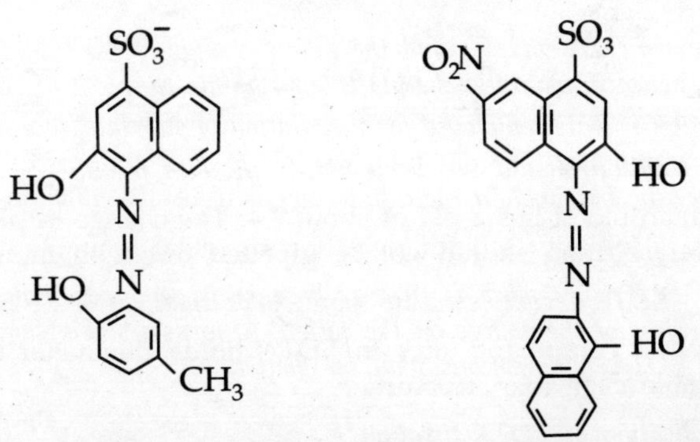

Calmagite Eriochrome Black T.

Both are week diprotic acids

8. EBT is pH sensitive indicator - comment.

Ans. *The ionized form of Eriochrome Black T is red in colour. The phenolic group on EBT may also get ionized at higher pH values. The indicator is usually abbreviated as H_3In. Its dissociation is represented as:*

$$\underset{\text{(Black)}}{H_3In} + H_2O \rightarrow \underset{\text{(Red)}}{H_2In^-} + H_3O^+$$

$$\underset{\text{(Red)}}{H_2In^-} + H_2O \rightarrow \underset{\text{(Blue)}}{HIn^{2-}} + H_3O^+$$

$$\underset{\text{(Blue)}}{HIn^{2-}} + H_2O \rightarrow \underset{\text{(Orange)}}{In^{3-}} + H_3O^+$$

Thus, it follows that indicator would exhibit different colours, depending upon its dissociation, which in turn depends upon the pH of the solution i.e. it is a pH sensitive indicator.

At $pH < 6$ *the indicator exists as* H_2In^- *(red)*

$pH > 7$ *the indicator exists as* HIn^{2-} *(blue)*

$pH > 10$ *the indicator exists as* In^{3-} *(orange)*

Thus, the pH of the titrating solution has to be maintained by adding a suitable buffer solution.

9. *Why EDTA titrations are performed in buffers of PH-8-10.*

Ans. *The EDTA titrations are performed in buffers of pH 8 -10, the range in which the predominant form of the indicator is blue* (HIn^{2-})*. When the metallic salts are added, the colour of the indicator changes from blue to red as indicator complexes with metal ion. At the end point in EDTA titration, the indicator will be set free because of the formation of a stable metal EDTA complex. The colour change will be red to blue.*

$$\underset{\substack{\text{(Indicator)}\\ \text{Blue}}}{HIn^{2-}} + M^{2+} \rightarrow \underset{\substack{\text{Red}\\ \text{Complex}}}{MIn^-} + H^+$$

$$MIn^- + H_2Y^{2-} \rightarrow \underset{\text{(Colour less)}}{MY^{2-}} + \underset{\text{(Blue)}}{HIn^{2-}} + H^+$$

Thus the metal-EDTA complex and the colour change of the indicator are sensitive to pH changes, hence to maintain the pH to appropriate range (8 to 10) ammoniacal buffer solution is added.

10. *Why buffer solution should not be pipetted out with mouth?*

Ans. Normal human blood has a pH of about 7.4. The change in pH may prove fatal. Hence buffer solution should not be pipetted out with mouth.

11. *Why metal - EDTA - complexes are more stable than metal EBT complex?*

Ans. The number of compexing sites in EDTA holds the metal ion more strongly giving a stable cage like structure.

12. *Give an application of EDTA titration.*

Ans. Calcium is an essential ingredient in the diet and is important for the development and preservation of strong bones and teeth. The loss of bone density and strenght with age is called *osteoporosis*. Thus, many people take calcium in tablet form, typically as calcium carbonate, to supplement their diettary intake.

A standard method of analysis for calcium involves *complexometric titration* with a standard solution of ethylenediaminetetraacetic acid (EDTA).

Chelation therapy may be used to remove any of several toxic elements from the body involving the use of EDTA.

Information about chemicals used

Ethylene Diamine Tetra Acetic Acid (EDTA)

Ethylenediamine Tetraacetic Acid is a white, odorless, crystalline synthetic amino acid, first used in the 1940's for treatment of heavy metal poisoning. It is used as a drug, food additive, in cosmetics, household and textile articles, in pharmaceutical products and in biochemicals.

Uses

- Sodium or calcium EDTA binds to metals, such as nickel, copper, and iron, making them unavailable to react with other ingredients in a product, or with compounds in the human body.
- Compounds that act in this way are called *sequestering* agents, or *chelating* agents (after the word for the claw of a crab - *chelae*).
- EDTA is used to treat lead and mercury poisoning, as it can lock up these metals so they can do no harm in the body.
- EDTA also sequesters calcium and magnesium from hard water, preventing them from forming insoluble soap films (scum) with soaps and detergents.

It is widely recognized as effective for that use as well as certain others, including emergency treatment of hypercalcemia and the control of ventricular arrhythmias associated with digitals toxicity.

Safety concern

Inhalation: No adverse health effects via inhalation.

Ingestion: Substance has low toxicity by ingestion. Large amounts may cause gastric upset due to osmotic imbalance through the sequestering of metal ions.

Skin contact: May cause irritation with redness and pain.

Eye contact: May cause irritation, redness and pain.

Chronic Exposure: No adverse health effects expected.

First Aid Measures

If large amounts are swallowed, give water to drink and get medical advice.

In case of skin contact, immediately flush skin with plenty of water for at least 5 minutes. Remove contaminated clothing and shoes. Wash clothing before reuse. Thoroughly clean shoes before reuse. Get medical attentiaon if irritation develops.

2

Water Analysis

The planet earth is abundantly rich in water, which constitute 75% of its surface. Besides, being an essential ingredient of animal and plant life, water forms the basis of human life. It constitutes about 65% of human body. Water is essential not only for survival of human beings, animals, plants and all other living beings but is also utilized in a large amount in industries. Out of the total water consumed by human beings, more than 50% of it is consumed for industrial activity and only a small proportion is used for drinking purposes.

Most of the water which is available is located in oceans (97.02%). The high concentration of salts make it virtually unusable as a source of water for municipal, agricultural or most industrial needs. Hence we are left with only 2.98% of water, out which, considerable amount of water is present in the form of polar ice caps and may remain locked as such for thousands of years. A small portion of it is also present in rock crevices and minerals, lying 800 meters deep below the earth's crust, which is again very expensive to pump out. Thus, the conclusion is that only 0.3 percent of world's water resources is available for possible human use.

Quantity of water is not the only concern. The second major problem is the **quality** of water. Overuse has resulted in the progressive deterioration of water quality. Even the water which is used for industries has to be treated prior to its use. The treatment to which it is subjected, depends upon the purpose, for which it would be used. Thus, both the problems **water quantity and quality** have to be treated with equal importance.

The impurities present in water vary greatly from one place to another. Treatment of water obtained from different sources is therefore essential so that it may be used safely for industrial or municipal purposes.

IMPURITIES IN WATER

Water from a water supply scheme may contain a number of impurities which it takes up from

- the ground or soil with which it comes in contact. For e.g. soil particles, minerals, garbage etc.

- the sewage and industrial waste. For eg. chemical wastes, acids, colours, heavy metals etc.
- the decomposition of plants and animals.
- the growth of bacterial, algae, viruses etc.
- the atmosphere in the form of dissolved gases.

Type of Impurities: The impurities present in water may be broadly classified as given below

IMPURITIES

Dissolved impurities	(*a*) Salts: Carbonates, bicarbonates, sulphates and chlorides of calcium, magnesium, iron, sodium and potassium. (*b*) Gases: CO_2, O_2, N_2, H_2S, NH_3 etc.
Suspended Impurities	(*a*) Inorganic – sand, clay (*b*) Organic – Vegetable and animal matter
Colloidal Impurities	Finely divided silica and clay, organic waste products, colouring matter etc.
Micro Organisms	Bacteria, Algae, Fungi, Virus etc.

Effect of Impurities: The various types of impurities present in water impart certain properties to water. Colour, decaying aquatic organisms and certain chemicals may impart odour, suspended matter may impart turbidity to the water. Presence of dissolved salts make the water hard; whereas excess quantities of metals and dissolved gases makes it corrosive. Bacteriological impurities due to pathogenic bacteria make water unfit for drinking or domestic purposes.

Hence, the water from any source has to be treated before it is used. The treatment to which it is subjected depends upon its use. For e.g.; if it has to be used for drinking purposes, the treatment would include removal of objectionable colour, taste and pathogenic microorganisms, whereas the water for industrial use require, removal of dissolved salts if it is used for steam generation, removal of Ca, Mg, Fe and Mn salts for its use in laundries. Similarly, the water used in alcoholic breweries requires the high purity and the presence of a few micro organisms.

Thus the water has to be treated before it can be used for various purposes. The treatment depends on the nature and extent of contamination which can be determined by analysing the water.

ANALYSIS OF WATER

Analysis of water includes the determination of different constituents in water to ascertain the suitability of water for a specific use, and the treatment to which it should be subjected prior to its use. The following characteristics of water are usually determined:

- Hardness
- Alkalinity
- Dissolved Oxygen (BOD, COD)
- Dissolved Carbondioxide
- Free Chlorine
- Chlorides

Each of these is discussed in the foregoing sections.

HARDNESS OF WATER

Hardness is one of the most important characteristic of water. This term was applied to the waters which make cleansing action of the soap difficult. The water which **does not form lather with soap** and is **hard** to wash in, is termed as hard water. Hardness of a water sample is a measure of its capacity to precipitate soap or its soap consuming capacity.

On the other hand, the water which forms copius lather with soap is termed as soft water.

CAUSES OF HARDNESS

Hardness of water is due to the presence of calcium and magnesium salts. Generally, soaps consist of sodium salts of higher fatty acids such as stearic, oleic, palmitic etc. which are highly soluble in water and thus exert their cleansing action. If calcium and magnesium salts are present in water, these react with soluble sodium soap to form insoluble salts of calcium and magnesium. For example,

$$2C_{17}H_{35}COONa + CaSO_4 \longrightarrow (C_{17}H_{35}COO)_2\,Ca + Na_2SO_4$$

(Sodium Stearate) (Insoluble salt)

$$2C_{17}H_{35}COONa + MgCl_2 \longrightarrow (C_{17}H_3COO)_2\,Mg + 2NaCl$$

Thus, a white scum or precipitate of insoluble soap of calcium and magnesium is formed. Lather is not produced until the cations, Ca^{2+} and Mg^{2+} are completely precipitated out in the form of insoluble salts. As a result, a large quantity of soap is required to produce lather with hard water. Thus hardness is rightly defined *as soap consuming capacity of water*.

Types of Hardness: Hardness of water is of two types:

(a) Temporary Hardness (b) Permanent Hardness

(a) Temporary Hardness: Hardness due to the presence of bicarbonate ions of calcium and magnesium is called temporary hardness or carbonate hardness. This hardness can be removed by mere boiling of water, which converts the bicarbonates into insoluble carbonates or hydroxides. The insoluble carbonates or *hydroxides* can be removed by filteration.

$$Ca(HCO_3)_2 \xrightarrow{\text{boil}} CaCO_3 \downarrow + H_2O + CO_2$$

$$Mg(HCO_3)_2 \xrightarrow{\text{boil}} MgCO_3 \downarrow + H_2O + CO_2.$$

Temporary hardness can also be removed by addition of calculated amount of lime $(Ca\ (OH)_2)$

$$Ca(HCO_3)_2 + Ca(OH)_2 \rightarrow 2Ca\ CO_3\downarrow + 2H_2O$$

$$Mg\ (HCO_3)_2 + Ca\ (OH)_2 \rightarrow MgCO_3\downarrow + CaCO_3\downarrow + 2H_2O$$

(b) Permanent Hardness. This type of hardness is also known as non-carbonate hardness. It is due to the presence of sulphates and chlorides of calcium and magnesium. Permanent hardness cannot be removed by boiling. Special methods are used for removal of permanent hardness.

HARDNESS IN TERMS OF $CaCO_3$ EQUIVALENTS

Hardness caused by different salts can be expressed in terms of single salt.

It is customary to express hardness in terms of equivalents of $CaCO_3$. The reason for choosing $CaCO_3$ as the standard for calculating hardness of water is due to-

- its molecular weight is excatly 100, which makes mathematical calculations easier.
- it is the most insoluble salt, thus can be easily precipitated in water treatment processes.

The $CaCO_3$ equivalents of hardness causing salts are

$$100g \text{ of } CaCO_3 \equiv 111g \text{ of } CaCl_2 \equiv 136g \text{ of } CaSO_4 \equiv 95g \text{ of } MgCl_2$$

$$\equiv 120g \text{ of } MgSO_4 \equiv 162g \text{ of } Ca\ (HCO_3)_2$$

$$\equiv 146g \text{ of } Mg\ (HCO_3)_2$$

EXPERIMENT NO. 2.1

AIM **Determination of total hardness by Complexometric titration method.**

CHEMICALS 0.1 M EDTA, Eriochrome Black T, water sample, buffer solution.

THEORY The hardness of water can be determined by *complexometric titration. EDTA is used as complexing reagent. The Ca^{2+} and Mg^{2+} present in water are titrated with EDTA using Eriochrome Black T as indicator.

Estimation of hardness by EDTA method is based on the following principle

(i) First, the indicator Eriochrome Black-T, which is a blue coloured dye, forms an unstable complex with calcium or magnesium ions in hard water at a pH of 9 to 10. The complex is wine red in colour.

Mg^{2+}/Ca^{2+} + Eriochrome Black T → [Mg/Ca – Eriochrome Black T]
Unstable Complex (Wine red)

(ii) As this solution is titrated against EDTA, the free Ca^{2+} and Mg^{2+} ions in water form stable metal-ion EDTA, complex

[Mg^{2+}/Ca^{2+} + EDTA → [Mg/Ca – EDTA]
Stable Complex
(Colourless)

Once the free metal ions are complexed, the EDTA replaces Ca and Mg ions from the unstable indicator complex also, to form a stable complex, with the result, the indicator is set free. Since the free indicator is blue in colour at the above mentioned pH, the endpoint is the appearance of blue colour.

[Mg/Ca-Eriochrome Black T] + EDTA
→ [Mg/Ca – EDTA] + Eriochrome Black T
(Blue)

Thus the amount of EDTA used corresponds to the hardness of water.

The temporary hardness is removed by boiling and after the removal of precipitate by filteration, the permanent hardness in the filterate is determined by titration with EDTA as above.

Temporary hardness will be given by the difference of total hardness and permanent hardness. Hardness in water is generally expressed in $CaCO_3$ equivalent.

---

* Please refer to page 82, chapter 1, for details on complexometric titrations

PROCEDURE

1. Rinse and fill the burette with EDTA solution
2. Pipette out 25 ml of hard water into volumetric flask and add 20 ml of distilled water and 10 ml of buffer solution.
3. Add 2-3 drops of indicator Eriochrome Black T.
4. Titrate against 0.1 M EDTA solution till the colour of the solution changes from wine red to blue.
5. Note the final reading and repeat to get three concordant readings.

OBSERVATIONS

Volume of hard water taken for each titration = 25 ml

Volume of 0.01 M EDTA used (from titration) = V_1 ml

Observation table

S. No.	*Volume of the solution taken in the titration flask (ml)*	*Burette Readings*		*Volume of the titrant used (Final-Initial reading) (ml)*
		Initial Reading	*Final Reading*	
1.				
2.				
3.				

CALCULATIONS

EDTA forms 1 : 1 complex with Ca^{2+} and Mg^{2+},

$\therefore$ 1000 ml of 1M EDTA = 1 Mole of $CaCO_3$ = 100 gm of $CaCO_3$

1 ml of 0.01 M EDTA = $100 \times \frac{1}{1000} \times \frac{1}{100}$ gm of $CaCO_3$ = 1mg of Ca CO_3

V ml of 0.01 M EDTA = 1.00 × Vmg of $CaCO_3$ = x mg of $CaCO_3$.

25ml of hard water contains = x mg of $CaCO_3$

1000 ml of hard water contains = $\frac{x}{25} \times 1000$ mg of $CaCO_3$ = y mg/L

Note. For determining permanent hardness of water, take 50 ml of water and boil it for about 15- 20 minutes. Temporary hardness will precipitate out. Filter and make up the volume of the filterate to 50ml and titrate it, as described above. Calculate the permanent hardness in the similar way as total hardness is calculated. Subtract permanent hardness from total hardness to get temporary hardness.

RESULT The total hardness of water = y mg/L

PRECAUTIONS

Same as in experiment no. 1.12 in complexometric titrations, chapter-1.

Instructions for the Lab Assistant

Provide the following

1. **0.1 M EDTA solution.** Dissolve 37.225 *g* of EDTA (Disodium salt) in distilled water and make up the volume to 1L.
2. **Indicator solution:** Dissolve 0.5g of the dye in 100 ml of methanol.
3. **Buffer solution:** Dissolve 70g of NH_4Cl in 570 ml of conc. NH_3 and dilute the solution to one litre with distilled water.

EXERCISE

1. Define hardness of water?
2. What do you mean by temporary and permanent hardness?
3. What is the principle of EDTA titration?
4. Why is it necessary to add buffer solution to the water sample in determination of hardness by EDTA method.
5. What is the structure of Eriochrome black T?

Ans.

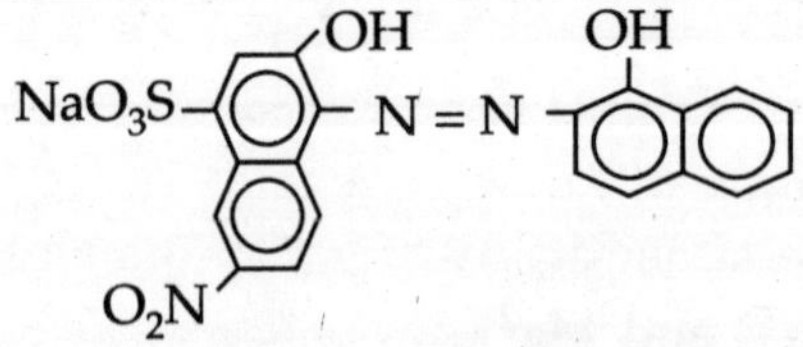

(a) Black (unionized form)

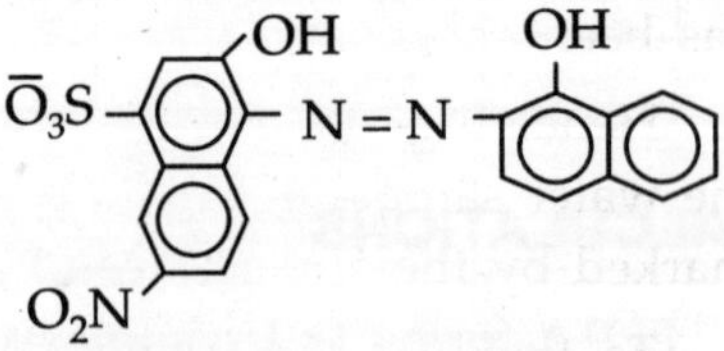

(b) Wine red (ionized form in aqueous medium)

Eriochrome Black T (sodium 1– (1–hydroxy–2–naphthylazo) 6 - nitro - 2 - naphthol - 4 -sulphonate

6. How would you find out permanent hardness of water?
7. What are the disadvantages of hard water.
8. If hard water sample does not contain Mg^{2+}, what will be the effect on the end point with EBT as indicator.

Ans. *The colour change from wine red to blue is not sharp if only Ca^{2+} ions are present. Therefore, Mg^{2+} ions have to be present to produce a sharp end point. For this a small amount of $MgCl_2$ is added to the EDTA solution before it is standardized.*

Note: Places consult Complexometric Titrations, Chapter 1, for more solved questions.

EXPERIMENT NO. 2.2

AIM **Determine the total hardness of a sample of water by soap solution method. (Clarke's Method)**

APPARATUS Burette, pipette, titration flask, beakers, measuring flask.

THEORY

Soap Titration Method: As was discussed previously that when hard water is treated with soap, it does not form lather. The calcium and magnesium ions present in water react with soap (sodium or potassium salt of higher fatty acid) to form insoluble precipitates of calcium or magnesium salt.

$$\underset{\text{(soap)}}{2\,C_{17}H_{35}COONa} + Ca(HCO_3)_2 \rightarrow \underset{\text{(Insoluble salt)}}{(C_{17}H_{35}COO)_2} \downarrow + 2NaHCO_3$$

$$2C_{17}H_{35}COONa + CaSO_4 \rightarrow (C_{17}H_{35}COO)_2\ Ca + Na_2SO_4$$

$$C_{17}H_{35}COONa + MgSO_4 \rightarrow (C_{17}H_{35}COO)_2\ Mg + Na_2SO_4$$

Lather is not produced with hard water until, Ca^{2+} and Mg^{2+} ions are completely precipitated out in the form of insoluble salts. *Thus, the amount of soap initially consumed before the formation of permanent lather is a measure of total hardness or the amount of calcium and magnesium salts present in the hard water.* A permanent lather is the lather which lasts for atleast one minute after the mixture has been shaken.

Hence the total hardness of a water sample can be determined by titrating the water sample against a soap solution of known strength. The end point is marked by the appearance of a stable lather at least for two minutes.

To find out the permanent hardness of the sample, the water sample is boiled for half an hour to remove the temporary hardness and the filtered water is treated with standard soap solution. The titre value corresponds to the permanent hardness of water.

Lather factor: It should be emphasized here that even if we use water of zero hardness, lather is not formed until some critical lower value of surface tension is reached. Thus initially, soap is consumed to reduce the surface tension to this critical value, after which lather formation takes place. The volume of soap solution required to form lather with distilled water (zero hardness) is known as lather factor. Thus, this volume of soap solution is subtracted from all the titre values obtained for hard water.

PROCEDURE

Preparation of Soap solution:

(1) Dissolve 100 g of pure dry caustic soap in 800 mL of alcohol and make up the volume to 1 L by adding distilled water.

Preparation of standard hard water: (SHW)

(2) Dissolve 1 g of pure dry $CaCO_3$ in dil HCl in a beaker. Boil the solution to dryness.

(3) Cool and dissolve the residue in distilled water and make up the volume to 1 L. This is the standard hard water containing 1 g of $CaCO_3$ in 1 L or 1 mg of $CaCO_3$ in 1 mL.

Determination of Lather Factor:

(4) Pipette out 50 mL of distilled water (zero hardness) in a titration flask.

(5) Titrate it with soap solution taken in the burette till the end point (formation of stable lather persisting for about 2 minutes).

(6) Note down the volume of the soap solution used. Let it be V ml.

Standardization of soap solution:

(7) Rinse and fill the burette with soap solution.

(8) Pipette out 50 mL of standard hard water prepared above in the titration flask.

(9) Titrate it against soap solution till the end point.

(10) Note the volume used. Let it be V_1 mL.

Determination of the total hardness:

(11) Now titrate 50 mL of the given sample water against soap solution as described above.

(12) Note down the volume used. Let it be V_2 mL.

OBSERVATIONS

(*a*) **Determination of lather factor**

S.No.	*Volume of distilled water taken*	*Burette readings*		*Volume of Soap solution used (V mL) (Initial - Final reading)*
		Initial reading	*Final reading*	

(*b*) **Standardization of Soap Solution:**

S. No.	*Volume of standard hard water taken*	*Burette readings*		*Volume of Soap solution used (V_1 mL) (Initial-Final reading)*
		Initial reading	*Final reading*	

(*c*) **Determination of Total Hardness**

S. No.	*Volume of the sample hard water taken*	*Burette readings*		*Volume of Soap solution used (V_2 mL) (Initial-Final reading)*
		Initial reading	*Final reading*	

CALCULATIONS

(*i*) **Strength of Soap Solution:**

Soap solution consumed by 50 mL of standard hard water = $(V_1 - V)$ mL

Soap solution consumed by 50 mg of $CaCO_3$ = $(V_1 - V)$ mL

($\because$ 1 mL of standard hard water corresponds to 1 mg of $CaCO_3$)

$(V_1 - V)$ mL of soap solution ≡ 50 mg of $CaCO_3$

1 mL of soap solution ≡ $\frac{50}{V_1 - V}$ mg of $CaCO_3$

(*ii*) **Total Hardness**

50 mL of sample hard water ≡ $(V_2 - V)$ mL of soap solution

$\therefore$ 50 mL of sample hard water ≡ $V_2 - V \times \frac{50}{V_1 - V}$ mg of $CaCO_3$

1000 ml of sample hard water contains

$$= V_2 - V \times \frac{50}{V_1 - V} \times \frac{1000}{50} \text{ mg of } CaCO_3 = \frac{V_2 - V}{V_1 - V} \times 1000 \text{ mg}$$

$$\therefore \text{ Total Hardness } = \frac{V_2 - V}{V_1 - V} \times 1000 \text{ mg/L.}$$

RESULT The total hardness of the given sample water ...... mg/L.

ALKALINITY OF WATER

Another important characteristic of water is its alkalinity, which is a measure of the ability of water to neutralize the acids. The alkalinity of water is normally due to the presence of bicarbonates, carbonates and hydroxides of sodium, potassium, calcium and magnesium. Some of the salts, not all, which cause alkalinity also cause hardness. However, presence of other bases such as borates, silicates, phosphates can also contribute to the total alkalinity, though to a small extent.

Depending upon the anion present, the alkalinity can be classified as:

(i) Caustic alkalinity (due to $\overline{O}H$ and CO_3^{2-} ions)

(ii) Bicarbonate alkalinity (due to HCO_3^- ions)

The alkalinity in water can be present due to the presence of the above mentioned ions present alone or in combination as

- hydroxide only ($\overline{O}H$)
- Carbonates only (CO_3^{2-})
- Bicarbonates only (HCO_3^-)
- Hydroxides and Carbonates ($\overline{O}H + CO_3^{2-}$)
- Carbonate and bicarbonates ($CO_3^{2-} + HCO_3^-$)

The presence of hydroxide and bicarbonates existing together is ruled out because they undergo the following reaction to form CO_3^{2-} ions.

$$\overline{O}H + HCO_3^- \rightarrow CO_3^{2-} + H_2O$$

And hence the possibility of all three ions existing together is also ruled out.

Significance

The knowledge of alkalinity in water is very important.

Highly alkaline waters are usually unpalatable. When alkaline waters are used in boilers for steam generation it leads to caustic embrittlement, deposition of scales and sludges, corrosion etc. Moreover, the knowledge of alkalinity is also important in calculating the amounts of lime and soda in water softening and in internal conditioning of boiler-feed waters.

EXPERIMENT NO. 2.3

AIM **To determine the alkalinity of a given water sample.**

APPARATUS Burette, pipette, conical flask, beakers.

CHEMICALS N/10 HCl, phenolphthalein and methyl orange.

THEORY Determination of alkalinity due to different ions is based on the titration of the water sample against a standard acid making selective use of indicators. The indicators used are–phenolphthalein and methyl orange*. The reactions taking place may be represented by the following equations:

$$\left.\begin{array}{l}\left.\begin{array}{l}\text{(a)}\ \overline{O}H + H^+ \rightarrow H_2O\\ \text{(b)}\ CO_2^{2-} + H^+ \rightarrow HCO_3^-\end{array}\right] P\\ \text{(c)}\ HCO_3^- + H^+ \rightarrow H_2CO_3\end{array}\right] M$$

- The volume of the acid used upto phenolphthalein end point corresponds to the reaction (a) and (b) i.e. complete neutralization of $\overline{O}H$ ions and neutralization of CO_3^{2-} ions upto HCO_3^- stage (half neutralization of CO_3^{2-} ions).
- The volume of the acid used upto methyl orange end point corresponds to the reaction (a), (b), (c) i.e. complete neutralization of $\overline{O}H$, CO_3^{2-} and HCO_3^- ions.

Thus from the respective volume of the acid used, the respective strengths of various ions can be determined.

A known volume of the sample is titrated against a standard acid using phenolphthalein as indicator and alkalinity is found out in terms of $CaCO_3$ equivalents by using normality equation. This alkalinity is called *phenolphthalein alkalinity (P)*. Similarly the sample is titrated against a standard acid using methyl orange as indicator and alkalinity is calculated in terms of $CaCO_3$ equivalents. This alkalinity is called *methyl orange alkalinity (M)* or total alkalinity of water sample.

From the measurement of phenolphthalein alkalinity and methyl orange alkalinity, it is possible to calculate the magnitude of various forms of alkalinity present in water sample,

(i) Alkalinity due to HCO^- only

(ii) Alkalinity due to CO_3^{2-} only

(iii) Alkalinity due to CO_3^{2-} and HCO_3^-

***Note.** (For the theory of acid base indicators refer to Acidimetry- Alkalimetry, chapter 1, page 9

(iv) Alkalinity due to CO_3^{2-} and $\overline{O}H$

(v) Alkalinity due to $\overline{O}H$ only.

Discussion.

Case (i) When phenolphthalein alkalinity = 0, this implies that volume of the acid used till phenolphthalein end point is 0. This can only happen if CO_3^{2-} and $\overline{O}H$ both are absent. Whatever alkalinity is present is due to HCO_3^- ions and can be detected using methyl orange as indicator.

Case (ii): When $P = \frac{1}{2}M$, this implies that only CO_3^{2-} ions are present. Neutralization using phenolphthalein as indicator reaches upto HCO_3^- stage. Same amount of acid will be further used when methyl orange is used as indicator to neutralize HCO_3^- to H_2O and CO_2 as represented in fig: 2-1 (a) The strength of CO_3^{2-} will be given by M or 2P

Case (iii) When $P < \frac{1}{2}$ M, this implies that besides CO_3^{2-} ions, HCO_3^- ions are also present. The acid used uptil phenolphthalein end point corresponds to half neutralization of CO_3^{2-} ions. Further titration using methyl orange as indicator corresponds to neutralization of HCO_3^- obtained from CO_3^{2-} ions and HCO_3^- originally present in the reaction mixture as shown in Fig. 2.1(b)

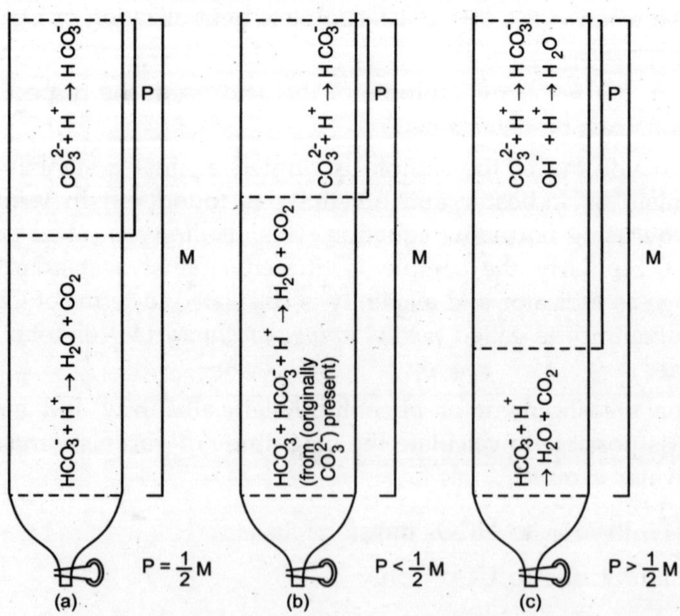

Fig. 2.1 Relative amounts of acid used, using phenolphthalein and methyl orange as indicator

Alkalinity due to CO_3^{2-} = 2P

Alkalinity due to HCO_3^- = (M – 2P)

Case (iv) When $P > \frac{1}{2}M$. This implies that besides CO_3^{2-} ions, $\overline{O}H$ ions are also present. $\overline{O}H$ will be completely neutralized by the acid whereas CO_3^{2-} will be neutralized upto the HCO_3^- stage using phenolphthalein as indicator. On further titration with the acid using methyl orange as indicator, the neutralization of HCO_3^- takes place as shown in Fig. 2.1(c)

Alkalinity due to CO_3^{2-} = 2 (M – P).

Alkalinity due to $\overline{O}H$ = M – 2 (M – P) = (2P – M)

Case (v): When P = M implies that only $\overline{O}H$ ions are present. Alkalinity due to $\overline{O}H$ = P = M.

Thus, on the basis of the analysis of water with respect to phenolphthalein alkalinity and methyl orange alkalinity, the alkalinity due to different ions can be calculated. The results are summarized in the following table.

Table 2.1 Computing the alkalinity due to different ions

Alkalinity	$\overline{O}H$ *(ppm)*	CO_3^{2-} *(ppm)*	HCO_3^- *(ppm)*
P = 0	NIL	NIL	M
$P = \frac{1}{2}M$	NIL	2P	NIL
$P < \frac{1}{2}M$	NIL	2P	(M – 2P)
$P > \frac{1}{2}M$	(2P – M)	2 (M – P)	NIL
P = M	P = M	NIL	NIL

INDICATORS Phenolphthalein and Methyl orange

END POINT Pink to colourless (Phenolphthalein), Yellow to red (Methyl orange)

PROCEDURE

1. Pipette out 20 ml of water sample into a conical flask. Add 1 – 2 drops of phenolphthalein.

2. Rinse and fill the burette with N/10 HCl.
3. Titrate the water sample in conical flask with N/10 HCl till the pink colour just disappears.
4. Note down the reading and repeat to get three concordant readings.
5. Again take 20ml of the water sample in conical flask and add methyl orange indicator to it.
6. Titrate it using N/10 HCl till a red colour is obtained.
7. Record the observation and repeat to get three concordant readings.

Observations

(a) Using Phenolphthalein

Normality of the acid used = N/10

S. No.	*Volume of the solution taken in the titration flask (ml)*	*Burette Readings*		*Volume of the titrant used (Final-Initial reading) (ml)*
		Initial Reading	*Final Reading*	
1. 2. 3.				

Volume of the acid used = Vml

(b) Using Methyl Orange

Normality of the acid used = N/10

S. No.	*Volume of the solution taken in the titration flask (ml)*	*Burette Readings*		*Volume of the titrant used (Final-Initial reading) ml*
		Initial Reading	*Final Reading*	
1. 2. 3.				

Notes 1. If no colour develops on addition of phenolphthalein to the water sample, it means that phenolphthalein alkalinity is zero and hence do not titrate the sample for phenolphthalein alkalinity.

2. Methyl orange can also be added to the solution in which phenolphthalein was added, after the phenolphthalein end point is reached. In that case the total volume of the acid used to methyl orange end point would be, the volume of acid used upto phenolphthalein end point plus the volume of acid used upto methyl orange end point.

CALCULATIONS

Phenolphthalein Alkalinity in terms of $CaCO_3$ equivalents:

$$\begin{array}{cc} \text{(Acid)} & \text{(Water sample)} \end{array}$$

$$N_1V_1 = N_2V_2$$

$$\frac{1}{10} \times V_1 = N_2 \times 20$$

$$N_2 = \frac{1}{10} \times \frac{V_1}{20} = \frac{V_1}{200}$$

Strength in terms of $CaCO_3$ equiv. = $N_2 \times$ Eq.wt of $CaCO_3$

$$= \left(\frac{V_1}{200}\right) \times 50\,g/L = x\,g/L$$

$\therefore$ Phenolphthalein alkalinity (P) = $x \times 1000$ mg/L = $x \times 1000$ ppm.

(II) Methyl orange Alkalinity in terms of $CaCO_3$ equivalents:

$$\begin{array}{cc} \text{Acid} & \text{water sample} \end{array}$$

$$N'_1V'_1 = N'_2V'_2$$

$$\frac{1}{10} \times V'_1 = N'_2 \times 20$$

$$N'_2 = \frac{1}{200} \times V'_1$$

Strength in terms of $CaCO_3$ equiv. = $N'_2 \times 50$ g/L

$$= \left(\frac{V'_1}{200} \times 50\right) g/L = y\,g/L$$

Methyl orange alkalinity or total alkalinity of water sample = $y \times 1000$ mg/L

$$= y \times 1000 \text{ ppm}$$

RESULT

Phenolphthalein alkalinity = ... ppm of $CaCO_3$

Mehtyl orange alkalinity = ... ppm of $CaCO_3$.

To find out the alkalinity in terms of individual ions, we find out to which case the values of P and M falls and from table 2.1, we calculate the amount of individual ions.

Examples

1. If P = 0, the alkalinity is due to bicarbonate ions.

   ∴ Bicarbonate Alkalinity in terms of $CaCO_3$ eq. = M.

   Bicarbonate ion concentration = $\frac{M \times 61}{50}$ mg/L.

   (The equiv. weight of HCO_3^- is 61 and of $CaCO_3$ is 50)

2. If P = $\frac{1}{2}$M, then alkalinity is only due to CO_3^{2-}. Carbonate alkalinity in terms of $CaCO_3$ equiv = 2P or M

   Carbonate ion concentration = 2P × $\frac{30}{50}$ mg/L.

   (The equiv. weight of CO_3^{2-} = 30)

3. If P < $\frac{1}{2}$ M, then the alkalinity is due to CO_3^{2-} and HCO_3^-

   Carbonate alkalinity in terms of $CaCO_3$ equiv = 2P

   Carbonate ion concentration = 2P × $\frac{30}{50}$ mg/L

   Bicarbonate alkalinity in terms of $CaCO_3$ equiv = (M – 2P)

   Bicarbonate concentration = (M – 2P) × $\frac{61}{50}$ mg/L

4. P > $\frac{1}{2}$ M, then the alkalinity is due to $\overline{O}H$ and CO_3^{2-}

   Hydroxide alkalinity in terms of $CaCO_3$ equiv = (2P – M)

   Hydroxide concentration = (2P – M) × $\frac{17}{50}$ mg/L

   (Equiv. weight of $\overline{O}H$ = 17)

   CO_3^{2-} alkalinity in terms of $CaCO_3$ equiv. = 2 (M – P)

   CO_3^{2-} concentration = 2 (M – P) × $\frac{30}{50}$ mg/L

5. If P = M, the alkalinity is due to hydroxide ions

   ∴ Hydroxide alkalinity in terms of $CaCO_3$ eq. = P or M

   Hydroxide ion concentration = P × $\frac{17}{50}$ mg/L

Note. P and M values in the above calculations are taken in mg/L.

EXERCISE

1. Alkalinity of water cannot be due to the simultaneous presence of $\overline{O}H$, CO_3^{2-} and HCO_3^-. Explain.
2. Write the structural formula of phenolphthalein. In what forms does it exist in acidic and alkaline medium.
3. What is the source of alkalinity in water.
4. A 200 ml of a water sample required 20 ml. of N/50 H_2SO_4 using methyl orange as indicator but did not give any colouration with phenolphthalein . What type of alkalinity is present. Express in mg/L.

ACIDITY OF WATER

Acidity in water is caused by the presence of mineral acids, free carbon dioxide, sulphates of iron and aluminium etc in waters. Some salts like $MgCl_2$ etc also liberate HCl due to hydrolysis.

$$MgCl_2 + 2H_2 \rightarrow Mg(OH)_2 + 2HCl$$

Acidic water reacts with iron in the pipes or in the boilers and hence cause corrosion. Moreover, in the lime soda process of water softening, the amounts of lime and soda to be added depend on the amount of mineral acids also, apart from other factors. Thus determination of acidity is important.

EXPERIMENT NO. 2.4

AIM **To determine the acidity of water sample.**

APPARATUS Burette, pipette, conical flask, beakers etc.

CHEMICALS N/50 NaOH, Methyl orange indicator, phenolphthalein indicator and N/10 $Na_2S_2O_3$ solution.

THEORY Acidity of water sample is determined by titrating it against a standard alkali solution using phenolphthalein as indicator. The acidity thus determined is known as total acidity as it measures the acidity due to mineral acids, organic acids and free CO_2.

The acidity due to the presence of free mineral acids is determined by titrating the given sample water with a standard alkali using methyl orange as indicator (*i.e.* the titration is carried to a pH of about 4.5). The results are expressed as equivalents of $CaCO_3$ per million parts of water.

INDICATOR Phenolphthalein, Methyl Orange

END POINT Appearance of pink colour (phenolphthalein) red to yellow (Methyl orange)

PROCEDURE ***Phenolphthalein Acidity***

1. Rinse and fill the burette with N/50 NaOH solution
2. Pipette out 100ml of the water sample in a conical flask and add drop of N/10 $Na_2S_2O_2$ solution to destroy any residual chlorine
3. Add 2-3 drops of phenolphthalein
4. Titrate the solution with N/50 NaOH solution till the appearance of pink colour.
5. Note the reading and repeat to get atleast three concordant readings.

Methyl orange Acidity

1. Pipette out 100 ml of the water sample in a conical flask and add one drop of N/10 $Na_2S_2O_3$.
2. Add a few drops of methyl orange and titrate the solution with N/50 NaOH solution from the burette till the solution turns yellow.
3. Record the reading and repeat to get three concordant readings.

OBSERVATIONS

Volume of water sample taken for each titration = 100ml

Normality of NaOH solution = N/50

Phenolphthalein as indicator

S. No.	*Volume of the solution taken in the titration flask (ml)*	*Burette Readings*		*Volume of the titrant used (Final-Initial reading) (ml)*
		Initial Reading	*Final Reading*	
1.				
2.				
3.				

Methyl Orange as indicator

S. No.	*Volume of the solution taken in the titration flask (ml)*	*Burette Readings*		*Volume of the titrant used (Final-Initial reading) (ml)*
		Initial Reading	*Final Reading*	
1.				
2.				
3.				

CALCULATIONS

Let the volume of the NaOH used in presence of phenolphthalein as indicator = V_1 ml

Let the volume of the NaOH used in presence of methyl orange as indicator = V_1' ml.

Phenolphthalein Acidity

$$\text{NaOH} \quad \text{Water sample}$$

$$N_1V_1 = N_2V_2$$

$$\frac{1}{50} \times V_1 = N_2 \times 100$$

$$N_2 = \frac{V_1}{5000}$$

Phenolphthalen Acidity in terms of $CaCO_3$ equiv.

$$= N_2 \times \text{eq. wt of } CaCO_3$$

$$= \frac{V_1}{5000} \times 50 = \frac{V_1}{100}\ \text{g/L} = \frac{V_1}{100} \times 1000\ \text{mg/L}$$

$$= 10\ V_1\ \text{mg/L} = 10\ V_1\ \text{ppm}$$

Methyl Orange Acidity

$$\text{NaOH} \quad \text{water sample}$$

$$N'_1V'_1 = N'_2V'_2$$

$$\frac{1}{50} \times V'_1 = N'_2 \times 100$$

$$N'_2 = \frac{V'_1}{5000}$$

Methyl Orange acidity in terms of $CaCO_3$ equiv.

$$= \frac{V'_1}{5000} \times 50 \text{ g/L}$$

$$= 10V'_1 \text{ mg/L}$$

$$= \mathbf{10V'_1 \text{ ppm.}}$$

RESULT The phenolphthalein acidity =

The methyl orange acidity =

Instructions to the Lab Assistant

Provide the following

1. N/50 NaOH solution: Dissolve 0.8g of NaOH in distilled water and make up the volume to 1L. Standardize with oxalic acid.
2. Phenolphthalein
3. Methyl Orange

FREE CARBONDIOXIDE

Varying amounts of CO_2 are present in water. CO_2 can be picked up from the atmosphere or it is generated at the bottom of the lake due to the decay of organic matter. Rivers receiving acid waters may also show high CO_2 content.

CO_2 content of a water sample contributes to corrosion and imparts unacceptable taste to waters. Thus the determination of free CO_2 is important.

EXPERIMENT NO. 2.5

AIM **Determination of free CO_2 in given sample of water.**

APPARATUS Burette, pipette, conical flask, beakers etc.

CHEMICALS N/50 Na_2CO_3 (or N/50 NaOH), phenolphthalein

THEORY Free CO_2 is determined by titrating the sample water with N/10 NaOH or N/10 Na_2CO_3 solution using phenolphthalein as indicator. The CO_2 reacts with NaOH or Na_2CO_3 to form HCO_3^-.

$$CO_2 + \overline{O}H \rightarrow HCO_3^-$$

$$CO_2 + CO_3^{2-} + H_2O \rightarrow 2HCO_3^-$$

The end point is appearance of pink colour.

PROCEDURE

1. Rinse and fill the burette with N/50 Na_2CO_3 solution
2. Pipette out 100 ml of water sample in a conical flask (there should be no air bubbles) Add a few drops of phenolphthalein.
3. Titrate the solution with N/50 Na_2CO_3 from the burette, till the pink colour persists for atleast 30 seconds.
4. Note the reading and repeat to get three concordant readings

OBSERVATIONS

Volume of water sample taken for each titration = 100 ml.

Normality of Na_2CO_3 used = N/50.

S. No.	*Volume of the solution taken in the titration flask (ml)*	*Burette Readings*		*Volume of the titrant used (Final-Initial reading) (ml)*
		Initial Reading	*Final Reading*	
1.				
2.				
3.				

CALCULATIONS

Water sample (Na_2CO_3)

$$N_1V_1 = N_2V_2$$

$$N_1 \times 100 = \frac{1}{50} \times V_2$$

$$N_1 = \frac{1}{5000} \times V_2$$

Strength of $CO_2 = N_1 \times$ Eq. wt

$$\text{Strength of } CO_2 = \frac{V_2}{5000} \times 22 \text{ g/L}$$

$$= \frac{V_2}{5000} \times 22 \times 1000 \text{ mg/L} = \frac{22}{5} V_2 \text{ mg/L}.$$

RESULT The amount of free CO_2 present in water sample = ...ppm.

PRECAUTIONS

1. The reaction mixture should be stirred very gently during titration.
2. Same amount of indicator should be added while taking different readings.

Instructions to the Lab Assistant

Provide the following

1. N/50 Na_2CO_3 solution: Dissolve 1.06 g of Na_2CO_3 in distilled water and make up the volume to 1 litre.
2. Phenolphthalein.

EXERCISE

1. What is the source of acidity in water?
2. How the acidity of water is determined?
3. What are the disadvantages of acidic water?
4. What are the disadvantages of dissolved CO_2 present in water.
5. How can you determine the amount of dissolved CO_2 in water.

DISSOLVED OXYGEN (D.O) AND OXYGEN DEMAND

One of the most important water quality parameters is the amount of dissolved oxygen (DO) present. Oxygen is poorly soluble in water. The amount of oxygen in water depends on physical, chemical and biochemical activities taking place in water. The solubility of dissolved oxygen in water at saturation at any temperature and pressure is given by Henry's law. The saturation value of DO in water is of the order of 8 to 15 mg/L, depending upon temperature. Minimum amounts required for aquatic life or for a healthy fish population is 5-8 mg/L.

Significance of the determination of DO

- In aquatic life, as DO drops, fish and other specie are threatened and may get killed.
- Fall in DO levels causes undesirable odours, tastes and reduce the acceptability of water for domestic use.
- In steam generation, DO is one of the most important factors causing corrosion of the boiler material.

There are several measures of oxygen demand commonly used. Two of these are:

(i) Biochemical oxygen demand (BOD)

(ii) Chemical oxygen demand (COD)

BIOCHEMICAL OXYGEN DEMAND

When biodegradable organic matter is released into water, the micro organisms feed on the wastes and break it down into simpler organic or inorganic substances. If the decomposition takes place in the presence of oxygen i.e *aerobically* the non objectionable, stable end products are formed.

$$\text{Organic Matter} + O_2 \xrightarrow{\text{MICRO ORGANISMS}} CO_2 + H_2O + \text{Other Stable Products.}$$

On the other hand if insufficient oxygen is available, the decomposition takes place *anaerobically*. The micro organisms causing the decomposition of the organic matter without the presence of oxygen are entirely different from aerobic bacteria and produce highly objectionable end products including H_2S, NH_3 and CH_4.

$$\text{Organic Matter} \xrightarrow{\text{MICRO ORGANISMS}} CO_2 + CH_2 + \text{Unstable Products}$$

The amount of oxygen required by micro-organisms to oxidize organic wastes aerobically is called biochemical oxygen demand (BOD).

Actually, when sufficient oxygen is present, that is in aerobic systems, oxygen is the oxidizing agent to convert organic waste into CO_2 and H_2O. Similarly inorganic materials are also oxidized. The products obtained are not harmful.

$$\text{Fatty acids (organic waste)} \xrightarrow{\text{oxidation}} CH_3COO^- \xrightarrow{O_2 + \text{Bacteria}} CO_2 + H_2O$$

$$S^{2-} \xrightarrow{O_2 + \text{Bacteria}} SO_4^{2-}$$

$$NH_3 \xrightarrow{O_2 + \text{Bacteria}} NO_3^- + H^+ + H_2O$$

However, if sufficient oxygen is not present (anaerobic conditions), organic matter is oxidized by nitrate to form undesirable products.

$$5\,CH_3\,COO^- + 8\,NO_3^- + 8H^+ \xrightarrow{\text{Bacteria}} 5CO_2 + 5HCO_3^- + 4N_2 + 9\,H_2O$$

$$CH_3\,COO^- + NO_3^- \xrightarrow{\text{Bacteria}} CO_2 + HCO_3^- + NH_2^-$$

The products in this case like NH_2^- etc give bad odour to water. Example NH_2^- give smell of rotten fish. Similarly under anaerobic conditions, SO_4^{2-}, PO_4^{2-}, CO_2 are reduced to H_2S, HS^-, phosphine, methane etc. giving rotten egg smell or suffocating smell.

Therefore oxygen demanding wastes are termed as water pollutants because they deplete water from its dissolved oxygen. Thus sufficient oxygen may be present or supplied to maintain aerobic conditions and hence prevent bad odour, taste and colour.

CHEMICAL OXYGEN DEMAND

Although BOD test is applicable to organic wastes, there are many drawbacks. Certain organic materials are not biodegredable and hence can give wrong conclusions that less organic matter is present because BOD of such water samples will be low. In this case COD, Chemical Oxygen Demand reveals the real organic content present. Here the oxidation of organic substances present in water is done chemically.

COD, is the amount of oxygen required by organic matter in a sample of water for its oxidation by a strong chemical oxidizing agent such as $K_2Cr_2O_7$.

Since in COD determination, the organic matter, both biologically oxidizable (glucose) and biologically inert (cellulose) is completely oxidized to CO_2 and H_2O, COD values are greater than BOD values.

EXPERIMENT NO. 2.6

AIM **Determination of dissolved oxygen in the given sample of water.**

APPARATUS Burette, pipette, conical flask, beakers etc.

CHEMICALS $Na_2S_2O_3$ (N/40), Mn SO_4 solution, KI, starch, conc. H_2SO_4.

THEORY Oxygen itself is not a pollutant in water but its deficiency is an indicator of several types of pollution in water.

Dissolved oxygen (DO) is determined by Winkler's method or iodometric titration.* The dissolved oxygen in water oxidizes KI and an equivalent amount of iodine is liberated. This iodine is titrated against a standard hypo solution. However since dissolved oxygen in water is in molecular state and is not capable of reacting with KI, therefore an oxygen carrier such as manganese hydroxide is used.

*__Note:__ For details on iodometric titrations refer to chapter 1, page 64.

The method involves introducing a conc. solution of Mn SO_4, NaOH and potassium iodide, azide reagent, into the water sample. The white precipitate of Mn $(OH)_2$ which is formed, is oxidized by oxygen in water sample to give a brown precipitate of basic manganic oxide $MnO(OH)_2$. This MnO $(OH)_2$, in acidic medium dissolves and liberates free iodine from the added KI in a equivalent amount of dissolved oxygen in water sample. This liberated I_2 is then titrated against $Na_2S_2O_3$ solution using starch as indicator. The reactions involved are:

$$MnSO_4 + 2NaOH \rightarrow Mn\,(OH)_2 + Na_2\,SO_4$$

$$4\,Mn\,(OH)_2 + O_2 \rightarrow 2MnO\,(OH)_2$$

Basic Manganic oxide

$$2MnO\,(OH)_2 + H_2SO_4 \rightarrow Mn\,SO_4 + 2H_2O + O$$

$$2KI + H_2SO_4 + O \rightarrow K_2SO_4 + H_2O + I_2$$

$$2Na_2\,S_2O_3 + I_2 \rightarrow Na_2S_4O_6 + 2NaI$$

The nitrites present in water, interfere with the titration as these can also liberate I_2 from KI

$$2HNO_2 + H_2SO_4 + 2KI \rightarrow 2NO + K_2SO_4 + 2H_2O + I_2$$

Thus to destroy nitrite, sodium azide is used.

$$2\,NaN_3 + H_2\,SO_4 \rightarrow 2HN_3 + Na_2SO_4$$

(Hydrazoic acid)

$$HNO_2 + HN_3 \rightarrow NO_2 + N_2 + H_2O$$

PROCEDURE

1. A known amount of sample water (say 250 ml) is taken in a stoppered bottle avoiding contact with air.
2. Add 0.2 ml of $MnSO_4$ solution to it by means of a pipette, dipping the end well below the surface of water. Also add 2 ml of alkaline iodide–azide solution to it.
3. Stopper the bottle and shake thoroughly. Allow the brown precipitates of MnO $(OH)_2$ formed, to settle down
4. When some portion of the liquid below the stopper is clear, add 2 ml of conc. H_2SO_4 with the help of pipette. Stopper and mix till the precipitate is completely dissolved. The characteristic brown colour of iodine is produced.
5. Transfer 100 ml of the above solution in a 250 ml flask with a pipette. Titrate the liberated I_2 with standardized sodium thiosulphate solution until the sample solution becomes pale yellow.

6. Add 2 ml of starch solution, the solution will turn blue.
7. Continue titration till the blue colour disappears.
8. Repeat to get another reading.

OBSERVATIONS

Volume of the water sample taken for titration = 100 ml

Observation Table:

S. No.	*Volume of the solution taken in the titration flask (ml)*	*Burette Readings*		*Volume of the titrant used (Final-Initial reading) (ml)*
		Initial Reading	*Final Reading*	
1.				
2.				
3.				

CALCULATIONS

$$\text{Hypo} \quad O_2 \text{ in water}$$

$$N_1V_1 = N_2V_2$$

$$\frac{1}{100} \times V_1 = N_2 \times 100$$

$$N_2 = \frac{V_1}{10000}$$

Strength of dissolved oxygen $= N_2 \times \text{Eq. wt}$

$$= \frac{V_1}{10000} \times 8\,g/L$$

$$= \frac{V_1}{10000} \times 8 \times 1000\,mg/L$$

$$= 0.8\ V_1 \text{ ppm.}$$

RESULT The amount of dissolved oxygen in water = . . . ppm.

PRECAUTIONS

1. The water should be taken in the stoppered bottle very carefully without trapping air bubbles which could raise oxygen level by aerating the sample.

Note. Sodium azide is poisonous. It may also explode if exposed to heat.

2. $MnSO_4$ and alkaline iodine -azide solutions are added to the water sample just below the surface of water.
3. Whole of the precipitate of $MnO(OH)_2$ should be dissolved in H_2SO_4.

Instructions For the Lab Assistant:

1. $MnSO_4$ sulphate: Dissolve 50 g of $MnSO_4$ in distilled water and make up the volume to 1L.
2. Alkaline potassium iodide azide reagent: Dissolve 40 g of NaOH, 20 g of KI in distilled water, add 0.5 g of reagent grade sodium azide to the cooled solution and make up the volume to 100 ml.
3. N/40 $Na_2S_2O_3$ solution: Dissolve 6.20 g of $Na_2S_2O_3$ solution in water and make up this volume to 1L. Standardize with N/40 $K_2Cr_2O_7$ solution.
4. Starch solution.

EXERCISE

1. Which method is used to determine the amount of dissolved oxygen in water and why?

Ans. ***Winkler's method.*** *Oxygen is present in small amounts in water, at concentrations often less than 3×10^{-4} M. In contrast, air is roughtly 21% oxygen. Any technique used to measure dissolved oxygen must exclude atmospheric oxygen from the analysis or the sample is highly contaminated. Also, the technique must be highly precise, so that small differences in oxygen content of natural samples can be detected. In most of its versions, the Winkler method meets these two requirements.*

2. What is Dissolved Oxygen?

Ans. *Oxygen gas that is dissolved into water is dissolved oxygen, or DO. DO is essential for rivers, streams, lakes, and aquatic life. It is a positive sign to have DO in water. If there is an absence of oxygen in water, it can be a sign of severe pollution. Water ranges from very high levels of DO to very low levels of DO. Sometimes the DO is so low, there is hardly any aquatic life.*

3. Where does dissolved oxygen come from?

Ans. *Most DO in water comes from the atmosphere or plants. The oxygen comes from tumbling water on fast moving rivers that mix atmospheric oxygen with water. Photosynthesis is another way of oxygen entering water. DO levels are often highest in the afternoon because of high levels sunlight that cause plants and algae to photosynthesize rapidly. Because photosynthesis doesn't occur at night. DO levels exhibit diurnal and nocturnal cycles.*

4. What are the factrors which contribute to the DO?

Ans. *Many factors contribute to how high or low a DO level may be. Water temperature is one of the causes. Oxygen and other gases dissolve more easily is cooler water than in warmer water. Certain factors affect the water temperature. These factors are seasons of the year, time of day, and water depth. DO is highest just before dark because the* *plants have photosynthesized all day. At night, levels drop because the plants aren't photosynthesizing and are conducting respiration.*

5. What is eutrophication?

Ans. *Eutrophication is an increase in nutrients that can lead to an overgrowth of algae. The overgrowth doesn't allow the plants to photosynthesize. If the plants don't photosynthesize, the levels of oxygen decrease possibly causing oxygen sensitive organisms to die.*

6. What causes low DO?

Ans. *If BOD (Biochemical Oxygen Demand) is too high, the DO content of the water becomes depleted, and the DO content of the water becomes too low to support all the life in the water. If a body of water has large amounts of decaying vegetation, like leaves or aquatic microorganisms, then the amount of DO is reduced. Sometimes people dump raw sewage, garbage, grass, and other decay-able substances in the water. As these things decay, the amount of DO falls. When bacteria decompose food, leaves, and faeces, they require a great amount of oxygen, lowering the DO level. Urban and agricultural runoff, such as fertilizers, makes algae grow. The algae block the sunlight from the other plants, and the plants die so bacteria again decompose this organic matter.*

7. How does DO affect water supplies?

Ans. *High DO levels in a commonly water supply are good because it makes drinking water taste better. On the other hand, high DO levels speed up corrosion in water pipes. Industries use water with the least amount of DO. In boiler waters it is important to remove ALL oxygen. Slightly acid water containing both carbon dioxide (CO_2) and oxygen (O_2) will be corrosive. The acid water condition and CO_2 initiate corrosion, but O_2 is necessary for it to continue. Salt water holds less O_2 than fresh water.*

EXPERIMENT NO. 2.7

AIM **To determine Biological Oxygen Demand (BOD) of a given sample of water.**

APPARATUS Burette, pipette, conical flasks, beakers

CHEMICALS $MgSO_4$ solution (22.5 g/L) $CaCl_2$ solution (27.5 g/L) $FeCl_3$ solution (0.25 g/L), phosphate buffer solution, BOD bottles, $Na_2S_2O_3$.

THEORY The BOD in a water sample is directly related to the extent of pollution by sewage or other oxygen demanding waste. If the oxygen in water is *less than* Biological Oxygen Demand, the decomposition of microorganisms takes place anaerobically (in absence of oxygen) and hence produce highly objectionable end products including H_2S, NH_3 and CH_4.

Therefore sufficient oxygen must be present or supplied externally to maintain aerobic conditions so that organic matter decomposes into acceptable products. To know the amount to oxygen required by the given sample, the determination of BOD is necessary.

Polluted waters will continue to absorb oxygen for many months, till the oxidation gets completed and hence it is not practical to determine this ultimate oxygen demand which would require an extended period of time. As a result, it has become standard practice to simply measure the oxygen over a shorter period of 5 days. This is known as 5-day BOD test. **BOD_5 is the total amount of oxygen consumed by micro organisms during the first five days of biodegradation.**

In its simplest form BOD_5 would involve diluting a known volume of sample water with a known volume of pure water, whose oxygen content is already known. Let it be D_1. This is kept for 5 days at 20°C in a stoppered bottle, away from light. The dissolved oxygen is measured after the period of incubation. Let it be D_2. The difference between the original oxygen content and residual oxygen content will indicate the oxygen consumed by the water sample in five days in oxidizing the organic matter present in water sample. Thus BOD_5 is given by:

$$BOD_5 = D_1 - D_2 \times \frac{\text{Volume of sample after dilution}}{\text{Volume of sample before dilution}}$$

where D_1 = D.O of the sample in mg/L at the start of the experiment

D_2 = D.O of the sample in mg/L after 5 days.

= Loss of oxygen in mg/L × Dilution Factor.

The dilution of the waste water sample with pure water is necessary because waste water contains more of oxygen demanding material and hence the dissolved oxygen may not be sufficient to decompose it aerobically.

PROCEDURE

Preparation of dilution water:

1. Take 1000 ml of distilled water in a bottle. Add 1 ml each of phosphate buffer (pH = 7.2), $CaCl_2$ (27.5g/L), $MgSO_4$ (22.5g/L) and $FeCl_3$ (0.25g/L)
2. Water is aerated by passing compressed air through a diffusion tube untill it is completely saturated.

Determination.

3. Take about 5ml of the sample and dilute it to 300 ml with dilution water prepared above. (or any appropriate dilution)
4. The diluted sample is taken in two bottles filled upto the neck.
5. The dissolved oxygen in one bottle is determined immediately and in the other bottle after five days of incubation. The procedure for determining the DO is the same as given in experiment number: 5.

OBSERVATION

(i) Determination of DO in bottle No - 1

S. No.	*Volume of the solution taken in the titration flask (ml)*	*Burette Readings*		*Volume of the titrant used (Final-Initial reading) (ml)*
		Initial Reading	*Final Reading*	
1.				
2.				
3.				

(ii) Determination of DO in bottle No-2 (after incubation for 5 days

S. No.	*Volume of the solution taken in the titration flask (ml)*	*Burette Readings*		*Volume of the titrant used (Final-Initial reading) (ml)*
		Initial Reading	*Final Reading*	
1.				
2.				
3.				

Volume of sample taken for each titration = 100 ml

Normality of $Na_2S_2O_3$ used = N/100

CALCULATIONS

Volume of the sample before dilution = 5 ml.

Volume of the sample after dilution = 300 ml.

Note. The dilution depends upon the nature of the sample. For example .1 – 1% for industrial wastes, 1 – 5% for raw and settled sewage, 25 – 100% for polluted water, 2 – 5% for domestic waste water)

* (i) D.O in bottle no.– 1.

$$\underset{\text{(Hypo)}}{N_1V_1} = \underset{(O_2 \text{ in sample})}{N_2V_2}$$

$$\frac{1}{100} \times V_1 = N_2 \times 100$$

$$N_2 = \frac{V_1}{10000}$$

Strength of dissolved oxygen = N_2 × Eq. wt

$$(D_1) = \frac{V_1}{10,000} \times 0.8 \text{ g/L.}$$

$$= 0.8\ V_1 \text{ ppm.}$$

* (ii) D.O in bottle– 2.

Similarly find D.O in bottle-No-2

Let it be D_2

(iii) $$\text{BOD} = D_1 - D_2 \times \frac{\text{Volume of sample after dilution}}{\text{Volume of sample before dilution}} \text{ mg/L}$$

RESULT BOD of the given water sample = ...ppm

Instructions For The Lab Assistant.

Prepare the solutions as in the case of determination of dissolved oxygen in Experiment No.: 2.6

EXERCISE

1. What do you means by B.O.D?
2. What is BOD_5 test?
3. What is the significance of B.O.D. determination?
4. Why bad odour is developed in water if insufficient dissolved oxygen is present in water?
6. How is dilution water prepared?

Ans. *It is prepared from distilled water by seeding with a small amount of domestic waste water to provide a mixed population of bacteria or it can be prepared as given in the experiment.*

7. Why, while preparing dilution water is added, one ml each of phosphate, $FeCl_3$, $CaCl_2$, $MgSO_4$ etc.?

Ans. *Phosphate is added to maintain a pH of 7 and it also serves as a nutrient. Salt like $FeCl_3$ is added to supply nutrients to bacteria and Ca^{2+} Mg^{2+} is added to serve as trace elements needed for growth of bacteria.*

Note. The amounts of the solutions taken in the conical flasks for the determination of D.O is same as taken in Experiment 2.6, page - 118.

EXPERIMENT NO. 2.8

AIM **To find the chemical oxygen demand (COD) of a waste water sample using $K_2 Cr_2 O_7$.**

APPARATUS Burette, pipette, conical flask, beakers, round bottomed flask, condenser, water bath.

CHEMICALS N/4 $K_2Cr_2O_7$, N/4 Mohr's salt solution, ferroin indicator, sample solution, $HgSO_4$, silver sulphate reagent.

THEORY The chemical oxygen demand is defined as the amount of oxygen used while oxidizing the organic matter content of a sample with a strong chemical oxidant under acidic conditions.

A known amount of $K_2Cr_2O_7$ is added to the measured amount of the sample and mixture is boiled with conc. H_2SO_4. The organic matter gets completely oxidized to produce CO_2 and H_2O

$$C_xH_yO_z + \left(x + \frac{y}{4} - \frac{z}{2}\right) O_2 \rightarrow x\,CO_2 + \frac{y}{2} H_2O$$

The excess of $K_2Cr_2 O_7$ remained after the reaction is titrated against standard Mohr's salt solution using ferrion as indicator. The dichromate consumed is calculated which gives the oxygen required for the oxidation of organic matter.

INDICATOR Ferroin

END POINT Blue to wine red or (colourless to red)

PROCEDURE

1. Take 50 ml of the sample in a round bottomed flask and add 1 g of $HgSO_4$ and some broken porcelain pieces.
2. Immerse the flask in cold water and slowly add 75 ml silver sulphate reagents with continuous shaking. (As this is an exothermic process that is why the flask is immersed in cold water)

Notes 1. For details on $K_2Cr_2O_7$ titrations, refer to chapter – 1, page 57.

2. $HgSO_4$ is used to remove the chloride ions which may interfere because of their oxidation under the experimental conditions.
3. Ag_2SO_4 is used as a catalyst to achieve the oxidation of straight-chain aliphatic hydrocarbons and low molecular weight fatty acids.

3. Now add 25ml of $K_2Cr_2O_7$ (0.25N) to this solution and mix the contents of the flask.
4. Attach the reflux condenser and reflux for 2 hours.
5. Wash the condenser with distilled water into the flask. Cool and dilute to about 300ml by distilled water.
6. Add 2-3 drops of ferroin as indicator and titrate against 0.25N Mohr's salt solution till the end point *i.e. from blue to wine red.*
7. Record the volume of Mohr's salt solution used. Let it be X ml.
8. Perform a blank titration using distilled water in place of sample solution. For this, take 50ml of distilled water in a round bottomed flask, add the same amounts of reagents and reflux for two hours. Titrate it the same way as is done with the sample water.
9. Record the volume of Mohr's salt solution used. Let it be Y ml.

OBSERVATIONS

Volume of the sample taken = 50 ml.

Observation Table

(a) For sample solution

S. No.	*Volume of the solution taken in the titration flask (ml)*	*Burette Readings*		*Volume of the titrant used (Final-Initial reading) (ml)*
		Initial Reading	*Final Reading*	
1.				
2.				
3.				

For distilled water (blank titration)

S. No.	*Volume of the solution taken in the titration flask (ml)*	*Burette Readings*		*Volume of the titrant used (Final-Initial reading) (ml)*
		Initial Reading	*Final Reading*	
1.				
2.				
3.				

Volume of Mohr's salt solution used in case of sample solution = X ml (say)

Volume of Mohr's salt solution used in case of distilled water = Y ml (say)

$$\begin{bmatrix}\text{Volume of } K_2Cr_2O_7 \text{ used}\\ \text{to oxidize the organic}\\ \text{matter present in water sample}\end{bmatrix} = \begin{bmatrix}\text{Volume of Mohr's salt}\\ \text{solution equivalent to}\\ K_2Cr_2O_7 \text{ used}\end{bmatrix} = [Y - X]\text{ ml}$$

$$\underset{\text{(sample)}}{N_1V_1} = \underset{\text{(Mohr's salt solution)}}{N_2V_2}$$

$$N_1 \times 50 = \frac{1}{4}(Y - X)$$

$$N_1 = \frac{1}{4 \times 50}(Y - X)$$

Therefore COD = $N_1 \times$ Eq.wt of oxygen = $N_1 \times 8$ g/L

RESULT The COD of the given sample is = $N_1 \times 8 \times 1000$ mg/L

PRECAUTIONS

1. The addition of $Ag_2SO_4 - H_2SO_4$ to the sample should be done slowly with shaking and cooling during mixing.

2. It should always be added through the open end of the condenser.

3. Smaller volume of the sample should be taken.

Instructions to the Lab Assistant

Provide the following:

1. $\frac{N}{4}$ $K_2Cr_2O_7$ solution: Dissolve 12.25g of $K_2Cr_2O_7$ in distilled water and make up the volume to 1000 ml.

2. $\frac{N}{4}$ Ferrous ammonium sulphate (Mohr's salt solution): Weigh 98 g of Mohr's salt and transfer in a flask having a few drops of conc. H_2SO_4. Add distilled water to dissolve it. Make up the volume to 1000 ml.

3. Ferroin indicator: It can be purchased from the market as such or can be prepared in the lab. Dissolve 3.7125 g of O-phenanthroline monohydrate with 1.6375 g of $FeSO_4 \cdot 7H_2O$ in distilled water and make up the volume to 250 ml

4. $HgSO_4$ (solid)

5. Silver sulphate reagent: Dissolve 5g of Ag_2SO_4 in 500 ml of conc. H_2SO_4.

EXERCISE

1. Define COD?
2. What is the significance of COD determination?

Ans. *The COD is widely used for measuring the polluting properties of domestic and industrial wastes.*

(3) What is the structure of ferroin indicator?

Ans. *Ferroin is a ferrous complex of the organic compound 1, 10-phenanthroline*

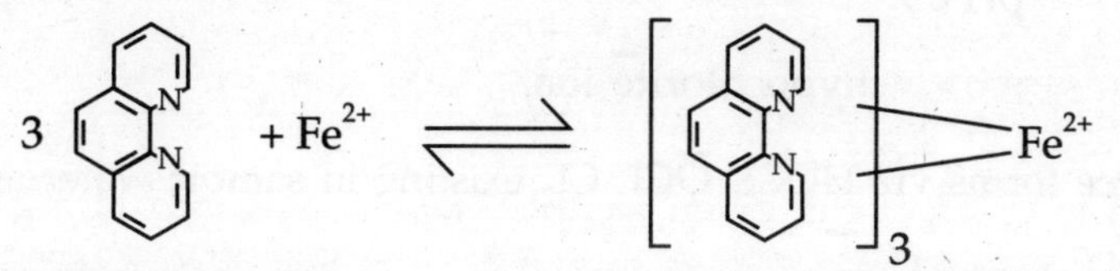

1, 10-phenanthroline Ferrous –1, 10- phenanthroline (Ferroin)

4. What is the colour of oxidized and reduced form of the indicator?

Ans. $(phen)_3\ Fe^{3+} + e^- \rightarrow (phen)_3\ Fe^{2+}$

pale blue *(Appears colourless sometimes)* — *Red* *(Reduced form)*

5. Why is $HgSO_4$ added to the sample water?

Ans. *The chloride ions if present, may interfere because of their oxidation under the experimental conditions. Thus $HgSO_4$ i.e. added to eliminate chloride ions by forming complexes.*

6. What is the use of adding silver sulphate in the sample to be refluxed?

Ans. *The straight-chain aliphatic compounds, aromatic compounds are oxidized more effectively if Ag_2SO_4 is used as a catalyst.*

RESIDUAL CHLORINE

It has been known that contaminated water is responsible for the spread of many contagious diseases, number of epidemics such as typhoid or cholera are water born. Water borne diseases are those acquired by ingestion of pathogens (disease causing bacteria) not only in drinking water, but also from water that makes it into persons mouth through washing food, utensils and hands.

The process of destroying these pathogenic bacteria and micro organisms is known as sterilization or disinfection. The chemicals used for the above purpose are called disinfectants. The various methods used for disinfection of water include, boiling, addition of bleaching powder, chlorine, U.V radiations etc. However chlorination of water is widely used for disinfecting water supply since chlorine is a powerful oxidizing agent and is cheaply available.

Chlorine is applied to water in its elemental form or as hypochlorite (bleaching powder)

Disinfecting Action of Chlorine. Cl_2 reacts with H_2O to form hypochlorous acid or hypochlorite ions which have immediate disastrous effect on microrganisms

$$Cl_2 + H_2O \xrightarrow{pH>5} HOCl + HCl.$$

The hypochlorous acid is unstable and may dissociate as

$$HOCl \underset{pH<7}{\overset{pH>8}{\rightleftharpoons}} H^+ + \overline{O}Cl$$

hypochlorite ion

All the three forms viz HOCl, $\overline{O}Cl$, Cl_2 existing in sample water are termed as *free chlorine.*

When HOCl react with NH_3 likely to be present in water, it forms chloramines.

$$NH_3 + HOCl \rightarrow NH_2Cl + H_2O$$

$$NH_2Cl + HOCl \rightarrow NHCl_2 + H_2O$$

$$NHCl_2 + HOCl \rightarrow NCl_3 + H_2O$$

These chloramines also have disinfecting properties and act as chlorine reserves. The chlorine in this form is called *combined chlorine.*

The chlorine is added to water upto the break point. The break point chlorination of water is defined as chlorination to such an extent that microorganisms as well as bad tastes and odours present in water are also destroyed.

However excess of free chlorine in drinking water is undesirable. An overdose of chlorine imparts an unpleasant taste to water and is also injurious to health.

Hence chlorine determination is necessary in those samples of water that have been treated with chlorine or hypochlorites (bleaching powder) so that necessary adjustments in dose rate can be made accordingly.

EXPERIMENT NO. 2.9

AIM **Determination of total residual chlorine in water**

APPARATUS Burette, pipette, conical flask and beakers.

CHEMICALS N/50 $Na_2S_2O_3$, Glacial acetic acid, KI, conc. HCl.

THEORY The estimation of residual chlorine is based on oxidation of KI by residual chlorine i.e by iodometric titration*.

When the water sample containing residual chlorine is treated with KI, residual chlorine (both free and combined) oxidizes KI and liberate I_2 in equivalent amount.

$$Cl_2 + 2KI \rightarrow 2KCl + I_2$$

$$\overline{O}Cl + 2KI \rightarrow 2KCl + I_2 + H_2O$$

$$HOCl + 2KI + HCl \rightarrow 2KCl + I_2 + NH_3$$

$$NHCl_2 + 2KI + 2HCl \rightarrow 2KCl + 2I_2 + NH_3.$$

The liberated I_2 is titrated against hypo solution, using starch as indicator

$$I_2 + 2Na_2S_2O_3 \rightarrow 2NaI + Na_2S_4O_6$$

$$I_2 + \text{Starch} \rightarrow \text{Deep blue complex.}$$

The end point is disappearance of blue colour.

PROCEDURE

1. Pipette out 100 ml of the given water sample in conical flask and add 2g of KI solution and about 3ml of glacial acetic acid to maintain pH upto 3-4.
2. Cover the flask and shake it well to mix the solution.
3. Using a wash bottle rinse the sides of the flask.
4. Titrate it with N/50 $Na_2S_2O_3$ solution from the burette till the solution becomes straw yellow.
5. Add 2ml of the starch solution. The solution will turn blue.
6. Continue titration with $Na_2S_2O_3$ till the blue colour disappears.
7. Note the final reading and repeat to get three concordant readings.

---

* For details on iodometric titrations, please refer to chapter 1, page 64.

OBSERVATIONS

Normality of the $Na_2S_2O_3$ solution = N/50

S. No.	*Volume of the solution taken in the titration flask (ml)*	*Burette Readings*		*Volume of the titrant used (Final-Initial reading) (ml)*
		Initial Reading	*Final Reading*	
1.				
2.				
3.				

CALCULATIONS

- Sample $Na_2S_2O_3$

$$N_1V_1 = N_2V_2$$

$$N_1 \times 100 = \frac{1}{50} \times V_2$$

$$N_1 = \frac{V_2}{5000}$$

$$\text{Total chlorine residuals} = N_1 \times \text{Eq. wt}$$

$$= N_1 \times 35.5 \text{g/L}$$

$$= N_1 \times 35.5 \times 1000 \text{ mg/L}$$

RESULT Amount of total residual chlorine in given water sample = ...mg/L

PRECAUTIONS

1. Chlorine vapours are harmful so the solution should not be sucked into the pipette with mouth.
2. The titration should be completed rapidly in order to avoid atmospheric oxidation of iodide.
3. First disappearance of blue colour may be taken as end point

Instructions to the Lab Assistant

Provide the following

1. N/50 $Na_2S_2O_3$ solution. Dissolve 4.96 *g* of $Na_2S_2O_2$ in distilled water and make up the volume to 1L. Standardize with N/50 $K_2Cr_2O_7$
2. Glacial acetic acid
3. KI (solid)
4. Starch solution.

EXERCISE

1. Why chlorine is added to water?
2. What are the disadvantages of excess of chlorine present in water.
3. What is the action of chlorine on pathogens?
4. Why and how pH is adjusted between 3 and 4.

Ans. pH is adjusted between 3-4 using acetic acid. At neutral pH, some combined chlorine residuals do not liberate I_2 by reacting with KI.

EXPERIMENT NO. 2.10

AIM **To determine the percentage of available chlorine in the supplied sample of bleaching powder.**

APPARATUS Burette, pipette, conical flask, beakers etc.

CHEMICALS N/10 $Na_2S_2O_3$, KI, dil acetic acid, starch solution, bleaching powder solution.

THEORY Bleaching powder is a mixture of calcium hypochlorite Ca $(OCl)_2$ and the basic chloride, $CaCl_2 \cdot Ca(OH)_2 \cdot H_2O$ Free slaked lime is also present in addition. *Available chlorine in bleaching powder is the amount of chlorine liberated by the action of dilute acids on bleaching powder according to the equation,*

$$OCl^- + Cl^- + 2H^+ \rightarrow \underset{\text{Available chlorine}}{Cl_2} + H_2O$$

The available chlorine is expressed as the percentage by weight of the bleaching powder. It is measured by iodometric titration*. The bleaching powder solution or suspension is treated with an excess of KI and then acidified with acetic acid.

$$OCl^- + 2I^- + 2H^+ \rightarrow Cl^- + I_2 + H_2O$$

The liberated I_2 is treated with $Na_2S_2O_3$ solution using starch as indicator near the end point.

$$I_2 + 2S_2O_3^{2-} \rightarrow 2I^- + S_4O_6^{2-}$$

---

* For details of iodometric titrations refer to chapter 1 page 64.

PROCEDURE

1. Rinse and fill the burette with N/10 $Na_2S_2O_3$ solution.
2. Pipette out 25 ml of bleaching powder solution in a conical flask.
3. Add 25 ml of distilled water, followed by 1 – 2g of solid KI and 10 ml of glacial acetic acid.
4. Keep the solution in dark for 3-4 minutes.
5. Titrate this solution with N/10 $Na_2S_2O_3$ till the brown colour of the solution changes to pale yellow.
6. Add 1-2ml of starch. The solution will turn blue.
7. Continue titration, till the blue colour disappears.
8. Repeat to get three concordant readings.

OBSERVATIONS

Normality of $Na_2S_2O_3$ solution $N_1 = 0.1N$

Observation Table.

S. No.	*Volume of the solution taken in the titration flask (ml)*	*Burette Readings*		*Volume of the titrant used (Final-Initial reading) (ml)*
		Initial Reading	*Final Reading*	
1.				
2.				
3.				

Volume of $Na_2S_2O_3$ used $V_2 = . . .$ ml.

CALCULATIONS

$$N_1V_1 = N_2V_2$$

(Bleaching powder) $\quad$ ($Na_2S_2O_3$)

$$N_1 \times 25 = \frac{1}{10} \times V_2$$

or

$$N_1 = \frac{1}{10} \times \frac{V_2}{25}$$

(Amount) Strength of available $Cl_2 = N_1 \times$ Eq.wt

$$= \left(\frac{1}{10} \times \frac{V_2}{25}\right) \times 35.5 g/L = zg/L$$

$$\% \text{ Available Chlorine} = z \times \frac{250}{1000} \times \frac{100}{w}$$

where w is the weight of bleaching powder (given) dissolved in 250 ml of distilled water for preparing the solution of bleaching powder

RESULT The available chlorine in the given sample is = ...%

PRECAUTIONS

1. The solution of bleaching powder is very unstable and hence should be titrated immediately after its preparation.
2. The solution should not be sucked into the pipette with mouth as chlorine vapours are harmful.
3. The bleaching powder solution should not be acidified in the absence of KI because some chlorine may be lost by doing so.

Instructions For The Lab Assistant

Provide the following:

1. **Solution of bleaching powder:** Weigh accurately 5g of bleaching powders into a watch glass. Add a little distilled water to form a paste. Transfer it in 250 ml flask with the help of water. Make up the volume to 250 ml with distilled water. A turbid solution will be obtained.
2. N/10 $Na_2S_2O_3$ solution
3. Starch solution
4. KI (Solid)
5. Glacial acetic acid.

DISSOLVED CHLORIDES

Chlorides are generally present in water in the form of NaCl, and Mg Cl_2 and may be due to leaching of marine sedimentary deposits, pollution from sea water, industrial or domestic wastes etc. Their concentrations above 250 mg/L produce noticeable salty taste in drinking water and are thus objectionable. Further, the presence of high quantity of chloride in river or stream waters may indicate pollution of water due to sewage and industrial wastes.

The presence of chlorides in the boiler water leads to problems like priming and foaming. Salts like $MgCl_2$ may undergo hydrolysis under the high pressure and prevailing high temperatures in boiler, leading to the formation of HCl. HCl leads to corrosion in boilers. Chlorides in the form of $CaCl_2$ and $MgCl_2$ cause hardness of water

The chloride concentration of raw waters should therefore regularly be tested.

EXPERIMENT NO. 2.11

AIM **To determine chloride ion in a given water sample by Argentometric method (Mohr's method)**

APPARATUS Burette, Pipette, conical flask, beakers.

CHEMICALS N/50 $AgNO_3$ solution, K_2CrO_4 indicator.

THEORY Chlorides ions can be determined by titration of the water sample against a standard solution of $AgNO_3$ using potassium chromate as indicator, (Argentometric titration) in a neutral or faintly alkaline medium.

This determination is based on precipitation titrations*. When $AgNO_3$ solution is added to the water sample in presence of K_2CrO_4, the chlorides present in it are precipitated as AgCl. As soon as all the chlorides are precipitated out, even a drop of $AgNO_3$ added in excess gives a red precipitate of silver chromate. This indicates the end point.

$$AgNO_3 + NaCl \rightarrow AgCl \downarrow + NaNO_3$$

$$AgNO_3 + K_2CrO_4 \rightarrow Ag_2\,CrO_4 + 2KNO_3$$

The pH of the solution should be in between 7 to 8 because at higher pH Ag^+ ion is precipitated as Ag OH

$$Ag^+ + \overline{O}H \rightarrow AgOH$$

If the pH is lower than 7 to 8 then the chromate ion is converted to dichromate ion

$$2CrO_4^{2-} + 2H^+ \rightleftharpoons Cr_2O_7^{2-} + H_2O$$

The required pH range can be achieved easily by adding a pinch of pure $CaCO_3$. Excess $CaCO_3$, being insoluble, does not interfere.

$$2H^+ + CaCO_3 \rightarrow Ca^{2+} + CO_2 + H_2O$$

PROCEDURE

Blank Correction

1. Transfer 50 ml of distilled water in a conical flask and add 3-4 drops of K_2CrO_4 indicator.

---

* **Note.** For details on precipitation titrations, refer to chapter 1, page 75.

2. Titrate it against N/50 $AgNO_3$ solution till the brick red colour appears in the solution.

3. Note the reading and repeat to get three concordant readings. Let it be X ml. This volume has to be deducted from the volume of the $AgNO_3$ used in titrating the sample water as below.

Sample Titration

4. Pipette out 50 ml of water sample and add a drop of methyl orange. If red or pink colour appears, add a pinch of $CaCO_3$, so that red or pink colour vanishes.
5. Add 1 ml of K_2CrO_4 indicator solution and titrate against N/50 $AgNO_3$ taken in the burette until the end point i.e a permanent tinge of red colour appears.
6. Note the reading and repeat to get three concordant readings. Let it be Y ml

OBSERVATIONS

(a) For Blank Correction

S. No.	*Volume of the solution taken in the titration flask (ml)*	*Burette Readings*		*Volume of the titrant used (Final-Initial reading) (ml)*
		Initial Reading	*Final Reading*	
1.				
2.				
3.				

Let it be X ml.

(b) For titration of the sample

S. No.	*Volume of the solution taken in the titration flask (ml)*	*Burette Readings*		*Volume of the titrant used (Final-Initial reading) (ml)*
		Initial Reading	*Final Reading*	
1.				
2.				
3.				

Let it be Y ml.

Normality of $AgNO_3$ solution used = N/50.

Volume of N/50 $AgNO_3$ used for the titration of sample

$$\text{water} = Y - X \text{ ml} = V_1 \text{ ml.}$$

$$\underset{(Ag\,NO_3)}{N_1V_1} = \underset{(Cl^-\text{ ion})}{N_2V_2}$$

$$\frac{1}{50} \times V_1 = N_2 \times 50$$

$$N_2 = \frac{1}{50 \times 50} \times V_1$$

$$\begin{aligned}\text{Strength of } Cl^- \text{ ions} &= N_2 \times \text{Eq.wt of } Cl^- \\ &= N_2 \times 35.5\text{g/L} \\ &= N_2 \times 35.5 \times 1000 \text{ mg/L.}\end{aligned}$$

RESULT The chloride ion concentration in the given sample is ....mg/L.

PRECAUTIONS

1. The glass apparatus should be washed with distilled water.
2. Volume of indicator should be same in all the titrations.
3. The end point should be noted when a tinge of brownish red colour persists.

Instructions to the Lab Assistant

Provide the following:

1. N/50 $AgNO_3$ solution: Dissolve 3.398 g of $AgNO_3$ in distilled water and make up the volume to IL
2. K_2CrO_4 indicator solution

EXERCISE

1. What is the significance of Cl estimation?
2. Why a pH of 7 to 8 is maintained?
3. What is the principle of argentometric titrations.
4. How the solution is neutralized if it is acidic?

❑❑❑

3

Ores and Alloy Analysis

Volumeretric analysis is widely employed for ores and alloy analysis. The solution of the ore is prepared in the appropriate solvent and the ore is estimated by employing a suitable titration.

Determination of Fe^{2+} through $K_2Cr_2O_7$ Titrations

EXPERIMENT NO. 3.1

AIM **Determination of percentage of Fe in an iron ore/alloy by redox titration using N-phenyl anthranilic acid as internal indicator.**

THEORY The estimation of iron in an iron ore or alloy can be done by potassium dichromate titrations.* The common iron ores are: Haematite (Fe_2O_3), Magnetite (Fe_3O_4), Siderite ($FeCO_3$). When these ores or alloys are dissolved, the iron is usually obtained in the form of Fe^{3+} ions. The Fe^{3+} ions present in the ore solution are converted to Fe^{2+} ions and then the resulting solution is titrated against standard $K_2Cr_2O_7$ solution in the presence of acid. The reducing agent used for converting Fe^{3+} to Fe^{2+} is metallic zinc or stannous chloride.

(*i*) Reduction of Fe^{3+} to Fe^{2+}

$$2\,FeCl_3 + SnCl_2 \rightarrow 2FeCl_2 + SnCl_4.$$

$$2\,HgCl_2 + \underset{\text{(excess)}}{SnCl_2} \rightarrow Hg_2Cl_2 + SnCl_4.$$

$HgCl_2$ is added to remove excess of $SnCl_2$.

(ii) Titration

$$Cr_2O_7^{2-} + 6Fe^{2+} + 14H^+ \rightarrow 2Cr^{3+} + 6Fe^{3+} + 7H_2O$$

* **Note.** For details on $K_2Cr_2O_7$ titrations refer to chapter-1, Volumetric analysis, page - 57.

When all the Fe^{2+} ions are oxidized to Fe^{3+} by $K_2Cr_2O_7$, an extra drop of $K_2Cr_2O_7$ oxidizes the indicator, N-phenyl anthranilic acid which turn the solution purple red because the indicator appears purple red in oxidized form.

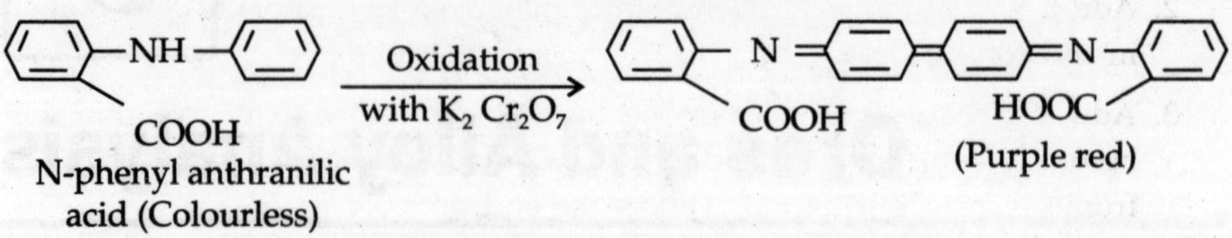

APPARATUS Burette, pipette, conical flask, beakers etc.

CHEMICALS $K_2Cr_2O_7$, 5% $SnCl_2$ solution, N-phenyl anthranilic acid, conc. H_2SO_4 conc. HCl, $HgCl_2$ solution (saturated).

INDICATOR N-phenyl anthranilic acid.

END POINT Green to light purple red.

PROCEDURE.

(I) Preparation of the ore/alloy solution

1. Weigh out accurately 2g of ground dried ore on a watch glass.
2. Transfer it carefully into a 100 ml conical flask and add about 10-15 ml of conc. HCl.
3. Warm the contents gently until the residue is not having any coloured material or only a whitish silica precipitate remains. (Boiling should be avoided).
4. Transfer the residue and the washings in 250 ml measuring flask and make up the solution to 250 ml.

(II) Reduction of Fe^{3+} to Fe^{2+} in the ore solution

1. Pipette out 20 ml of the above solution in the conical flask.
2. Add 2-3 ml of conc. HCl and heat the solution to boiling for 1-2 minutes.
3. Add $SnCl_2$ solution to the hot solution dropwise, till the yellow colour of the solution (due to $FeCl_3$) gets discharged. Cool the solution under tap water and dilute to 100 ml with distilled water.
4. Add 1-2 ml of saturated $HgCl_2$ solution in one lot to oxidize excess of $SnCl_2$ solution. A white turbidity is obtained.

(III) Titration with $K_2Cr_2O_7$

1. Rinse and fill the burette with 0.1 $K_2Cr_2O_7$.
2. Add a few drops of N-phenyl anthranilic acid (indicator) to the solution in the conical flask.
3. Add $K_2Cr_2O_7$ solution from the burette very slowly (the solution will become green). Keep on adding $K_2Cr_2O_7$ untill the solution changes colour from green to purple red. This is the end point.
4. Take three concordant readings.

OBSERVATIONS

Weight of iron ore taken = wg.

Volume of the iron ore solution prepared = 250 ml

$$\therefore \text{ Weight of ore presents in 1000 ml} = \frac{1000}{250} \times \omega = 4\ \omega g.$$

Observation Table.

S. No.	*Volume of the solution taken in the titration flask (ml)*	*Burette Readings*		*Volume of the titrant used (Final-Initial reading) (ml)*
		Initial Reading	*Final Reading*	
1.				
2.				
3.				

CALCULATION

$$(Fe^{2+}) \quad (K_2Cr_2O_7)$$

$$N_1V_1 = N_2V_2$$

$$N_1 \times 20 = \frac{1}{10} \times V_2$$

$$N_1 = \frac{V_2}{10 \times 20} = \frac{V_2}{200}$$

$$\text{Strength of } Fe^{2+} \text{ in the solution} = N_1 \times \text{Eq. wt} = \frac{V_2}{200} \times 56\ g/L$$

$$\text{Hence, percentage of iron in the ore} = \frac{56\ V_2}{200} \times \frac{100}{x}$$

Note. The same experiment can be done using diphenylamine as indicator. In that case along with the indicator add 5 ml of syrupy phosphoric acid also. The end point will be from green to violet blue

RESULT The percentage of iron in iron ore =...%

PRECAUTIONS.

1. $SnCl_2$ solution should be added dropwise and the solution should be shaken after each addition.
2. The addition of $HgCl_2$ should be in one lot.
3. If on reduction, no precipitate appears or if grey precipitate of metallic mercury are formed then reject the solution and repeat the process of reduction

Instruction for the Lab Assistant

Provide the following:

1. N/10 $K_2Cr_2O_7$	:	Dissolve 4.9 g of dry solid $K_2Cr_2O_7$ in distilled water and make up the volume to 1L
2. $SnCl_2$ solution	:	Dissolve 5g of $SnCl_2$ in 100 ml water.
3. $HgCl_2$ solution	:	Saturated solution
4. N-phenylanthranilic acid	:	Dissolve 0.25 gm of solid in 12 ml of N/10 NaOH and dilute to 250 ml with water.
or		
Diphenylamine	:	Dissolve 1g of diphenylamine in 100 ml of conc. H_2SO_4

EXPERIMENT NO. 3.2

AIM **To determine the ferrous content in the supplied sample of iron ore by titrimetric analysis against standard $K_2Cr_2O_7$ solution using potassium ferricyanide [K_3 Fe $(CN)_6$] as external indicator.**

THEORY The iron ore is dissolved in a suitable acid, the iron obtained is generally in the form of Fe^{3+} ions. These Fe^{3+} ions are reduced to Fe^{2+} ions by adding $SnCl_2$ and the resulting solution is titrated against $K_2Cr_2O_7$ solution in the presence of acid.

Note. For details of $K_2Cr_2O_7$ titrations using external indicator, refer to chapter-1, page - 57.

(*i*) Reduction of Fe^{3+} to Fe^{2+}

$$2\ FeCl_3 + SnCl_2 \rightarrow 2FeCl_2 + SnCl_4.$$

$$2\ HgCl_2 + SnCl_2 \rightarrow Hg_2Cl_2 + SnCl_4.$$

$HgCl_2$ is added to remove excess of $SnCl_2$.

(ii) Titration

$$Cr_2O_7^{2-} + 6Fe^{2+} + 14H^+ \rightarrow 2Cr^{3+} + 6Fe^{3+} + 7H_2O$$

Potassium ferricyanide is used as external indicator which gives a greenish blue colour with Fe^{2+} ions due to the formation of ferrous-ferricyanide

$$3Fe^{2+} + 2\ [Fe\ (CN)_6]^{3-} \rightarrow Fe_3\ [Fe\ (CN)_6]_2$$

When all the Fe^{2+} are converted to Fe^{3+} ions at the end point, the indicator do not give blue colour when treated with a drop of titration mixture.

INDICATOR Potassium ferricyanide $[K_3Fe\ (CN)_6]$

END POINT Blue to colourless

PROCEDURE

(I) Preparation of the ore solution. Prepare in the same way as given in experiment 3.1.

(II) Reduction of Fe^{3+} to Fe^{2+} in the ore solution.

1. Pipette out 20 ml of the above solution in the conical flask.
2. Add 2-3 ml of conc. HCl and heat the solution to boiling for 1-2 minutes.
3. Add $SnCl_2$ solution to the hot solution dropwise, till the yellow colour of the solution (due to $FeCl_3$) gets discharged. Cool the solution under tap water and dilute to 100 ml with distilled water.
4. Add 1-2 ml of saturated $HgCl_2$ solution in one lot to oxidize excess of $SnCl_2$ solution. A white turbidity is obtained.

(III) Titration with $K_2Cr_2O_7$

1. Rinse and fill the burette with 0.1 N $K_2Cr_2O_7$.
2. Place small drops of potassium ferricyanide solution (1%) on a dry white tile by means of a thin glass rod.
3. Titrate the solution in the conical flask (obtained in step II) with $K_2Cr_2O_7$ solution.

4. After adding one ml of $K_2Cr_2O_7$ solution in the conical flask, withdraw a drop of the solution from the flask with a thin glass rod and mix it with one of the drop of potassium ferricyanide on the tile. Appearance of blue colour indicates that end point has not yet reached.
5. Continue the titration with $K_2Cr_2O_7$ and repeat the process of withdrawing a drop of solution and mixing it with $[K_3Fe(CN)_6)]$ on the tile. When the indicator-drop does not give blue colour, the end point is reached. Note the reading.
6. Repeat the titration to get the exact end point.
7. Take three concordant readings.

OBSERVATIONS

Weight of iron ore taken = ω.g.

Volume of the solution prepared = 250 ml

$$\therefore \text{ Weight of ore present in 1000 ml} = \frac{1000}{250} \times \omega = 4\ \omega g.$$

Observation Table

S. No.	*Volume of the solution taken in the titration flask (ml)*	*Burette Readings*		*Volume of the titrant used (Final-Initial reading) (ml)*
		Initial Reading	*Final Reading*	
1.				
2.				
3.				

CALCULATIONS

$$(Fe^{2+})\ (K_2Cr_2O_7)$$

$$N_1V_1 = N_2V_2$$

$$N_1 \times 20 = \frac{1}{10} \times V_2$$

$$N_1 = \frac{V_2}{10 \times 20} = \frac{V_2}{200}$$

$$\text{Strength of } Fe^{2+} \text{ in the solution} = N_1 \times \text{eq. wt} = \frac{V_2}{200} \times 56 \text{ g/L}$$

$$\text{Hence, percentage of iron in iron ore} = \frac{56\ V_2}{200} \times \frac{100}{x}$$

RESULT The percentage of iron in iron ore = ...

PRECAUTIONS Similar to previous experiment.

Instructions to Lab Assistant

Provide all the solutions as in the case of previous experiment, except the indicator.

Indicator solution, [K_3 Fe $(CN)_6$]. Take small crystal of pure [$K_3Fe(CN)_6$] in a clean test tube and wash it repeatedly with distilled water to remove any potassium ferrocyanide adhering on the surface. Reject the washings and dissolve the crystal in 1.5ml of water. (Indicator solution should always be prepared fresh as $K_3Fe(CN)_6$ solution slowly changes to $K_4Fe(CN)_6$.

EXERCISE

1. Why only a very small excess of $SnCl_2$ is added during the reduction of Fe^{3+})?

Ans. *If large excess of $SnCl_2$ is added then a much larger amount of $HgCl_2$ will have to be added to remove the excess of $SnCl_2$.*

2. Why the solution is cooled after reduction is complete?

Ans. *To avoid oxidation of Fe^{2+} to Fe^{3+} which takes place in hot solution.*

3 What is the function of adding H_3PO_4 during the titration involving diphenylamine as indicator?

Ans. *Refer to chapter-1, page 61*

4 What will happen if $K_3[Fe(CN)_6]$ is added to the titration flask?

Ans. *It will produce deep blue colour forming complex with Fe^{2+} ions which will not be available for titration with $K_2Cr_2O_7$ and also the end point cannot be detected.*

DETERMINATION OF Cu^{2+} THROUGH IODOMETRIC TITRATION

EXPERIMENT NO. 3.3

AIM **Determine the amount of Cu in the copper ore solution provided hypo solution (0.1 N)**

APPARATUS Burette, pipette, beakers, titration flask.

THEORY This experiment comes under the category of iodometry* which is used widely in the analysis of ores, alloys etc.

When an excess of KI is added to the solution containing Cu^{2+} in neutral or slightly acidic medium, quantitative liberation of iodine takes place.

$$2\,CuSO_4\;5H_2O + 4KI \rightarrow 2\,CuI_2 + 2CuSO_4 + 5H_2O$$

$$2\,CuI_2 \rightarrow Cu_2I_2 + I_2$$

This liberated iodine is then titrated against standard $Na_2S_2O_3$ solution using starch solution as indicator near the end point.

$$2Na_2\,S_2O_3 + I_2 \rightarrow Na_2S_4O_6 + 2NaI$$

(Sodium Tetrathionate)

Ionic equation will be:-

$$2\,Cu^{2+} + 2I^- \rightarrow Cu_2^{2+} + I_2 + 2e^-$$

$$I_2 + 2S_2O_3^{2-} + 2e^- \rightarrow S_4O_6^{2-} + 2I^-$$

From the above equation it is evident,

$$2Cu^{2+} \equiv I_2 \equiv 2S_2O_3^{2-} \equiv 2e^-$$

The equivalent weight of Cu^{2+} will be one half of twice the molecular weight since the reaction involves two electrons per two moles of Cu^{2+}.

$$\text{Equivalent weight of } Cu^{2+} = \frac{2\times 63.5}{2} = 63.5.$$

From the normality equation, normality of Cu^{2+} and hence the strength of Cu^{2+} in hydrated $CuSO_4$ crystals is determined.

Note. For details on iodometric titrations, please refer to chapter 1-iodometric titration, page 64

The titration fails when any mineral acid is present in the solution and therefore before commencing the titration the acid present should be neutralized. This is done by dropwise addition of a solution of Na_2CO_3 until a slight precipitate or turbidity appears.

$$2H^+ + CO_3^{2-} \rightarrow CO_2 + H_2O$$

$$Cu^{2+} + CO_3^{2-} \rightarrow \underset{\text{(Turbidity)}}{CuCO_3}$$

The turbidity is removed by dropwise addition of dilute CH_3COOH.

$$CuCO_3 + 2CH_3COOH \rightarrow Cu^{2+} + CO_2 + 2CH_3COO^- + H_2O$$

Moreover, the precipitate of Cu_2I_2 absorbs I_2 from the solution and releases it slowly making the detection of sharp end point difficult, so a small amount of NH_4SCN is added near the end point to displace the absorbed iodine from Cu_2I_2 precipitate

INDICATOR Starch solution

END POINT Blue to Colourless

PROCEDURE

Preparation of Solution of ore

1. Take about 1 g of the finely ground copper ore in a beaker and warm with 15-20 ml of conc. HNO_3 until the copper passes into solution.
2. Evaporate to a volume of 5 ml and add 10 ml of conc. HCl and equal volume of 1:1 H_2SO_4. Again evaporate till dense fumes of SO_3 appear indicating complete removal of nitrogen oxides.
3. Heat the mixture with 15 ml of bromine solution and an equal volume of water to complete oxidation of any As and Sb to the pentavalent state.
4. Boil off excess of bromine and dilute the resulting solution to 100 ml in a measuring flask.

Titration.

1. Rinse and fill the burette with 0.1 N $Na_2S_2O_3$ solution
2. Measure 20 ml of the ore solution into a conical flask.
3. Neutralize any free acid present by adding Na_2 CO_3 solution drop by drop till a faint permanent precipitate remains even on shaking.

4. Add dil. acetic acid drop wise until precipitate just dissolves
5. Add 1 gm of solid Kl, cover the mouth of conical flask by filter paper and allow the mixture to stand for 2 to 5 minutes in the dark The solution becomes brown due to liberated iodine.
6. Titrate the liberated iodine with hypo solution added from the burette, the brown colour of iodine becomes fainter as the addition by hypo solution proceeds and when only a faint yellow colour remains add 2ml starch solution. This immediately form a deep blue iodo-starch complex
7. Now add hypo solution further, drop by drop when only milky white colour remains.

Observation Table

S. No.	*Volume of the solution taken in the titration flask (ml)*	*Burette Readings*		*Volume of the titrant used (Final-Initial reading) (ml)*
		Initial Reading	*Final Reading*	
1.				
2.				
3.				

CALCULATIONS.

$$\begin{aligned} &\text{(Ore Soln.)} \quad (Na_2S_2O_3) \\ N_1V_1 &= N_2V_2 \\ N_1 \times 20 &= 0.1 \times V_2 \\ N_1 &= \frac{0.1 \times V_2}{20} \\ N\,(Cu^{2+}) &= \frac{V_2}{200} \\ \text{Strength of } Cu^{2+} &= \frac{V_2}{200} \times 63.5\, g/L \\ \text{Percentage of Cu in ore solution} &= \frac{V_2}{200} \times \frac{63.5}{\text{Strength of ore Solution}} \times 100 \end{aligned}$$

RESULT. The amount of Cu present in copper ore solution is.......

---

* **Note.** If in the ore solution Fe^{2+} is present as an impurity, then instead of $NaHCO_3$, a 1:1 solution of ammonia is added dropwise.

PRECAUTIONS

1. The solution of $Na_2S_2O_3$ is always taken in the burette in iodometric titrations.
2. Sufficient amount of KI solution is to be added for the reason explained earlier.
3. The indicator should be added just before the end point.
4. NH_4SCN should be added near the end point to displace the adsorbed iodine and to get a sharp end point.

EXERCISE

Refer to Chapter-I, iodometric titrations.

❑❑❑

4

Potentiometry and pH metric Measurements

There are several analytical methods that are based upon oxidation and reduction reactions. These include potentiometry, colometry, voltametry etc. Some of the fundamentals that are necessary to understand the theory of these measurements are as follows:

(i)**REDOX REACTIONS** Various chemical reactions occur through redistribution of electrons among the reacting substances. Any substance that loses electrons is said to be oxidized and the one which gains electrons is said to be reduced.

However, in a chemical reaction, a substance can lose electrons only if there is present another substance which can gain electrons. This implies that oxidation can take place only if reduction also occurs at the same time or vice-versa. These reactions in which reduction and oxidation take place are called redox reactions.

$$Zn(s) + Cu^{2+}_{(aq)} \rightarrow Zn^{2+}_{(aq)} + Cu_{(s)}$$

In this $Zn_{(s)}$ loses two electrons whereas Cu^{2+} gains two electrons, and hence this is a redox reaction.

The redox reactions are of two types:

(a) ***Direct redox reactions.*** The redox reactions in which oxidation and reduction take place in the same vessel are called direct redox reactions. Example, displacement of copper from $CuSO_4$ solution when a zinc rod is dipped in it.

$$Zn + CuSO_4 \rightarrow ZnSO_4 + Cu_{(s)}$$

or $$Zn_{(s)} + Cu^{2+} \rightarrow Zn^{2+} + Cu_{(s)}$$

The above reaction can be split into two halves,

$Zn \rightarrow Zn^{2+} + 2e^-$ (oxidation half reaction)

$Cu^{2+} + 2e^- \rightarrow Cu$ (reduction half reaction)

(b) ***Indirect redox reaction.*** If redox reaction is allowed to take place in such a way that oxidation reaction takes place in one beaker and the reduction half reaction in another beaker, the electrons given out by former will be taken by latter and a current will flow. These indirect redox reactions form the basis of electro chemical cells.

(ii) **ELECTRODE POTENTIAL** A metal rod or plate in contact with its own ions is called electrode. When a metal is in contact with the solution of its own ions, it can undergo oxidation or reduction. *The tendency of an electrode to lose or gain electrons is called electrode potential.* Since each electrode represents a half cell, therefore electrode potential is also called potential for half cell.

Consider a copper rod dipped in the solution of its own ions, i.e. Cu^{2+} ions. The copper rod has a tendency to lose electrons to form Cu^{2+} ions which pass into the solution. Thus, copper rod acquires a net negative charge because electrons which are released get accumulated around the Cu rod. These electrons will attract Cu^{2+} ions from the solution. This indicates that copper rod will not keep on losing electrons and a equilibrium will soon be established.

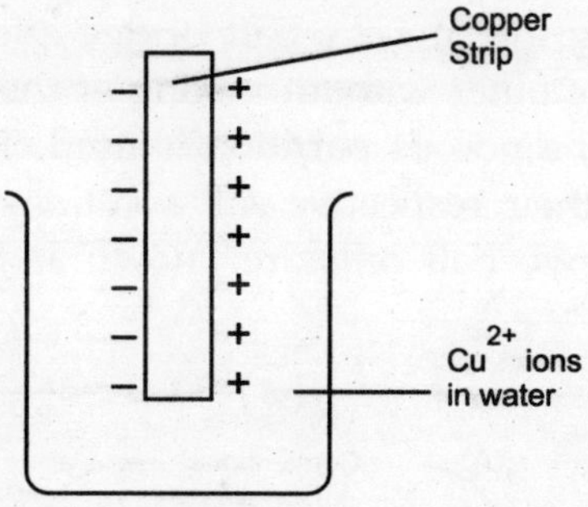

Fig.4.1 *Development of a potential difference between metal and its ions.*

$$Cu(s) \rightleftharpoons Cu^{2+}_{(aq)} + 2e^-$$

When such a equilibrium is reached, a separation of positive and negative charges would occur. This separation creates a potential difference between metal rod and the solution. This potential difference is called the **electrode potential**. Its magnitude depends upon-

(a) The nature of the metal and ions

(b) Concentration of ions

(c) Temperature

Electrode potential measured under standard conditions i.e. 1 molar concentration of metal ions and a temperature of 298 K are called standard electrode potentials and are denoted by E^o. A single electrode potential cannot be measured directly. It can be indirectly measured by coupling the electrode with another electrode called reference electrode.

(iii) ELECTROCHEMICAL CELL A device used to convert the chemical energy produced in a redox reaction into electrical energy is called an electrochemical cell or simply a chemical cell. These are also called galvanic cells or voltaic cell.

A simple voltaic cell in which the reaction

$$Zn\,(s) + CuSO_{4\,(aq)} \rightarrow ZnSO_{4\,(aq)} + Cu_{(s)}$$

takes place is shown in (Fig.4.2)

The anode in voltaic cell is zinc rod immersed in zinc sulphate solution while the cathode is copper rod immersed in copper sulphate solution placed in two separate beakers. The metallic rods in the two beakers are connected to a galvanometer by means of an insulated wire. The solution in two beakers are connected by an inverted U-tube known as salt bridge. The tube is filled with a solution of electrolyte such a KNO_3, KCl or $NH_4\,NO_3$ which does not undergo chemical change during the process. Salt bridge allows the movement of ions from one solution to the other without mixing of the two solutions. Sometimes, instead of a salt bridge, a porous partition is used. The two beakers containing metal rods dipped in their respective salt solutions are called **half cells**. The reactions occurring in two half cells are known as **half reactions**.

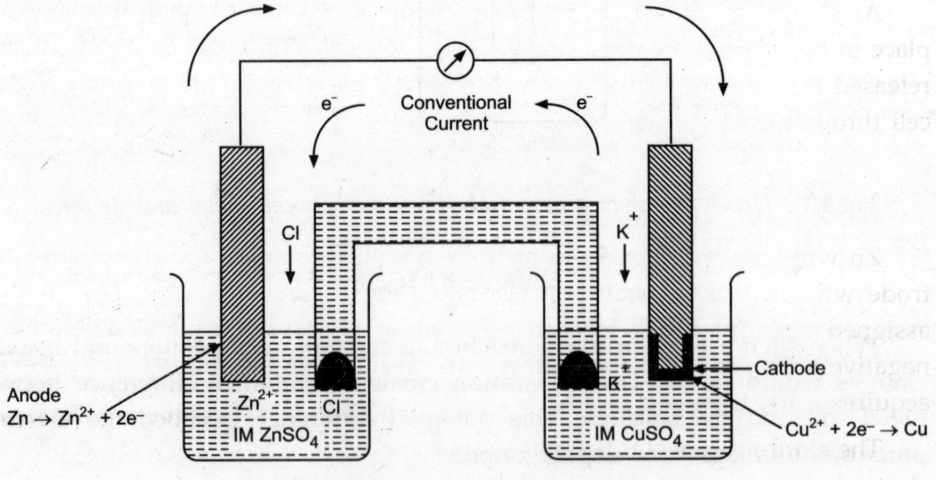

Fig. 4.2 *Electrochemical cell*

(iv) HALF REACTIONS AND ELECTRODES A redox reaction is a reaction in which there is a transfer of electrons from one substance to another. Any redox reactions may be expressed as the sum of the two half reactions. One of the half reaction shows loss of electrons and other shows the gain of electrons. For example, the reaction occurring in Galvanic cell

$$Zn_{(s)} + Cu^{2+}_{(aq)} \rightarrow Zn^{2+}_{(aq)} + Cu_{(s)}$$

can be expressed as the sum of the following two half reactions.

$$Cu^{2+}_{(aq)} + 2e^- \rightarrow Cu(s) \qquad \text{(Reduction of } Cu^{2+}\text{)}$$

$$Zn(s) \rightarrow Zn^{2+}_{(aq)} + 2e^- \qquad \text{(Oxidation of Zn)}$$

The half cell in which oxidation takes place is called **oxidation half cell** while the other in which reduction takes place is called **reduction half cell**. In common practice, both the half reactions are written as reduction reactions and then the overall reaction is the difference of the two

$$Cu^{2+}_{(aq)} + 2e^- \rightarrow Cu(s) \qquad \text{(Reduction of } Cu^{2+}\text{)}$$

$$Zn^{2+}_{(aq)} + 2e^- \rightarrow Zn(s) \qquad \text{(Reduction of } Zn^{2+}\text{)}$$

$$Cu^{2+}_{(aq)} + Zn_{(s)} \rightarrow Cu_{(s)} + Zn^{2+}_{(aq)} \qquad \text{(Over all)}$$

The reduced and oxidized substances in a half reaction form a redox couple, denoted as Ox/Red. In this reaction Cu^{2+}/Cu and Zn^{2+}/Zn are the redox couple.

As shown in Fig. 4.2, in an electrochemical cell, the two half reactions take place in two different compartments. When the reaction takes place, the electron released in one compartment travels through the external circuit and enter the cell through the other electrode. For example,

$$Zn(s) \rightarrow Zn^{2+}_{(aq)} + 2e^- \qquad \text{(oxidation)}$$

Zn will lose the electrons which are travelled through the circuit. The electrode where oxidation half reaction takes place is called **anode**. The anode is assigned negative charge, because oxidation is accompanied by liberation of negatively charged electrons. These are given to the electrode which, thereby acquires a negative charge.

The electrons given at the anode are travelled through the circuit to the other electrode where the reduction half reaction will take place by gain of electrons.

In galvanic cell, the electrons lost by anode are accepted by the copper ions to form copper metal.

$$Cu^{2+}_{(aq)} + 2e^- \rightarrow Cu(s)$$

The electrode at which reduction takes place is called **cathode**. Since the copper electrode requires the electrons for reduction, therefore, the electrode is electron deficient and pulls the electrons from the external circuit. Thus the cathode is assigned positive charge.

When the electrons flow from anode (zinc rod) towards cathode (copper rod), the electrical energy is produced. The direction of conventional current is opposite to that of flow of electrons. *Thus, an electrochemical cell converts the chemical energy into electrical energy.*

Representation of Electrochemical Cell

The electrochemical cell is represented as

Anode	Salt bridge	Cathode
Zn/Zn^{2+}	KNO_3	Cu^{2+}/Cu

The electrode at which oxidation takes places (anode) is written on the left hand side. The electrode at which reduction takes place (cathode) is written on the right hand side of the salt bridge.

In some galvanic cells, the same electrolyte is used in two half cells, In such cases, no separation is shown between two electrolytes of two electrodes For example

$$Pt, H_2 \text{ (1 atm); HCl (0.1M); } Cl_2, Pt$$

Here HCl is an electrolyte and provides H^+ ions to the hydrogen electrode and Cl^- ion to the chlorine electrode.

Cell Potential (E.M.F. of the Cell) An electrochemical cell is made up of two electrodes, one of these electrodes must have a higher electrode potential (higher tendency to lose electrons) than the other electrode. As a result of this potential difference, the electrons flow from the electrode at higher potential to the electrode at lower potential. A cell that has not reached equilibrium, can do the electrical work till a transfer of electron takes place. *The electron transfer depends upon the potential difference between two electrodes. This potential difference is called the cell potential* and measured in *volts*. This potential is also known as *electromotive force of the cell*. When the cell potential is large, a given number of electron travelling between the electrodes can do a large amount of electrical work. When the cell potential is small, the work done by the same number of electrons will be small. When the cell potential is zero, the reaction is at equilibrium and no work is done.

The maximum electrical work done by a system is given by value of ΔG

$$W_{max} = \Delta G \quad \text{at the constant temperature and pressure.}$$

Therefore, the e.m.f. of a cell or cell potential is a measure of overall tendency which determines the spontaneity of a reaction to take place.

(v) MEASUREMENT OF ELECTRODE POTENTIAL A galvanic cell consists of the conjuction of two electrodes, and each making a characteristic contribution to the cell potential. Although it is not possible to determine the absolute value of the electrode potential of a single electrode, but it can be determined with respect to a reference electrode. The reference electrode used is the standard hydrogen electrode (SHE) which is assigned the value zero.

$$Pt \mid H_2(g) \mid H^+_{(aq)} \quad E^0 = 0 \text{ at all temperature.}$$

Thus, an electrode potential is the potential of a cell that has a standard hydrogen electrode as the anode.

Cell potential can be measured through potentiometric analysis. The cell for this analysis can be depicted as:

Reference electrode / Salt Bridge / Analyte solution / Indicator electrode

The **reference electrode** is the electrode with an accurately known electrode potential independent of the concentration of the solution under study. The **indicator electrode** is the electrode which responds rapidly to the changes in the concentration of the solution under study.

The various electrodes are discussed below:

1. REFERENCE ELECTRODES

(a) Standard Hydrogen Electrode (SHE) Standard hydrogen electrode contains hydrogen gas in contact with H^+ ions of 1 M concentration at 298 K and at 1 atmospheric pressure (Fig. 4.3)

In hydrogen electrode, hydrogen gas is bubbled through a solution of hydrogen ions. The inert metal (which is often Pt) act as a source or sink of electrons, but takes no other part in the reaction other than acting as a catalyst for it. This electrode is denoted as

$Pt \mid H_{2(aq)} \mid H^+ \, (aq)$

The hydrogen electrode may act as either a cathode or an anode, depending on the other electrode in the cell and the spontaneous direction of the overall reaction. When the electrode is acting as *cathode*, the reaction is

$$2H^+_{(aq)} + 2e^- \rightarrow H_{2\,(g)} \qquad \text{(reduction)}$$

when the hydrogen electrode is acting as *anode*

$$H_{2\,(g)} \rightarrow 2H^+\,(aq) + 2e^- \qquad \text{(oxidation)}$$

For determining the electrode potential of the given electrode, the electrode is coupled with **SHE**. The **E.M.F.** of the cell is then determined with the help of a potentiometer. The electrode potential of **SHE** is arbitrarily taken as zero, the **E.M.F.** of cell gives the value of electrode potential of the given electrode.

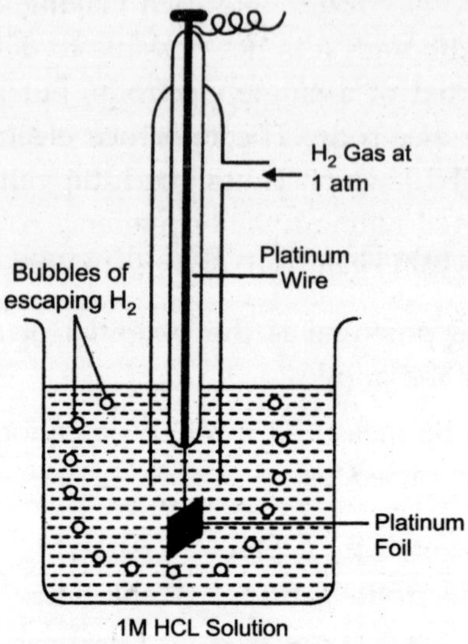

Fig.4.3 *Standard hydrogen electrode*

(b) **Calomel Electrode.** It is very convenient secondary reference electrode and is used in place of hydrogen electrode which is not easy to prepare. It consists of mercury in contact with a solution of potassium chloride saturated with mercurous chloride (calomel). The electrode is represented as

$$Hg \,/\, Hg_2\,Cl_{2\,(s)},\ KCl_{(aq)}$$

It must be emphasized here that the essential requirement of a reference electrode is that its potential through the potentiometric measurement must remain constant.

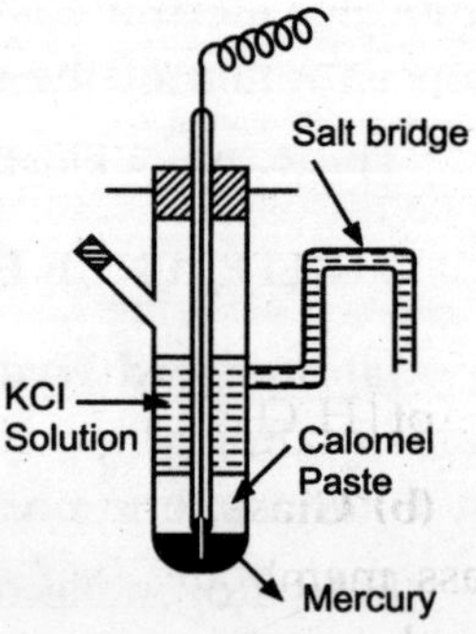

Fig. 4.4 *Calomel Electrode*

The cell reaction when calomel electrode acts as cathode is

$$\frac{1}{2}Hg_2Cl_{2(s)} + e^- \rightleftharpoons Hg + Cl^-$$

2. INDICATOR ELECTRODES

(a) **Quinhydrone electrode**. Quinhydrone is the name given to quinone-hydroquinone system involving the reaction,

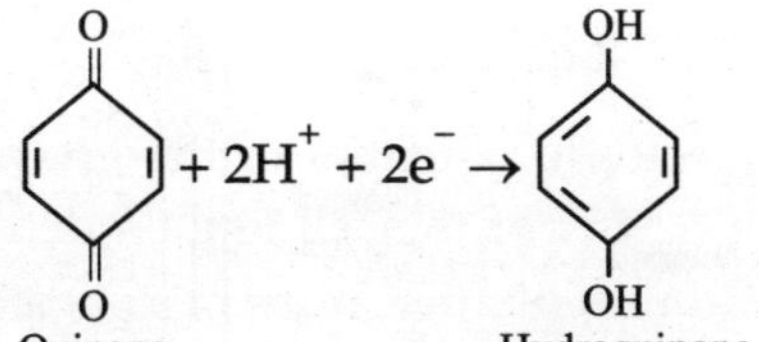

Quinhydrone electrode behaves as a form of hydrogen electrode and is used for measuring pH of solutions as an indicator electrode.

Quinhydrone is dissolved in the solution whose pH is to be determined until the solution becomes saturated and a platinum wire is dipped into the solution. The half cell so formed is connected with any other reference electrode, usually saturated calomel electrode and the EMF of the cell so formed is determined potentiometrically. The cell is represented as,

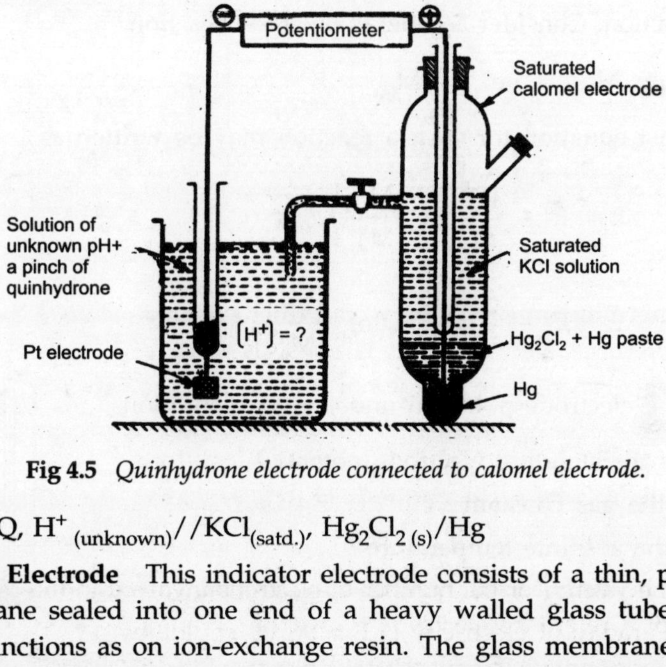

Fig 4.5 *Quinhydrone electrode connected to calomel electrode.*

$pt\,|\,H_2Q, Q, H^+_{(unknown)}//KCl_{(satd.)}, Hg_2Cl_{2\,(s)}/Hg$

(b) Glass Electrode This indicator electrode consists of a thin, pH sensitive glass membrane sealed into one end of a heavy walled glass tube. The glass membrane functions as on ion-exchange resin. The glass membrane exchanges its Na^+ ions with H^+ ions of the solution. A small volume of dilute HCl saturated with AgCl is contained in the tube. A silver wire in this solution forms a silver/silver chloride reference electrode, which inturn is connected to one of the terminals of potentiometer. To the other terminal of potentiometer, calomel electrode is connected. Thus glass electrode system contains *two* reference electrodes: viz., the external calomel electrode and the internal silver/silver chloride electrode. The glass electrode can be represented as:

$Ag/AgCl_{(s)}$, HCl (0.1 M)/Glass, 0.1 MHCl/ Glass

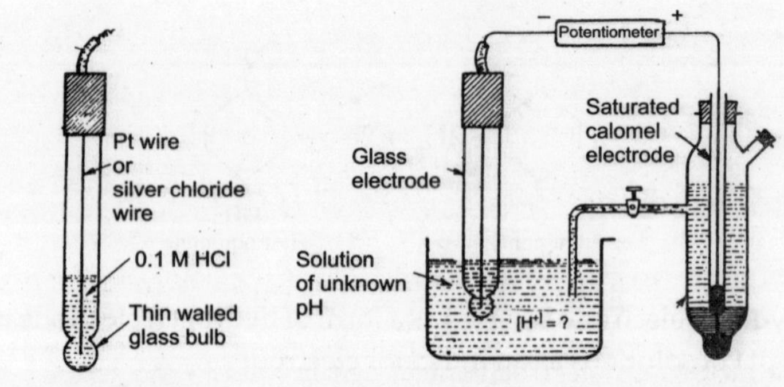

(a) Glass Electrode (b) Determination of pH using glass electrode

Fig. 4.6

(vi) NERNST EQUATION The relationship between EMF of a cell (a half-cell potential) and the concentration of species involved in cell reaction is given by Nernst equation. For this purpose, the electrode reaction is always written as reduction reaction. Consider a general electrode reaction.

$$M^{n+}_{(aq)} + ne^- \rightarrow M_{(s)}$$

The Nernst equation for such a reaction may be written as

$$E = E^o - \frac{RT}{nF} \text{ In } \frac{[M_{(s)}]}{[M^{n+}_{aq}]}$$

or

$$E = E^o - \frac{2.303\,RT}{nF} \log \frac{[M_{(s)}]}{[M^{n+}_{aq}]}$$

where E is the electrode potential under given conditions,

E ° is the standard electrode potential

R is the gas constant

T is the absolute temperature

n is the valency or the number of electrons involved in the redox reaction

F is Faraday of electricity (1 F = 96500 coulombs)

Now for pure solids or liquids or gases at one atmosphere pressure, the molar concentration is taken as unity i.e. $[M_{(s)}] = 1$

Thus, $$E = E^o - \frac{2.303\,RT}{nF} \log \frac{1}{[M^{n+}_{(aq)}]}$$

Substituting the values, R = 8.314 J K^{-1} mol^{-1}, T = 298 K, F = 96500 Coulombs in the above Nernst equation, we have,

$$E = \frac{E^\circ - 2.303 \times (8.314\, J\, K^{-1}\, mol^{-1}) \times (298\, K)}{n \times (96500\, C\, mol^{-1})} \log \frac{1}{[M^{n+}_{(aq)}]}$$

$$E = E^\circ - \frac{0.0591}{n} \log \frac{1}{[M^{n+}_{(aq)}]}$$

or
$$E = E^\circ + \frac{0.0591}{n} \log \left[M^{n+}_{(aq)}\right]$$

In general for any electrochemical cell involving the redox reaction,

$$aA + bB \rightleftharpoons cC + dD$$

the Nernst equation is given by,

$$E_{Cell} = E^\circ_{Cell} - \frac{2.303\, RT}{nF} \log \frac{[C]^c\,[D]^d}{[A]^a\,[B]^b}$$

At 298 K, the equation takes the form,

$$E_{Cell} = E^\circ_{Cell} - \frac{0.0591}{nF} \log \frac{[C]^c\,[D]^d}{[A]^a\,[B]^b}$$

MEASUREMENT OF E.M.F. OF A CELL (POTENTIOMETRY)

Potentiometry deals with the measurement of difference in potential between two electrodes which have been combined to form electrochemical cells. The cell potential measured nearly under zero or negligible current flow is called **electromotive force** (E.M.F.) of the cell. The difference between E.M.F. and potential difference may be summarized as follows:

	EMF	Potential Difference
1.	EMF is the potential difference of the cell when no current is flowing in the circuit i.e. an open circuit.	Potential difference is the difference between the electrode potentials of two electrodes under any condition.
2.	EMF can be measured by potentiometric method only.*	Potential difference can be measured by simple voltmeter
3.	EMF is the maximum voltage obtainable from the cell from the cell.	Potential difference is less than the maximum voltage obtainable

* The EMF sent by the potentiometric source is equal and opposite to that of the cell and there is no net current flowing in the circuit.

The EMF of a cell is measured with the help of Potentiometer. The experimental set up is as follows:

POTENTIOMETER

In potentiometer, the principle generally used is that of **Poggendorff compensation method**. This consists in balancing the unknown EMF against a known one in potentiometric circuit. In otherwords, we measure the external potential difference required to stop the passage of current in the cell.

Potentiometer consists of a uniform cross section wire AB. A battery S of constant EMF is connected to the two ends of the wire AB. The EMF of this battery or cell should be larger than the cell EMF to be measured. The Cell C whose EMF is to be measured is connected to A, with the poles in the same direction, as that of the cell C. The other terminal of the cell C is connected through a galvanometer to a sliding contact J, which can be moved along the wire AB.

Firstly the sliding contact is moved till at a certain point there is no deflection in the galvanometer i.e. no current flows through it. Say the point is Y. Thus the EMF of the cell E_C would be directly proportional to the distance AY.

$$E_c \propto AY$$

Then the cell C is replaced by the standard cell D and a point (say X) is found out when no current flows through the galvanometer. Then, EMF of this standard cell D, E_D will be given by

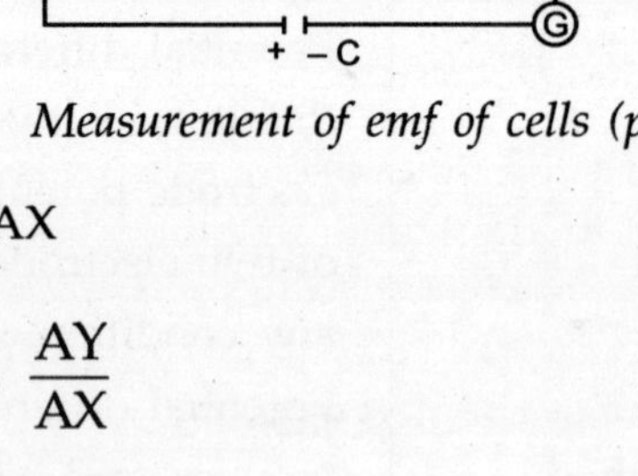

Fig 4.7 *Measurement of emf of cells (potentiometer).*

$$E_D \propto AX$$

Now $$\frac{E_C}{E_D} = \frac{AY}{AX}$$

Therefore, $$E_C = \frac{AY}{AX} \times E_D$$

Knowing the value of E_D, E_C can be calculated.

The commonly used standard cell for the above purpose is Western Cadmium cell (Cadmium amalgam cell).

Western Cadmium cell consists of an H shaped glass vessel with a platinum wire sealed through the bottom of each limb. One limb contains mercury which serves as positive electrode (anode). It is covered with a paste of sparingly soluble mercurous sulphate and cadmium sulphate crystals. The other limb i.e. the negative pole contains a cadmium amalgam containing 12.5 percent cadmium. It is covered with layer of cadmium sulphate crystals ($3\,CdSO_4 \cdot 8H_2O$). The electrolyte is a saturated solution of cadmium sulphate and Hg_2SO_4.

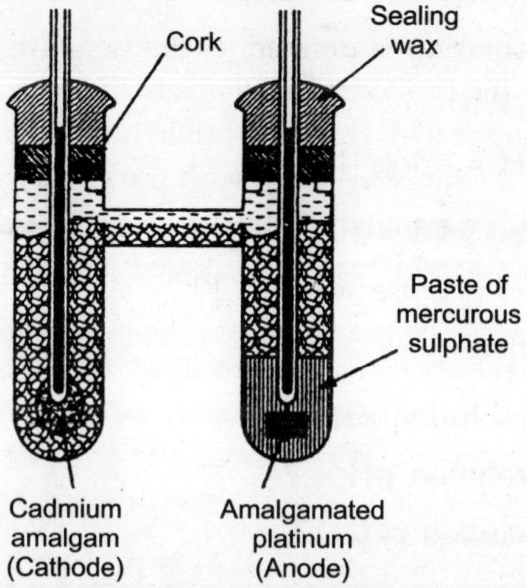

Fig. 4.8 *Standard cell.*

The cell is represented as

$$Cd - Hg/3\,CdSO_4 \cdot 8\,H_2O\ Hg_2SO_{4(s)}/Hg$$

Direct reading potentiometers. Now a days direct potentiometers are available in most of laboratories. In these the meter wire of a simple potentiometer is replaced by a series of coils of standard resistance (usually 2 ohms). Separate terminals are provided for connecting the working battery, standard cell, galvanometer and the EMF to be tested. By means of a selector switch, the potentiometer can be connected separately to the standardization and measurement of unknown EMF.

APPLICATION OF EMF MEASUREMENTS

pH Metric Measurements. An important application of potentiometry is the determination of pH of a solution. The technique is known pHmetry.

The term pH was introduced by Sorensen (1909) to express H^+ ion concentration i.e. the degree of acidity or alkalinity of a solution. As is well known that the product of H^+ and $\overline{O}H$ ions in water (ionic product) is constant and is equal to 10^{-14}. In neutral water therefore we have

$$[H^+] = [\overline{O}H] = 10^{-7}$$

The number of gram ions of hydrogen present in one litre of the solution is known as hydrogen ion concentration.

- If the hydrogen ion concentration is more than 10^{-7}, solution in **acidic**.
- If the hydrogen ion concentration is less than 10^{-7}, solution is **basic**.

Sorensen suggested the use of pH for expressing the acidity or alkalinity.

The pH of a solution is defined as the negative logarithm of the hydrogen ion concentration $[H^+]$

$$pH = -\log [H^+]$$

For pure water, $[H^+] = 10^{-7}$

$\therefore$ pH of pure water = 10^7

It is evident that

- For neutral solution pH = 7
- For acidic solution pH < 7
- For basic solution pH > 7

	pH	$[H^+]$ (mol L^{-1})	$[OH^-]$ (mol L^{-1})	
Acidic	0	10^0	10^{-14}	Acidic
	1	10^{-1}	10^{-13}	
	2	10^{-2}	10^{-12}	
	3	10^{-3}	10^{-11}	
	4	10^{-4}	10^{-10}	
	5	10^{-5}	10^{-9}	
	6	10^{-6}	10^{-8}	
Neutral →	7	10^{-7}	10^{-7}	← Neutral
Basic	8	10^{-8}	10^{-6}	Basic
	9	10^{-9}	10^{-5}	
	10	10^{-10}	10^{-4}	
	11	10^{-11}	10^{-3}	
	12	10^{-12}	10^{-2}	
	13	10^{-13}	10^{-1}	
	14	10^{-14}	10^{-0}	

Thus, lower the pH, more acidic is the solution and higher the pH more basic is the solution.

Determination of pH. For determining the pH of a solution, a number of methods are employed viz, potentiometry, cryoscopy etc. In the **potentiometric determination** of pH, first a cell is assembled in which the indicator electrode used is reversible to hydrogen ions and dips in the solution whose pH is to be determined. The reference electrode is usually a calomel electrode of known potential. The junction between the two is made through salt bridge or by immersing the reference electrode directly into the solution. By measuring the voltage of the cell so formed, pH of the solution can be determined using Nernst equation. A typical cell may be represented as follows:

Electrode reversible to H^+ ion / standard or unknown solution / KCl saturated solution $Hg_2Cl_{2\,(s)}$ / Hg

The emf of this cell, containing standard solution will be given by,

$$E = Eref - \frac{RT}{F} \text{ In } [H^+]$$

where R = gas constant, T = Absolute temperature and F = Faraday.

A number of indicator electrodes reversible to hydrogen ion are employed for pH measurements. These include hydrogen electrode, quinhydrone electrode and glass electrode. Standard calomel electrode (SCE) is used as reference electrode.

EXPERIMENT NO. 4.1

AIM **To find out the strength of the given hydrochloric acid solution (approximate strength N/10) by titrating it against sodium hydroxide using pH meter.**

APPARATUS pH meter with glass electrode, reference electrode, beaker, burette, stirrer etc.

CHEMICALS HCl, NaOH.

THEORY When an alkali is added to an acid solution, the pH of the solution increases slowly. But at the equivalence point, the rate of change of the solution is very rapid. A plot is drawn between volume of the alkali added and the pH of the solution. The sharp break in the curve gives the equivalence point, from which the strength can be calculated using normality equation.

INSTRUMENTATION In pH meter, the glass electrode is incorporated in an ordinary potentiometric circuit. The potentiometric pH meter differs from a simple potentiometer to the extent that the galvanometer is replaced by an electronic circuit that amplifies the current in the cell circuit by a factor of 10^9 or more.

Before using, the pH meter is first standardised with a buffer solution of known pH. Then the glass and reference electrodes are immersed in an unknown solution and the pH is read directly on pH scale.

PROCEDURE

1. Calibrate the pH meter with the glass electrode in the buffer solution of known pH (Details would be given in the manual supplied by the manufacturer).
2. Wash the glass electrode and reference electrode with distilled water and then rinse with the acid solution.
3. Take 5 ml of HCl solution in a 400 ml beaker. Add sufficient water so as reference and glass electrodes are completely dipped.
4. Note down the pH of the pure acid solution.
5. Now add 1 ml of N/10 NaOH from the burette in the beaker. Stir the contents and note the pH of the solution.
6. Continue adding NaOH solution from the burette and note down the pH after each addition.
7. Near the equivalence point the change in pH is much more rapid than in any other region.

OBSERVATIONS Volume of the acid taken = 5 ml

Volume of alkali added 0.0 1.0 2.0 3.0 4.0 5.0 6.0 7.0 8.0 9.0 10.0

CALCULATIONS Plot a curve with pH values as ordinate and the volume of alkali added as abscissa. The sharp break in the curve corresponds to the equivalence point

Volume of alkali added (ml)

Let the volume of alkali at equivalence point = x ml

$$\begin{array}{cc} \text{Acid} & \text{Alkali} \end{array}$$

$$N_1V_1 = N_2V_2$$

$$N_1 \times 5 = \frac{N}{10} \times x.$$

$$N_1 = \frac{N}{10} \times \frac{x}{5}$$

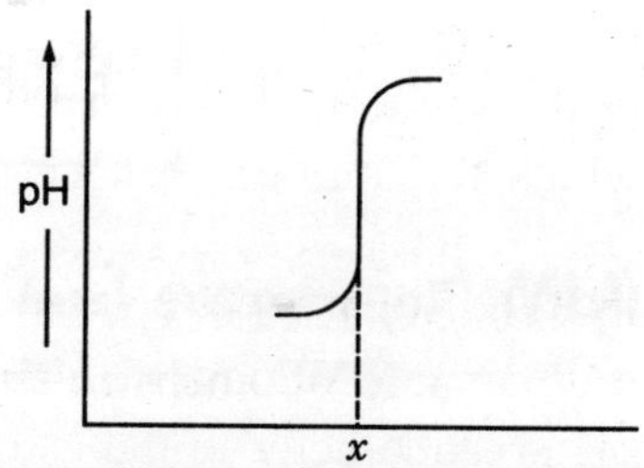

strength of HCl solution

$$= \frac{36.5 \times x}{10 \times 5}\ g/L$$

RESULT The strength of the given acid solution is .... g/litre.

PRECAUTIONS

1. The pH meter should be caliberated before use.
2. After the addition of the alkali, the solution should be thoroughly stirred.
3. Electrodes must be immersed properly in the solution.

EXERCISE

1. Define electrode potential.
2. What is electrochemical cell?
3. Voltmeter cannot be used in the precise measurement of EMF of galvanic cell. Give reason.
4. Why is hydrogen electrode generally not used in pH measurement

Ans. Because it is difficult to set up and cannot be used in presence of As, S, Fe^{3+}, $Cr_2O_3^-$ etc.

5. What do you mean by pH of a solution?
6. What is the effect of temperature on pH?
7. What are calomel and glass electrode?
8. How the equivalence point in acid base titrations is determined pH metrically?

EXPERIMENT NO. 4.2

AIM **To prepare and describe a titration curve for phosphoric acid - sodium hydroxide titration using pH-meter.**

APPARATUS pH meter, reference electrode, beaker, burette, stirrer etc.

CHEMICALS N/10 NaOH and phosphoric acid.

THEORY Neutralization of phosphoric acid takes place in steps.

$$H_3PO_4 + NaOH \rightarrow NaH_2PO_4 + H_2O$$

$$NaH_2PO_4 + NaOH \rightarrow Na_2HPO_4 + H_2O$$

$$Na_2HPO_4 + NaOH \rightarrow Na_3PO_4 + H_2O$$

Thus, on titrating the acid with NaOH pH metrically the following type of curve is obtained.

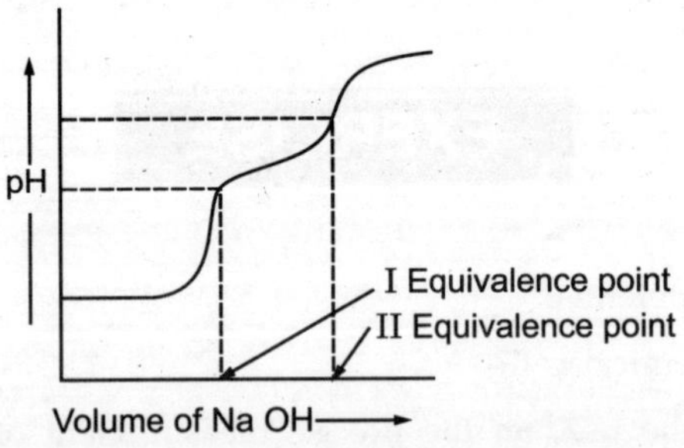

The solution remains acidic after the first neutralization step. After the second neutralization step it becomes mildly alkaline and strongly alkaline after the third neutralization step. If quinhydrone electrode is used as the indicator electrode, first and second inflexion points are observed in the curve but the third one will not be clearly seen.

PROCEDURE Proceed as the previous experiment taking H_3PO_4 in the beaker. Plot a graph between volume of NaOH used and pH of the solution.

(In the case of phosphoric acid, two end points would be observed)

❑❑❑

5

Conductometric Measurements

Conductometry deals with the measurement of electrical conductance of solutions containing electrolytes (acids, bases and salts in water). These solutions conduct electric current due to the movement of ions (cations and anions) towards oppositely charged electrodes.

Like metallic conductors, solutions of electrolytes also obey *Ohm's Law.* According to this law the current (I) passing through a given solution is directly proportional to the potential difference (E) between the two ends of the conductor through which the current is flowing. Thus, mathematically,

$$I \propto E \quad \text{or} \quad E \propto I$$

$$E = RI \qquad ...(1)$$

where I = Current strength

E = Potential difference

R = A proportionality constant and is characteristic of the conductor. It is called *resistance.*

$$R = \frac{E}{I} \qquad ...(2)$$

when E is measured in *volts,* current I is measured in *amperes,* the unit of resistance is known as *ohm* (Ω).

Thus the ratio E/I or R is the resistance of the conductor to the passage of electric current. The reciprocal of resistance R is called the *conductance* (C).

$$C = \frac{1}{R} \qquad ...(3)$$

The conductance is the ease of the flow of current through a solution of electrolyte. Since conductance is reciprocal of ohm its units are *ohm*$^{-1}$ or *mho.* Now a days this unit of conductance is

IS = 1 Ω^{-1} known as *seimen (S).*

SPECIFIC RESISTANCE AND SPECIFIC CONDUCTANCE

The resistance R of a conductor at a definite temperature, is directly proportional to its length (l cm) and inversely proportional to its area (a cm^2). Thus

$$R \propto \frac{l}{a} \quad \text{or} \quad R = \rho \cdot \frac{l}{a} \qquad ...(4)$$

where ρ is a constant and is known as specific resistance or resistivity of the material. Its value is characteristic of the material at a specified temperature. If l = 1 cm and a = 1cm^2 then,

$$R = \rho \qquad ...(5)$$

Thus specific resistance of a substance is defined as the resistance offered by a cube of unit length of it to the passage of electricity through it. The units of ρ are *ohm cm* or *ohm m*.

The reciprocal of specific resistance or resistivity is called specific conductance or conductivity. It is denoted by the symbol K (kappa). Thus

$$K = \frac{1}{\rho} \qquad ...(6)$$

The specific conductance of a conductor is defined as the conductance of a cube of unit length.

The specific conductance of a solution can be defined as the conductance of a solution contained between two parallel electrodes which have cross sectional area 1 cm^2 and which are kept 1cm apart .

Taking the inverse of equation (4)

$$\frac{1}{R} = \frac{1}{\rho} \cdot \frac{1}{l/a} \qquad ...(7)$$

Substituting the value of $\frac{1}{\rho}$ from equation (6) in equation (7)

$$\frac{1}{R} = K \frac{1}{(1/a)} \qquad ...(8)$$

or $$C = \frac{K}{(l/a)}, \text{ where C is the conductance} \qquad ...(9)$$

or $$K = C.\ (l/a) \qquad ...(10)$$

Units of K:

$K = C.\ l/a$, there fore its units are

$$K = mho \times \frac{cm}{cm^2} = \text{mho cm}^{-1} \text{ or siemens cm}^{-1}$$

Cell Constant. The ratio (l/a) in equation (10) is known as cell constant. It is constant for a given cell because it depends on the geometry of the cell i.e on the distance between the electrodes (l) and the cross sectional area of the electrodes (a). Thus.

$$\text{Cell Constant} = \frac{l}{a} \qquad ...(11)$$

Equation (10) can be written as,

Specific conductance = Measured conductance × cell constant or

$$\text{Cell constant} = \frac{\text{Specific conductance}}{\text{Measured conductance}}$$

The units of cell constant are cm^{-1}

Equivalent Conductance. It is defined as the conductance of a solution containing one gram equivalent of an electrolyte at any particular concentration, when placed between two large electrodes 1cm apart . Equivalent conductance of a solution is obtained by multiplying the specific conductance of the solution and the *volume in millilitres* which contains one gram equivalent of the solute. It is denoted by Λ_v, where v is the volume of the solution in millilitres

Equivalent Conductance, Λ_v = specific conductance (K) × Volume in $ml(v)$ per gram equivalent of the electrolyte ...(12)

The volume per gram equivalent of electrolyte is called the dilution factor of the solution.

It the concentration of the solution is c gram equivalent per litre then the volume containing 1 equivalent of the electrolyte will be

$$v = \frac{1000}{c} \text{ ml per gram equiv.}$$

As $\qquad \Lambda_v = K \times v$

$$\therefore \qquad \boxed{\Lambda_v = K \times \frac{1000}{c}} \qquad ...(13)$$

Units of $\Lambda_v = (ohm^{-1}\ cm^{-1}) \times cm^3\ g\text{-}equiv^{-1} = \boldsymbol{ohm^{-1}cm^2g - equiv^{-1}}$

Molar Conductance. Molar conductance of an electrolyte at the particular dilution is the conductivity of the solution containing 1 mole of the electrolyte when placed between two large electrodes one cm apart . It is represented by μ.

$$\mu = K \times V = K \times \frac{1000}{molarity} \qquad ...(14)$$

where V is the volume of the solution in ml. containing 1 mole of the electrolyte

Units of $\mu = ohm^{-1}\ cm^{-1} \times cm^3 mol^{-1} = ohm^{-1}\ cm^2 gmol^{-1}$

Effect of dilution on conductivity: The specific as well as equivalent conductance values of solutions of electrolytes vary with dilution. The *specific conductance values decrease* on diluting the solutions. This is mainly due to decrease in the concentrations of ions per unit volume. On the other hand, the equivalent or *molar conductance increases with dilution* and attains a limiting value at infinite dilution. This increase in conductance in case of weak electrolyte is attributed to the increased dissociation of the electrolyte on dilution. Whereas in case of strong electrolyte, this increase in conductance is due to the increased mobility of the ions. (Fig. 5.1)

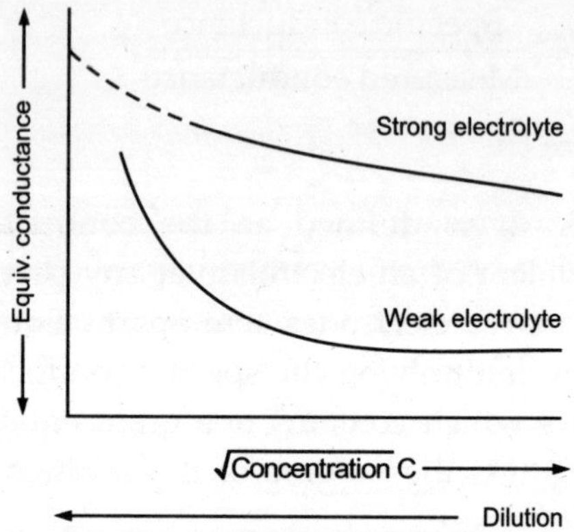

Fig. 5.1 *Variation of equivalent conductance with dilution.*

Effect of temperature on conductivity. The conductance of solution increases with increase in temperature. This is due to the.

- Increase in the velocity of ions,
- decrease in the viscosity of the medium
- decrease in the interaction between the ions.

In case of weak electrolytes, increase in the degree of ionization with increase in temperature also results in increased conductance. Generally, conductance of aqueous solutions increases by about 2% per degree rise in temperature.

Other factors influencing conductance

1. Number of free ions in the solution: More is the number of free ions, greater is the conductance.
2. Charge on the free ions: Greater the charge greater is the conductance.
3. Mobility of the ions: Greater the mobility greater is the conductance.

MEASUREMENT OF CONDUCTANCE

Conductance is the reciprocal of resistance which can be determined using the principle of wheat stone bridge circuit (Fig 5.2)

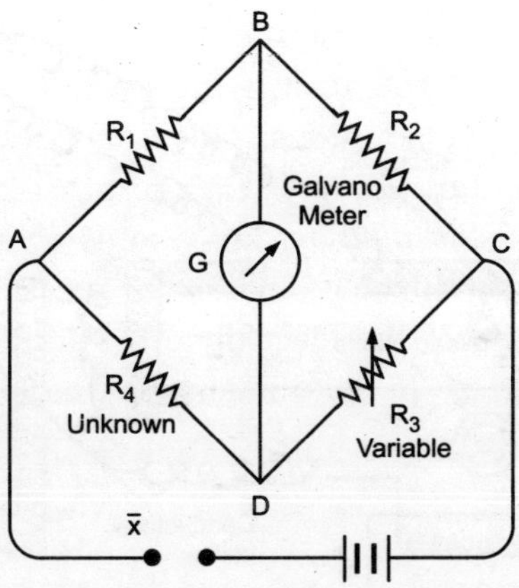

Fig. 5.2 *A Simple Wheatstone bridge circuit.*

In the figure R_1 and R_2 represent fixed resistance, R_3 is variable resistance ,and R_4 is unknown resistance. A small voltage is applied across the point AC and the resistance R_3 is varied till no current passes through the galvanometer connected across the points BD. Then according to wheat stone's principle,

$$\frac{R_1}{R_2} = \frac{R_4}{R_3}$$

Thus , the unknown resistance R_4 can be calculated , as all the other quantities are known. However, this method is faced with a number of difficulties when applied for determining the resistance of an electrolyte solution . The main difficulties are:

(i) Direct current (DC) cause electrolysis of the solution and this changes the concentration of the solution.

(ii) Products of the electrolysis accumulate at the electrodes resulting is polarisation effects i.e a back e.m.f is set up which opposes the current being passed.

The above difficulties are removed by using alternating current of strength 500-2000 cycles per second. The galvanometer becomes useless while using alternating current and hence it is replaced by head phone.

A few conductance bridges based on the above principle, with modifications are:

(i) Kohlrauch's Conductance bridge. It is the widely used circuit in the measurement of conductance. The arrangement of the apparatus is shown in Fig. 5.3

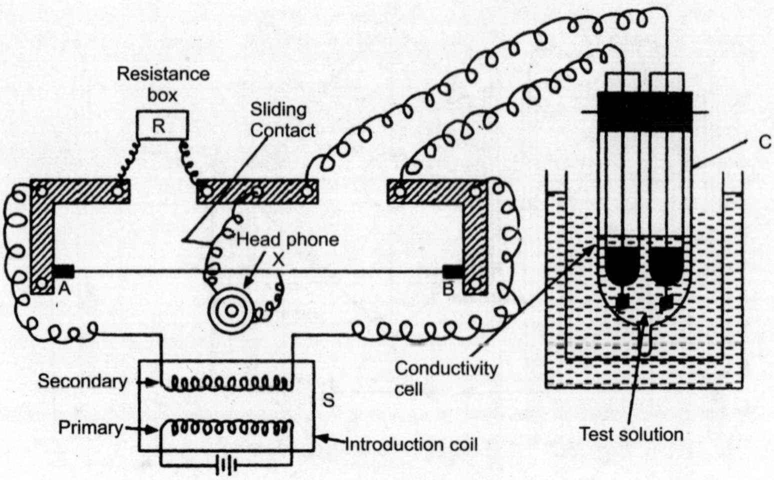

Fig. 5.3 *Measurement of conductivity*

The conductivity cell C is connected to the resistance box R on one side and to a long thin wire AB stretched along a scale on the other side.

An alternating current is passed and a point X at which the sound in the head phone P is minimum is located by moving the sliding contact along the wire AB. At every point other than X, current will flow through the head phone and a buzzing noise would be heard in the head phone. At point X when no sound is heard, no current flows through the head phone. At this point,

$$\frac{\text{Resistance of the solution}}{\text{Resistance R}} = \frac{\text{Resistance of wire AX}}{\text{Resistance of wire XB}} = \frac{\text{length of AX}}{\text{length of XB}}.$$

or $$\text{Resistance of solution} = \frac{AX}{XB} \cdot \text{Resistance R}$$

$$\therefore \text{Conductance of solution} = \frac{XB}{AX} \times \frac{1}{R}$$

(2) Direct Reading Conductance bridges. These days conductivity is determined by direct reading conductance bridges. In these conductivity bridges or conductometers, the head phone detector is replaced with visual devices such as electronic detector. The method of operation of an instrument depends on its make and the manual is supplied by the manufacturer. One such instrument is described under the experiment.

Conductivity Cells

The solution whose conductivity is to be determined is placed in a special type of cell known as conductivity cell. Conductivity cells are of various shapes and sizes. Two of these are shown is Fig. 5.4.

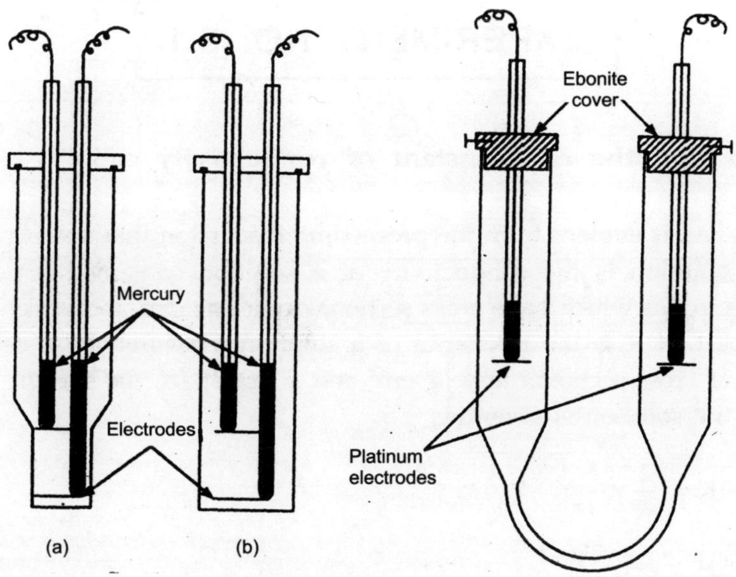

Fig. 5.4 *Conductivity cells*

The conductivity cells are made up of pyrex glass (or some other resistant glass), which is fitted with a pair of platinum electrodes. These electrodes are platinum plates fused into the glass tubes. The tubes are firmly supported by an ebonite cover. The cover along with the platinum electrodes is fitted into the pyrex glass vessel so that the distance between the electrodes may not change.

Conductivity Water As ordinary distilled water contains impurities like CO_2, metal silicates etc., it cannot be used for preparing solutions for conductivity measurements. Hence, water should be specially purified before it can be used. Such water is known as *conductivity water*. Conductivity water is obtained by distilling good quality distilled water after adding potassium permanganate and KOH in a pyrex glass distillation unit. However, conductivity water is used when conductance values of high accuracy are required. For all other purposes simple distilled water can be used as long as the conductivity of water is not more than 10^{-5}S cm^{-1}

EXPERIMENT NO. 5.1

AIM **To find the cell constant of conductivity cell.**

THEORY As is evident from the proceeding discussion that specific conductance of a solution is the conductivity of a solution contained between two parallel electrodes which have cross sectional area one cm^2 and which are kept one cm apart. If R is the resistance of a solution measured by a cell having electrodes of cross sectional area '*a*' cm^2 and '*l*' cm apart, the specific conductance K of the solution is given by,

$$K = \frac{1}{R} \times \frac{l}{a} \qquad \text{... (i)}$$

or

$$K = C \times \frac{l}{a} \qquad \text{...(ii)}$$

where C is the observed conductance and l/a is known as the cell constant. Thus to obtain the value of specific conductance, the observed conductance has to be multiplied by a certain factor known as cell constant. (The values of a and l are constant for a particular cell so l/a is constant and hence the name cell constant)

The cell constant i.e the ratio l/a cannot be determined from the geometrical dimensions of the cell because the exact area of cross section of electrode or exact distance between them cannot be measured accurately. Thus cell constant is determined by an indirect method. From equation (ii)

$$\frac{l}{a} = \frac{K}{C} \qquad \text{...(iii)}$$

or

$$\text{Cell constant} = \frac{\text{Specific conductance}}{\text{Observed conductance}}$$

For the determination of cell constant, a solution of known specific conductance is taken whose conductance is determined using the given cell. Thus cell constant can be determined using equation (iii). Solution of KCl of known concentration and known specific conductance are used for this purpose because,

- KCl is strong electrolyte
- The mobility of K^+ and Cl^- is quite comparable. Table-5.1 gives the specific conductance values of KCl (0.01 M) at different temperatures.

Table 5.1

Temp (°C)	25	26	27	28	29	30	31
Sp. Conductance (m mhocm–1)	1.408	1.458	1.487	1.531	1.561	1.642	1.683
Temp (°C)	32	33	34	35	36	38	
Sp. Conductance (m mhocm–1)	1.731	1.777	1.818	1.862	1.916	1.968	

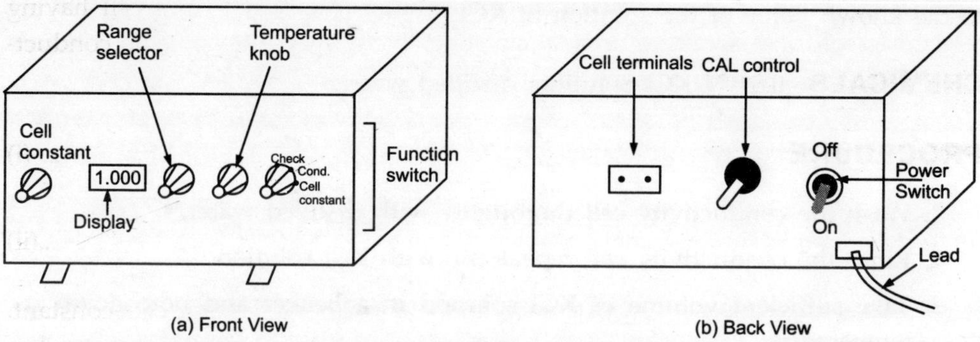

Fig. 5.5 *Conductometeric Bridge*

However these days the conductivity bridges have built in systems based on the above equation and hence the value of cell constant is read directly from the instrument.

APPARATUS. * Conductivity bridge (A.C mains operated), conductivity cell, thermometer, beaker, burette, pipette, funnel etc.

**** Description of Conductivity bridge**

The usual controls provided in these type of bridges (Fig. 5.5) are as follows.

(a) Power switch : This switch is on the back of the instrument. When it is switched on, the display on the front panel glows.

(b) Cell control : This is at the back of the instrument. The display is adjusted to 1.000 with the help of CAL control.

(c) Range Switch : This is on the front panel of the instrument. It has positions marked 0.2, 2, 20. These numbers stand for the full scale meter values in millisiemens. For example when the switch is at 2, the meter reads between 0 and 2 ms

* These days wheatstone bridges are replaced by automatic conductivity bridges which directly give the value of conductance and hence are more convenient to use. However the equations and principle remain the same.

** May vary according to the make.

(d) Temperature Knob. This is provided in the front panel of the instrument. This is to set the temperature control to the actual temperature of the solution (measured by the ordinary thermometer) under test.

(e) Function Switch. This control is also on the front panel of the instrument. It has positions marked *check, cell constant* and *conductivity*. This is used to set the instrument in any of the above positions.

(f) Display window. This window on the front panel displays the readings (cell constant or conductivity)

(g) Cell constant knob : This knob, provided at the front panel is adjusted to the known value of the solution of KCl.

CHEMICALS 0.01 N KCl Solution, distilled water.

PROCEDURE

1. Wash the conductivity cell thoroughly with distilled water.
2. Rinse the conductivity cell repeatedly with KCl solution.
3. Take sufficient volume of KCl solution in a beaker and note down its temperature.
4. Connect the instrument to the mains and switch on the instrument using the power switch. Connect the electrode leads in the input sockets at the rear of the instrument.
5. Set the 'Function' switch to 'check' position and adjust the display to 1.000 with CAL control at the back panel.
6. Dip the conductivity cell in KCl solution and adjust the temperature knob of the conductivity bridge at the temperature of the KCl Solution.
7. Move the 'Function' switch to 'conductance' position and range position to appropriate range.
8. Adjust the cell constant knob so that the display reads the known value of the KCl solution at that temperature
9. Bring the 'Function switch' to 'cell constant' position and read the value of the cell constant from the display window.

RESULT The cell constant of the given cell is ...cm^{-1}

Instructions for Lab Assistant

Provide the following

1. Distilled water
2. 0.01 N KCl solution : Dissolve 0.7456g of anhydrous KCl, dried at 180°C for one hour in distilled water and make up to 1000 ml.

CONDUCTOMETRIC TITRATION

Conductometric titration is the volumetric analysis based upon the measurement of the conductance during the course of titration. Conductance of an aqueous solution containing an electrolyte depends upon the-

- Number of free ions.
- Charge on the free ions
- Mobility of the ions.

When the solution of one electrolyte is added to another, a change in the conductance will result due to the change in number of free ions or substitution of ions of one mobility by the ions of another mobility. Let us illustrate this by taking specific cases:

(i) Titration of a strong acid with a strong base. Before NaOH solution is added from the burette, the acid solution has high conductivity due to the highly mobile H^+ ions. When NaOH is added to the acid the highly mobile H^+ ions are removed by combination with $\overline{O}H$ ions to form feebly ionized water and hence are replaced by less mobile sodium ions.

$$H^+Cl^- + Na^+ \overline{O}H \rightarrow Na^+Cl^- + \underset{\text{(feebly ionized)}}{H_2O}$$

Thus the conductance of the solution will decrease and continues decreasing with the addition of NaOH until the equivalence point is reached. On further addition of NaOH, the conductivity of the solution will increase due to addition of Na^+ ions and highly mobile hydroxyl ($\overline{O}H$) ions. If we plot the conductance of the solution against the volume of alkali added, two straight lines intersecting at a point will be obtained [Fig. 5.6 (*a*)]. This point of intersection corresponds to the equivalence point since at this point there is no excess of either acid or base.

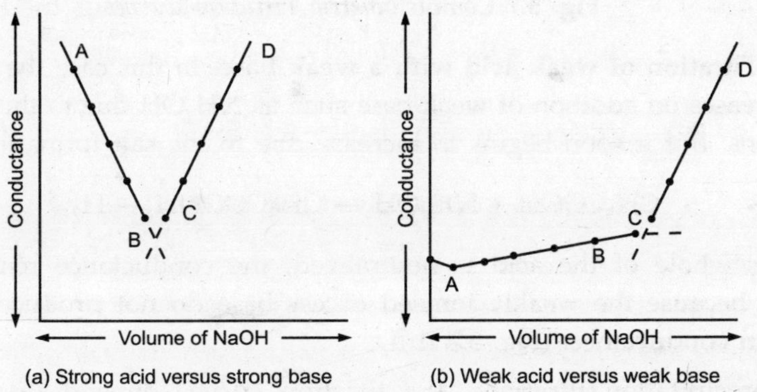

Fig.5.6 *Conductometric Titration*

(ii) Titration of a weak acid with a strong base: In case of weak acid, like CH_3COOH, against a strong base, the conductivity first decreases owing to the formation of the salt which suppresses the ionization of weak acid. Mostly this decrease is not observed unless one adds very small volumes of NaOH. As more of NaOH is added, highly ionized salt, sodium acetate is formed thereby increasing the conduction.

CH_3COOH	+	NaOH	→	$CH_3COO^-Na^+$	+	H_2O
Poorly conducting acid				Highly ionized salt		unionized.

When the acid is neutralized, further addition of NaOH causes a sharp rise in conductance due to addition of conducting $\overline{O}H$ ions. Thus the intersection of lines AB and CD in Fig 5.6(b) represents the equivalence point.

(iii) Titration of strong acid with a weak base. When a weak base like NH_4OH is added to a strong acid, the conductance first decreases due to the replacement of H^+ ions. It keeps on decreasing upto the equivalence point after which it remains practically constant. This is because the excess of weak base added is not appreciably ionized in presence of its salt. Fig 5.7(a)

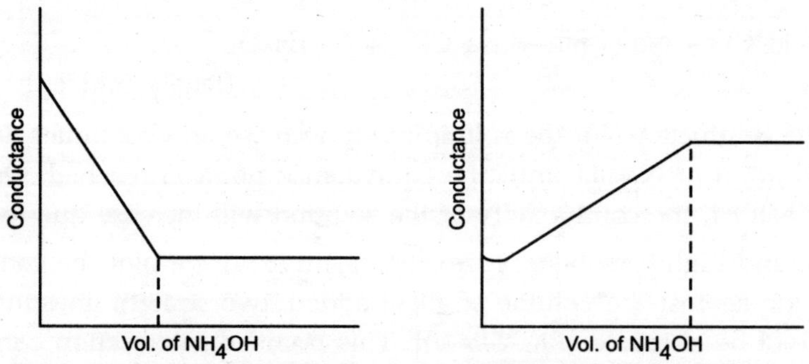

(a) Strong acid versus weak base (b) Weak acid versus weak base

Fig. 5.7 *Conductometric Titration Curves*

(iv) Titration of weak acid with a weak base. In this case the conductance first decreases on addition of weak base such as NH_4OH due to the replacement of H^+ ions. But it soon begins to increase due to the salt formation.

$$CH_3COOH + NH_4OH \rightarrow CH_3COONH_4 + H_2O$$

After whole of the acid is neutralized, the conductance remains almost constant because the weakly ionized excess base do not produce appreciable change in conductance. (Fig. 5.7 (b))

In precipitation titrations, the titration curves depend upon the ionic conductance of the various ions taking part in the reaction.

Advantages of Conductometric Titrations

1. This method can be employed to very dilute solutions (upto the order of 10^{-4} M solutions)
2. Give very accurate end points with an error of ± 0.5%
3. These titrations are very useful in case of coloured solutions which cannot be titrated by ordinary volumetric methods because the colour change of indicator is not clear.
4. Useful for titrating weak acids against weak bases, which otherwise do not give sharp end point.
5. No keen observation is necessary near the end point since it is detected graphically.

EXPERIMENT NO. 5.2

AIM **Determine the strength of hydrochloric acid solution by titrating it against standard sodium hydroxide solution conductometrically.**

THEORY As given on page 155 under the heading "titration of strong acid with a strong base."

APPARATUS Conductometer, conductivity cell, beaker, burette, pipette, funnel etc.

CHEMICALS 0.1N NaOH, standard conductivity reagent (KCl solution) and unknown HCl solution.

PROCEDURE

1. Determine the cell constant of the given conductivity cell as is done in the previous experiment (Caliberation of Conductivity cell)
2. Rinse the conductivity cell with the solution whose conductivity is to be measured.
3. Take 20ml of the given HCl Solution in a 100 ml beaker.
4. Wash the conductivity cell with distilled water and then rinse it with the given HCl solution. Dip the cell in the solution taken in the beaker.
5. Connect the conductivity cell to the conductometer.

6. Set the function switch to check position. Display must read 1.000. If it does not, set it with CAL control at the back panel.
7. Put the 'Function Switch' to 'cell constant' and set the value of cell constant determined in step-1 with the help of cell constant knob.
8. Set the temperature control to the actual temperature of the solution under test.
9. Set the 'Function Switch' to 'conductivity' and read the display. This will be the exact conductivity Note it down.
10. Take alkali (NaOH) in the burette and add 0.5 ml of it into the beaker containing HCl. Stir and determine the conductivity.
11. Repeat the procedure of addition of 0.5 ml of NaOH and noting down the conductivity in the observation table . Take 12-15 readings in this way. After each addition, stir the solution gently.

OBSERVATIONS.

Volume of unknown acid solution take V_1 = 20.0 ml

Observation Table.

S.No	Volume of NaOH added (ml)	Specific conductance (mho/cm)
1	0.0	
2	0.5	
3	1.0	
4	1.5	
5	2.0	
6	.....	.... and so on.

CALCULATIONS

1. Plot a graph between specific conductance (along y-axis) and volume of NaOH added (along *x* axis) A graph of the type shown in Fig. 5.6(a) is obtained.
2. The point of intersection will give the amount of alkali required for neutralization of the acid. Suppose the volume of the alkali used = V_2 ml Applying normality equation,

$$N_1V_1 = N_2V_2$$
(HCl) (NaOH)

$$N_1 = \frac{0.1 \times V_2}{20}$$

$$\text{Strength of HCl Solution} = \text{Eq.wt} \times \text{Normality}$$
$$= 36.5 \times N_1 \text{ g/litre}$$

RESULT Strength of given HCl solution is . . . g/liter

PRECAUTIONS

1. The conductivity cell should be handled very carefully as it is very delicate
2. In order to have a sharp break in conductance titration curve at the equivalence point, there should be no appreciable change in the volume during the course of titration. Therefore, the titrant solution should be 10 to 100 times stronger than the solution to be titrated.
3. Stirring should be done after each addition of the titrant.

Instructions for Lab Assistant

Provide the following

1. Distilled water
2. 0.01 N KCl : As is prepared in the previous experiment.
3. 0.01 N NaOH : Dissolve about 4 g of NaOH in distilled water and make up to 1000ml. Standardize it by titrating against exact. N/10 oxalic acid solution. It is then diluted accordingly by distilled water to get exact 0.1 N solution.
4. 0.01N HCl : Dissolve 0.83 ml of conc. HCl (12N) in distilled water and make up the volume to 1000 ml in a measuring flask.

   The titrations involving weak acid and strong base or strong acid and weak base or weak acid or weak base are carried out similarly. The equivalence points are determined through the graphs and hence corresponding strengths can be calculated.

EXERCISE

1. Define conductance, specific conductivity and equivalent conductivity. What are their units?
2. How specific conductance and equivalent conductance vary with dilution?
3. What is the effect of temperature on conductance ?
4. Define cell constant. How is it determined ?
5. What is conductivity water ?
6. What are the advantages of conductometric titrations ?
7. Why KCl is used as a standard for verification of cell constant ?
8. What is the relation between specific conductivity, conductance and cell constant?

9. Define ionic mobility.

Ans. *It is the rate of migration of ion under a potential gradient of 1 volt per cm.*

10. Out of Li^+, Na^+, K^+, Rb^+, Cs^+, which one has highest ionic mobility in water?

Ans. *Smaller the ion, greater is its mobility. Cs^+ ion has highest mobility in water. All these ions are hydrated in water. The hydrated sphere of Li^+ is largest due to its small size and hence its mobility is least among the ions of first group elements.*

11. H^+ ion$^+$ is small in size and should be hydrated to a greater extent and its ionic mobility should be even less than that of Li^+. But its mobility is highest in water among all the ions Explain why ?

Ans. *The mobility of H^+ and $\bar{O}H$ is exceptionally high in hydroxylic solvents such as water. This is due to the reason that the migration of H^+ through water takes place through different mechanism known as Grotthus-type mechanism. In water, hydrogen ion is strongly hydrated forming hydronium ion. The mobility of H_3O^+ is not entirely due to the actual migration of H_3O^+ ion through the solution but also by transfer of a proton from H_3O^+ to the neighbouring hydrogen bonded water molecules. Thus, the proton is transferred from one molecule of water to another and travels further along a series by a rearrangement of hydrogen bonds. The rate of proton transfer are much faster than the velocity of H_3O^+ and hence the mobility also becomes high.*

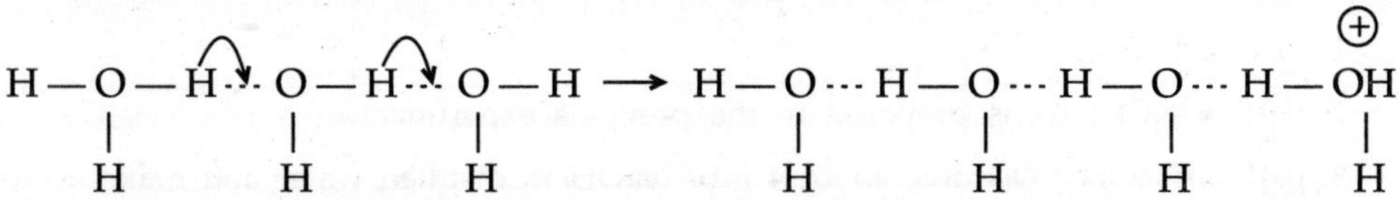

Similarly the high mobility of hydroxyl ion in water is also due to a proton transfer between the hydroxyl ion and water molecules.

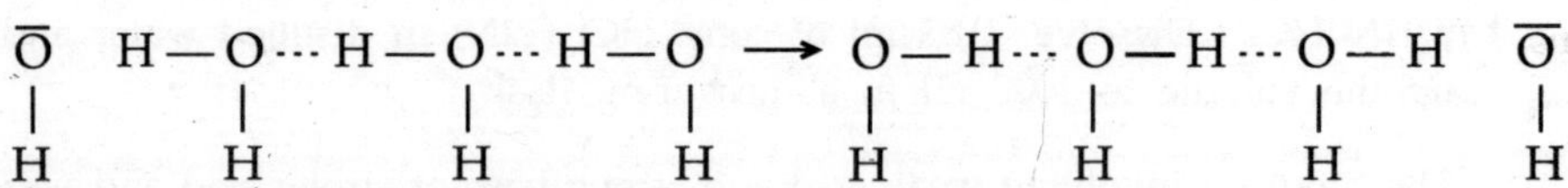

12. What is the effect of temperature on conductance ?

13. Why precipitation titrations cannot be carried out effectively as the acid base titrations, conductometrically?

Ans. *Acid base titrations are characterized by sharp breaks because both H^+ and $\overline{OH}$ ions have very high equivalent conductances.*

Precipitation titrations suffer from,

(i) *Slow separation of precipitate.*

(ii) *Adsorption of the ions on precipitate*

(iii) *Coprecipitation or post precipitation.*

These errors can be eliminated by performing precipitation conductometric titrations at elevated temperatures.

VIVA-VOCE

1. What are electrolytes?

Ans. *The substances which undergo decomposition when current is passed through them are called electrolytes.*

2. What do you mean by strong and weak electrolytes?

Ans. *Electrolytes are classified into two main categories as strong electrolytes and weak electrolytes. Strong electrolytes are substances like strong acids e.g., HCl, HNO_3, H_2SO_4, etc., strong bases e.g., NaOH and KOH etc., and most of the inorganic salts. These dissociate almost completely in the aqueous solution or in the molten state and hence conduct electricity to a large extent. Weak electrolytes are substances like weak acids e.g., CH_3COOH, HCN, H_2CO_3, H_3PO_4 etc. and weak bases e.g., NH_4OH, $Ca(OH)_2$, $Al(OH)_3$ etc. These have low degree of dissociation and hence conduct electricity to a small extent.*

3. What are the factors which affect electrolytic conduction?

Ans. ***Factors affecting electrolytic conduction.*** *The conductance of the solution of an electrolyte depends upon the following factors:*

(i) interionic attractions which depend upon the interactions between the ions of the solute and hence are called solute-solute interactions.

On this basis, the substances are classified as strong electrolytes and weak electrolytes, as already mentioned.

(ii) Solvation of ions which depends upon the interactions between the ions of the solute and the molecules of the solvent and hence are called solute-solvent interactions.

(iii) Viscosity of the solvent which depends upon the solvent-solvent interactions.

The effect of all these factors decreases with increase of temperature. Hence the electrolytic conduction increases with increase of temperature. On the other hand, the electrical conduction of the electronic conductors decreases with increase of temperature.

4. Define conductance. What are its units?

Ans. *The reciprocal of the electrical resistance is called the conductance. It is usually represented by C. Thus*

$$C = \frac{1}{R} \quad \text{where } I = \text{current}$$

$$R = \text{Resistance} = E/I$$

It is a common practice to speak of the conductance of a solution rather than its resistance.

Units. *The units of conductance are siemens (S) or reciprocal ohms written as ohm^{-1} or mho or r.o ($1S = 1\,\Omega^{-1}$).*

For example, if a solution has a resistance of 10 ohms, it is said to have a conductance of 1/10 ohm^{-1} or 1/10 mhos or 1/10 Ω^{-1} or 1/10 siemens.

5. Define specific conductivity, equivalent conductivity and molar conductivity?

Ans. **Specific conductivity** *of a solution is defined as the conductance of a solution of 1 cm length and having 1 sq. cm as the area of a cross section.* Alternatively, it may be defined as *"conductance of one centimeter cube of the solution of the electrolyte."*

Equivalent conductivity *of a solution at a dilution V is defined as the conductance of all the ions produced from one gram equivalent of the electrolyte dissolved in V cm^3 of the solution when the distance between the electrodes is one cm and the area of the electrodes is so large that whole of the solution is contained between them.* It is represented by $\wedge_{eq}$ *(lambda).*

The molar conductivity *of a solution at a dilution V is the conductance of all the ions produced from one mole of the electrolyte dissolved in V cm^3 of the solution when the electrodes are one cm apart and the area of the electrodes is so large that the whole of the solution is contained between them.* It is usually represented by $\wedge_m$.

6. What is the relation between molar and specific conductivity?

Ans. ***Relationship between molar conductivity and specific conductivity.*** *The molar conductivity is related to the specific conductivity as follows:*

$$\wedge_m = K_V \times V$$

or

$$\wedge_m = K_c \times \frac{100}{c} = K_c \times \frac{100}{\text{Molarity}}$$

where K is the specific conductivity and V is the volume of the solution containing one mole of the electrolyte and c is the molar concentration i.e., mol L^{-1} (or mol dm^{-3}):

7. What is the relation between equivalent and specific conductivity?

Ans. *In general, if the volume of the solution containing one gram equivalent of the electrolyte is V cm^3, we have*

Equivalent conductivity = Specific conductivity × V

$$\wedge_m = K_V \times V$$

In terms of concentration, if the solution has a concentration of c gram equivalent per litre i.e., c gram equivalents are present in 1000 cm^3 of the solution, then the volume of the solution containing one gram equivalent will be $\frac{100}{C}$. Hence the above expression may be written as

$$\wedge_m = K_c \times \frac{100}{C_{eq}} = K_c \times \frac{1000}{\text{Normality}}$$

8. Give units of specific conductivity, equivalent conductivity, and molar conductivity.

Ans. *The units of specific conductivity are mho cm^{-1} or siemens cm^{-1}. Equivalent conductivity is expressed as ohm^{-1} cm^2 g – equiv^{-1} and molar conductivity is expressed as ohm^{-1} cm^2 g mol^{-1}.*

9. What is the difference between metallic and electrolytic conductors

Ans.

Metallic Conductors	Electrolytic Conductors
1. Flow of electricity takes place without the decomposition of the substance.	1. Flow of electricity takes place accompanied by the decomposition of the substance (electrolyte).
2. Flow of electricity is due to the flow of electrons only *i.e.,* there is no flow of matter.	2. Flow of electricity is due to the movement of ions and hence there is flow of matter.

10. Define cell constant. What are its units?

Ans. ***Cell Constant.*** *The ratio (l/a) is known as cell constant. It is constant for a given cell because it depends on the geometry of the cell i.e. on the distance between the electrodes (l) and the cross sectional area of the electrodes (a). Thus:*

$$\text{Cell Constant} = \frac{1}{a}$$

Equation (10) can be written as,

Specific conductance = Measured conductance × cell constant or

$$\text{Cell constant} = \frac{\text{Specific conductance}}{\text{Measured conductance}}$$

The units of cell constant are cm^{-1}.

11. Discuss the effect of dilution on conductivity.

Ans. *Effect of dilution on conductivity: The specific as well as equivalent conductance values of solutions of electrolytes vary with dilution. The specific conductance values decrease on diluting the solutions. This is mainly due to decrease in the concentrations of ions per unit volume. On the other hand, the equivalent or molar conductance increases with dilution and attains a limiting value at infinite dilution. This increase in conductance in case of weak electrolyte is attributed to the increased dissociation of the electrolyte on dilution. Whereas in case of strong electrolyte, this increase in conductance is due to the increased mobility of the ions.*

12. Why conductance of a solution increase with increase in temperature?

Ans. *Effect of temperature on conductivity. The conductance of solution increase with increase in temperature. This is due to the.*

- *increase in the velocity of ions.*
- *decrease in the viscosity of the medium.*
- *decrease in the interaction between the ions.*

In case of weak electrolytes, increase in the degree of ionization with increase in temperature also results in increased conductance. Generally, conductance of aqueous solutions increases by about 2% per degree rise in temperature.

13. What are the factors influencing conductance?

Ans. 1. *Number of free ions in the solution: More is the number of free ions, greater is the conductance.*

2. *Charge on the free ions: Greater the charge greater is the conductance.*
3. *Mobility of the ions.*

14. Define ionic mobility.

Ans. *It is the rate of migration of ion under a potential gradient of 1 volt per cm.*

15. *Out of* Li^+, Na^+, K^+, Rb^+, Cs^+, *which one has highest ionic mobility in water?*

Ans. *Smaller the ion, greater is its mobility. Cs+ ion has highest mobility in water. All these ions are hydrated in water. The hydrated sphere of* Li^+ *is largest due to its small size and hence its mobility is least among the ions of first group elements.*

16. H^+ ion$^+$ is small in size and should be hydrated to a greater extent and its ionic mobility should be even less than that of Li^+. But its mobility is highest in water among all the ions Explain why?

Ans. *The mobility of* H^+ *and H is exceptionally high in hydroxylic solvents such as water. This is due to the reason that the migration of* H^+ *through water takes place through different mechanism known as Grotthus-type mechanism. In water, Hydrogen ion is* *strongly hydrated forming hydronium ion. The mobility of H_3O^+ is not entirely due to the actual migration of H_3O^+ ion through the solution but also by transfer of a proton from H_3O^+ to the neighbouring hydrogen bonded water molecules. Thus, the proton is transferred from one molecule of water to another and travels further along a series by a rearrangement of hydrogen bonds. The rate of proton transfer are much faster than the velocity of H_3O^+ and hence the mobility also becomes high.*

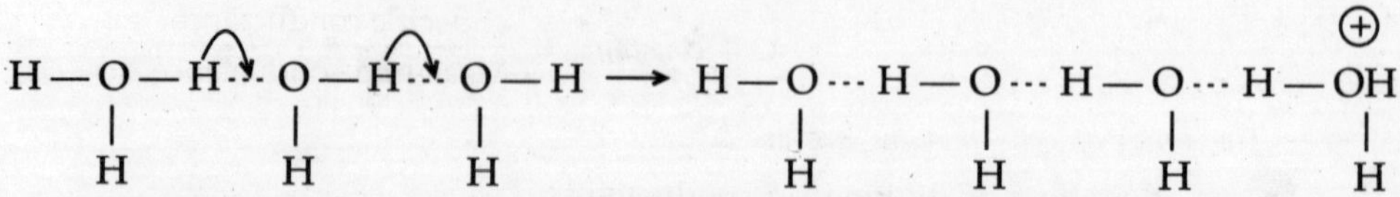

Similarly the high mobility of hydroxyl ion in water is also due to a proton transfer between the hydroxyl ion and water molecules.

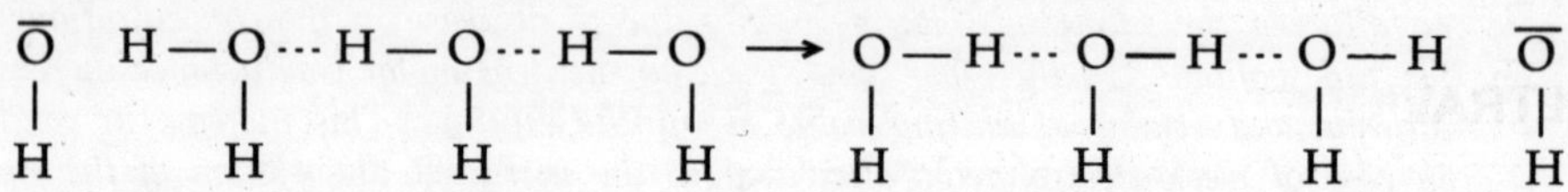

17. Why precipitation titrations cannot be carried out effectively as the acid base titrations, conductometrically?

Ans. *Acid base titrations are characterized by sharp breaks because both H^+ and ions have very high equivalent conductances.*

Precipitation titrations suffer from,

(i) Slow separation of precipitate.

(ii) Adsorption of the ions on precipitate

(iii) Coprecipitation or post precipitation.

These errors can be eliminated by performing precipitation conductometric titrations at elevated temperatures.

18. Why DC is not used in conductance measurements?

Ans. *Because this may lead to electrolysis of the solute thereby changing the concentration of the solution.*

❑❑❑

6

Colorimetric and Spectrophotometric Analysis

ULTRAVIOLET AND VISIBLE SPECTROMETRY

Spectroscopy is the branch of science dealing with the study of interaction of electromagnetic radiation with matter. It is the most powerful tool available in the analysis of wide range of samples.

Spectroscopic methods of analysis are based on measurements of the electromagnetic radiation emitted or absorbed by the solutes. When an electromagnetic radiation is passed through a sample (solid, liquid or gas), certain characteristic wavelengths are absorbed by the sample. (The extent of absorption depends upon the wavelength of the radiation, thickness to the medium and also on the concentration of the sample). Thus there is a decrease in the intensity of transmitted radiation. By measuring the decrease in intensity of radiation we can qualitatively and quantitatively analyze a given substance. The technique is known as *photometry*.

A group of methods by which the *concentration* of substance is determined by measuring the interaction of electromagnetic radiation with the sample is called *Photometric analytical methods*. The most common of these are:

(1) Colorimetry (2) Spectrophotometry.

Colorimetry Colorimetry is concerned with the determination of concentration of a substance by measurement of the relative absorption of light with respect to a known concentration of a substance. The absorption employed is generally in the *visible region*. Any coloured substance can be analyzed by this method. In otherwords, the variation of the colour of a system with a change in concentration of a component, forms the basis of colorimetric analysis. The instrument used in this technique is called a colorimeter.

Spectrophotometry Spectrophotometry is a method of chemical analysis through measurements of radiations after these are allowed to pass through a sample. It deals with ultraviolet (185 – 400 nm) visible (400 – 760 nm) and infrared (0.76–15 μm) region of spectrum. It employs a much narrower band of wavelengths as provided by a monochromator operating in UV, visible and infrared region. The instrument used in this method is called spectrophotometer.

The absorption of light by a substance is governed by certain laws. These are:

(i) Lambert's law. When a monochromatic radiation is passed through a solution, the decrease in intensity of radiation with the thickness of the solution is directly proportional to the intensity of the incident radiation. Let,

Intensity of incident radiation = I

∴ Rate of decrease in intensity of incident radiation with thickness dx of the

$$\text{medium} = \frac{-dI}{dx}$$

thus $$\frac{-dI}{dx} \propto I \qquad \text{...(i)}$$

or $$\frac{-dI}{dx} = kI \qquad \text{...(ii)}$$

where k is the proportionality constant.

Further $$\frac{-dI}{I} = k\,dx \qquad \text{...(iii)}$$

On integrating equation (iii) between limits $I = I_0$ at $x = 0$ and $I = I$ at $x = l$, we get

$$\ln \frac{I}{I_0} = -kl \qquad \text{...(iv)}$$

$$2.303 \log \frac{I}{I_0} = -kl \qquad \text{...(v)}$$

$$\log I/I_0 = \frac{-k}{2.303}\, l \qquad \text{...(vi)}$$

or $$\log I_0/I = \frac{k}{2.303}\, l \qquad \text{...(vii)}$$

where $\log I_0/I$ is called **'Absorbance'** and is represented as 'A'

and $$\frac{k}{2.303} = E \text{ is } \textbf{'Absorption coefficient'}$$

∴ $$\boxed{A = El} \qquad \text{... (viii)}$$

(2) Beer's Law. Beer's law takes into consideration the concentration of solution along with thickness of the solution. The law states that when a monochromatic light is passed through a solution, the decrease in intensity of radiation with the thickness of the solution is directly proportional to the incident light as well as concentration of the solution. Let a monochromatic light of intensity I pass through a solution of concentration C molar and thickness dx, then rate of decrease of intensity of radiation is given by,

$$\frac{-dI}{dx} \propto CI \qquad \text{...(ix)}$$

$$\frac{-dI}{dx} = k'CI$$

$$\frac{dI}{I} = k'C\,dx \qquad \text{...(x)}$$

On integrating the equation (x) between the limits

$I = I_0$ at $x = 0$ and $I = I$ at $x = l$ we get

$$\int_{I_0}^{I} \frac{dI}{I} = -k'C \int_0^l dx$$

$$\ln \frac{I}{I_0} = -k'Cl$$

$$2.303 \log \frac{I}{I_0} = -k'Cl$$

$$\log I_0/I = \frac{k}{2.303} Cl$$

$$\boxed{A = \varepsilon Cl} \qquad \text{...(xi)}$$

Where C is the concentration of solute expressed in mole/litre, l = length of the cell containing the sample and ε is ***molar extinction coefficient.***

Also transmittance $T = \frac{I}{I_0}$,

or $\quad -\log T = \log \frac{I_0}{I} = A$

thus, from equation (xi)

$$A = -\log T = \varepsilon Cl \qquad \text{...(xii)}$$

From equation (xi) it is seen that the absorbance, which is also called optical density (OD), of a solution in a container of fixed path length is directly proportional to the concentration of a solution. A plot between absorbance and concentration is expected to be linear.

INSTRUMENTATION. Abs]orption measurements are carried out with various instruments such as colorimeter, spectrometer. All of these have the following features in common:-

1. Source of electromagnetic radiation. The incident light of required wave lenght is provided by using appropriate source. The source of visible radiation is *incandescent tungsten filament*, whereas the source of UV radiation is *hydrogen discharge lamp.*

2. Filter or Monochromator. In colorimeter filters are used whereas in spectrophotometers, filters are replaced by *monochromators. The filters or monochromators convert polychromatic radiation from source into narrow bands i.e these are used to eliminate the radiant energy of certain wavelengths and radiation in only selected region is allowed to pass. The monochromators are capable of isolating a much narrower band of wavelength.

3. Sample Holder: The sample is placed in the sample container which are known as cuvettes or cells. For visible or Ultraviolet radiations, the cuvettes made of quartz are used.

4. Photodetectors: The main function of any detector is to absorb the energy of photons and convert it into measurable quantity such as electric current. Detectors generate an electrical signal, which activate meter or recorder. In UV-visible spectrophotometers photocells are used as detectors.

5. Amplification and Recording unit: The main function of amplifier is to amplify a signal so that it can be measured. It takes an input signal from circuit through electronic operations and produce an output signal which is several order of magnitude than input . The signal is recorded on a graph.

Most spectrophotometers record wavelength versus absorbance. A schematic diagram of the instrument design is given in Fig. 6.1

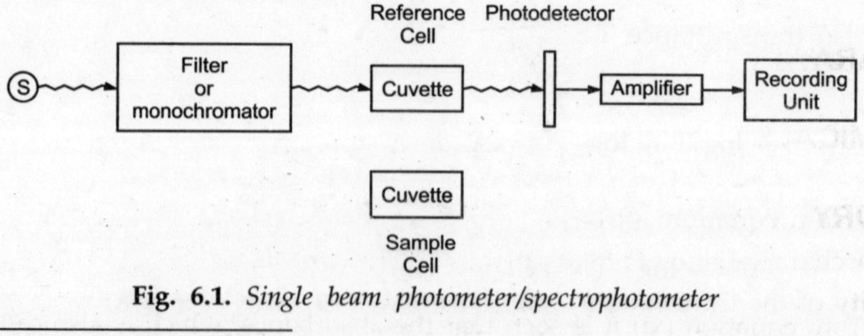

Fig. 6.1. *Single beam photometer/spectrophotometer*

* Monochromator may be regarded as an assembly consisting of a dispersing element, such as prism, differaction grating together with two narrow slits.

Generally two types of spectrophotometers/photometers are used.

(a) Single beam spectrophotometers

(b) Double beam spectrophotometers.

Single beam is described above whereas in double-beam spectrophotometers, a beam of light is split in to two identical beam known as sample beam and reference beam respectively. Sample beam passes through the sample and the reference beam is directed through an cuvette (Cell) which contains only the solvent. The function of a double beam operation is to measure the difference in intensities of the two beams at each wavelength. A schematic layout of double beam spectrophotometer is given in fig. 6.2

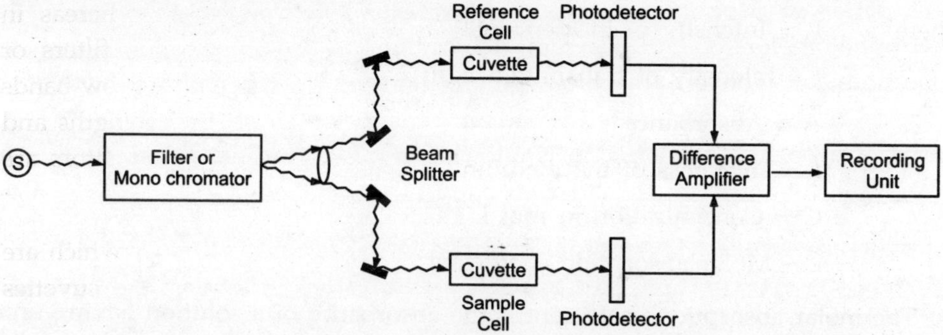

Fig. 6.2 *Double beam spectrophotometer*

EXPERIMENT NO. 6.1

AIM **(a) To determine λ_{max} (wave length of maximum absorption) of solution of $KMnO_4$ using a spectrophotometer**

(b) Verify Beer's law and apply it to find the concentration of the given unknown solution.

APPARATUS UV- visible spectrophotometer, beaker.

CHEMICALS Distilled water, standard solution of $KMnO_4$, tissue paper.

THEORY When an electromagnetic radiation is passed through a sample, certain characteristic wavelengths are absorbed by the sample. As a result the intensity of the transmitted light is decreased. The measurement of the decrease in intensity of radiation is the basis of spectrophotometry. Thus the spectrophotometer compares the intensity of the transmitted light with that of incident light.

The absorption of light by a substance is governed by certain laws (discussed in detail in the preceding section page 162)

According to the Beer Lambert's law the intensity of the incident light is proportional to the length of thickness of the absorbing medium and the concentration of the solution,

$$\log \frac{I_0}{I} = A = \varepsilon C l \qquad \text{...(i)}$$

where I_0 = Intensity of incident light

I = Intensity of transmitted light

A = Absorbance

l = Thickness of the medium

C = Concentration in mol L^{-1}

ε = Molar absorption coefficient

The molar absorption coefficient is the absorbance of a solution having unit concentration (C = 1 M) placed in a cell of unit thickness (l = 1 cm). Absorbance is also called Optical Density (OD)

From equation (xii), it is seen that the absorbance (OD) of a solution in a container of fixed path length is directly proportional to the concentration of a solution. i.e.

$$\boxed{A = \varepsilon Cl}$$

A plot between absorbance and concentration is expected to be linear. Such a straight line plot, passing through the origin, shows that Beer-Lambert's law is obeyed. This plot, known as caliberation curve can also be employed in finding the concentration of a given solution.

Spectrophotometer. A spectrophotometer is a device which detects the percentage transmittance of light radiation when light of certain intensity and frequency range is passed through the sample. Thus the instrument compares the intensity of the transmitted light with that of the incident light (A detailed discussion is given in the preceding section, page 164)

There are many spectrophotometers available for the visible range extending from 3800 – 7800 A°. One of there is represented in the Fig. 6.3

Fig. 6.3 *Spectrophotometer*

PROCEDURE

Setting of the instrument

1. Spectrophotometer should initially read zero on transmittance scale (T). If it does not read zero, set it mechanically with adjusting knob.
2. Connect the instrument to the mains and put on the power switch.
3. Adjust the wavelength knob to the required wavelength region on scale (approximately)
4. Choose the position of wavelength switch correspondingly either to 340 – 400 nm or 400-960 nm.
5. Adjust the meter needle on zero on Transmittance scale and 100 on O.D scale

Working

6. Open the lid of the cell compartment and insert a cuvette containing the blank solvent (distilled water). Close the lid.
7. Adjust the needle to 100% transmittance or zero optical density
8. Remove the cuvette and close the lid tightly again. Empty the cuvette and rinse it with the standard solution of $KMnO_4$ (0.01 IM). Fill it with standard solution.
9. Now place the cuvette containing the standard solution in the cell compartment. Note the O.D and Transmittance.

Determination of λ max

10. Now change the wavelength by 20 nm and note absorbance (OD) and transmittance for each wavelength.

    Imp. Transmittance through the pure solvent (blank) is always adjusted to 100% before each wavelength measurement. (For this repeat step 5 to 9). This is because the extinction coefficient of the blank also changes with the wavelength.

11. Plot a graph between wavelength on the x axis and Absorbance (OD) on the y axis

Verification of Beer's Law

12. Fix the wavelength at λ max position.
13. Prepare $KMnO_4$ solutions with concentrations 0.2%, 0.5%, 1.0%, 1.5%, 2.0%, 2.5%, and 3.0% etc. (20 ml each)
14. Note down the absorbance (OD) of series of solution of $KMnO_4$ prepared above by the method described above.
15. Plot a graph between O.D against concentration. (If a straight line is obtained Beer's law is verified)
16. Now find out the OD of the unknown solution of the $KMnO_4$. Find out the concentration of this solution from the graph.

OBSERVATIONS.

(i) *Determination of λmax*

Wavelength (nm)	Absorbance (OD)

(ii) *Verification of Beer's Law*

S.No.	*Concentration (C) (moles/L)*	*Absorbance (OD)*

Calculations.

(i) A curve is plotted between wavelength and absorbance (OD)

From the graph, the maximum value of A gives λmax.

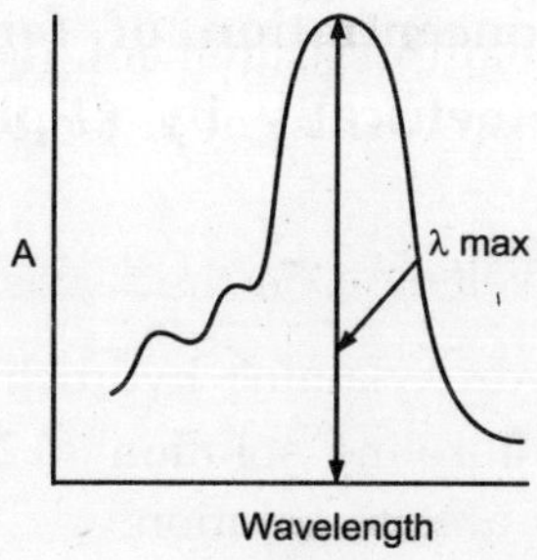

Fig. 6.4

(ii) A curve is plotted between O.D and concentration and if a straight line is obtained as shown by equation (i), Beer's law is verified

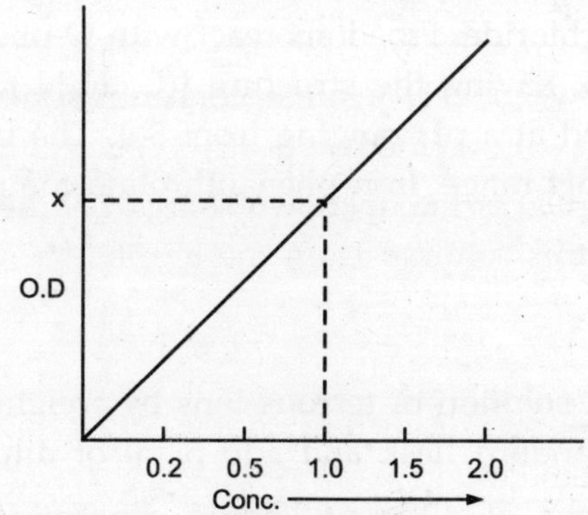

Fig. 6.5

(iii) From the graph of O.D Versus concentration, the concentration of the unknown solution can be found out. For example, in the Fig. 6.5 if x is the O.D of unknown solution then its concentration will be 1.0 %.

RESULTS

(i) λmax for $KMnO_4$ = . . nm

(ii) $KMnO_4$ solution obeys Beer's law

(iii) Concentration of the unknown solution = ...mg/L

PRECAUTIONS.

1. Always use dilute solutions for getting caliberation curve.
2. Cuvette should be cleaned properly and must be wiped with tissue paper.
3. Do not leave any finger marks on the cuvette.

EXPERIMENT NO. 6.2

AIM **To determine the concentration of ferrous ions in a given sample spectrophotometrically by O-phenanthroline method.**

APPARATUS Spectrophotometer,

CHEMICALS Conc. HCl, Hydroxylamine hydrochloride (10%), Ammonium acetate buffer solution, Phenanthroline solution (0.25%), stock iron solution, standard iron solution, Sodium acetate solution.

THEORY Similar to experiment No. 6.1 for discussion on spectrophotometry in general.

Fe^{3+} ions in the sample are reduced to Fe^{2+} ions by boiling with HCl and hydroxyl amine hydrochloride. Fe^{2+} ions react with O-phenanthroline and give an orange red complex having the structure $[(C_{12}H_8N_2)_3\ Fe]^{2+}$. The colour is stable for a long period at a pH ranging from 3-9. The intensity of the colour is independent of this pH range. Iron phenanthroline complex is then measured spectrophotometrically.

PROCEDURE

1. Prepare 1% stock solution of ferrous ions by weighing 0.3511g of Mohr's salt in 100 ml standard flask and add 5 ml of dilute H_2SO_4.
2. Prepare 5 solutions of different known concentrations by diluting the above stock solution in 5 different flasks.
3. Transfer 1ml of solution from first flask in a 50 ml standard flask. Now add 5 ml each of 10% hydroxylamine solution, sodium acetate solution and O-phenanthroline solution. Dilute the solution to 50 ml
4. Repeat the same process with the solutions of other concentrations prepared above and also with the unknown sample.

   Also prepare the blank solution by dissolving the specified amounts of each reagent except the standard iron solution.
5. Keep all these coloured solutions at rest for five minutes.
6. Note down the absorbance of these six solutions with spectrophotometer against blank solution, selecting λmax at 515 nm. (as discussed in experiment 6.1)

7. Prepare a caliberation curve by plotting absorbance against concentration.
8. From this, the concentration of the unknown solution corresponding to its absorbance can be found out.

OBSERVATIONS.

S.no.	Conc.	Absorbance
1		
2		
3		
4		
5		
6	Unknown	

RESULT The conc. of ferrous ions in a given sample solution is .. mg/*l*

Instruction to the lab assistant

Prepare the following

1. Ammonium acetate buffer (pH 5.0)- Dissolve 250g of ammonium acetate in 150 ml of distilled water and add 700 ml of glacial acetic acid.
2. Hydroxylamine hydrochloride - 100g of the compound is dissolved in 100 ml of distilled water and 2 drops of conc. HCl are added to it.
3. Phenanthroline solution : 100g of it is dissolved in 100 ml of distilled water.
4. Sodium acetate. Dissolve 2g of it in 1L of distilled water.

EXPERIMENT NO. 6.3

AIM Determine the concentration of iron in the given sample of water by spectrophotometric method (colorimetric method) using KCNS as colour developing agent.

APPARATUS Spectrophotometer.

CHEMICALS Stock solution of Fe^{3+} ions, conc. HCl, Potassium thiocyanate.

THEORY For general discussion on spectrophotometry refer to experiment No. 6.1.

The colorimetric determination of iron using KCNS as colour developing agent is based on the formation of a red coloured complex between Fe^{3+} and CNS^- ions

$$Fe^{3+} + 3CNS \rightleftharpoons \underset{\text{Red}}{[Fe\,(CNS)]^{2+}}$$

In the determination of iron in the water sample, a series of standard solutions having iron is treated with KCNS to get iron-thiocyanate complex. The absorbance of all the standard solution prepared is noted at λ_{max} 480 nm because iron-thiocyanate complex shows maximum absorbance at this wavelength. Now the absorbance for the unknown solution is also determined. A graph is plotted against the absorbance (OD) against the concentrations of the known solutions. From the graph the concentration of the unknown solution can be found out.

PROCEDURE

(1) Set the colorimeter (spectrophotometer) as discussed in experiment 6.1. Adjust the wavelength to 480 nm.

(2) Prepare the stock solution of Fe^{3+} ions by dissolving 2.0g of ferric ammonium sulphate in 50 ml of distilled water and 10 ml of conc. HCl. HCl is added to suppress the hydrolysis of ferric ammonium sulphate.

(3) Take the stock solution in the burette and prepare 5 solution of different known concentrations by diluting the above stock solution in five different flasks.

(4) Transfer 1 mL solution of standard stock solution to 50 mL measuring flask and add 10 mL of thiocyanate solution Dilute the solution to 50 mL.

(5) Repeat the process (step 4) with the solutions of other concentrations prepared above and also with the unknown sample.

(6) Also prepare the blank solution by dissolving 10 mL of thiocyanate solution in distilled water and making the volume to 50 mL.

(7) Keep the solution at rest for 5 minutes.

(8) Note down the absorbance of all the solution against blank solution, spectrophotometrically at 480 nm. (as discussed in experiment 6.1).

(9) Prepare a calibration curve by plotting absorbance against concentration.

(10) From the calibration curve, the concentration of the unknown solution corresponding to its absorbance can be found out.

OBSERVATIONS

S.No.	*Conc.*	*Absorbance*
1		
2		
3		
4		
5		
6	unknown	

CALIBRATION CURVE

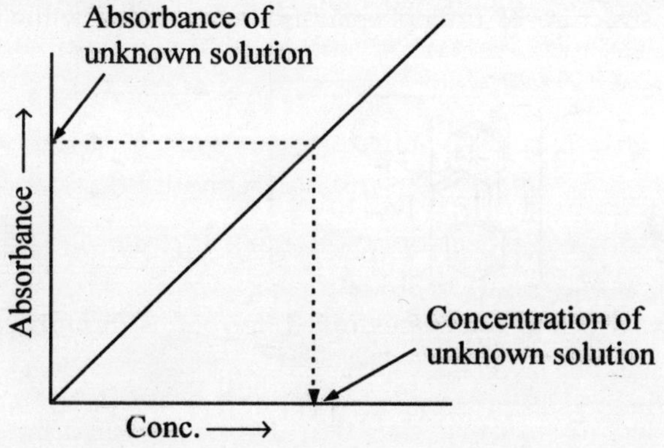

RESULT

The amount of Fe(III) in water was found to be = ... g/L.

PRECAUTIONS

Same as in experiment 6.2.

Instructions to the lab Assistant:

Provide the following:

1. 40% solution of KCNS
2. Conc. HCl
3. Ferric ammonium sulphate

EXERCISE

1. What is spectrophotometry?
2. Discuss Lambert's and Beer's law of absorbance.
3. What is colorimetry?
4. What is absorbance, transmittance, extinction coefficient.
5. How you can verify Beer's law?
6. How you can find out the conc. of the given unknown solution spectrophotometrically?
7. What is monochromatic light ?
8. What is the significance of spectrophotometry.
9. What is the structure of orthophenanthroline complex with iron (II).

Ans.

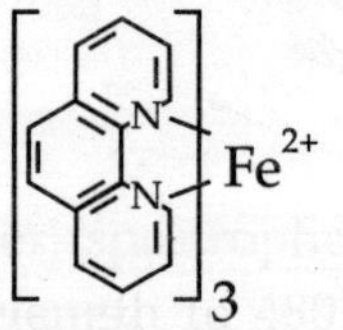

This complex, which is sometimes called 'ferroin' is formulated as $(phen)_3\ Fe^{2+}$.

❑❑❑

7

Thermochemistry

Thermochemistry is the branch of physical chemistry which deals with the heat changes accompanying chemical reactions or change of phase. It is based on the first law of thermodynamics.

Chemical reactions are almost always accompanied by energy changes. These changes appear in the form of evolution or absorption of heat. If a chemical reaction is accompanied by evolution of heat, it is called *exothermic reaction*. For example.

$$C_{(s)} + O_{2(g)} \rightarrow CO_{2(g)} + 393.5 \text{ kJ}$$

On the other hand, if the chemical reaction is accompanied by absorption of heat, it is called *endothermic reaction*. For example:

$$H_{2(g)} + I_{2(g)} \rightarrow 2HI_{(g)} - 52.5 \text{ kJ}$$

HEAT OF REACTION OR ENTHALPY OF REACTION

Heat of reaction is defined as the amount of heat absorbed or evolved in a reaction when the number of moles of reactants as represented by the balanced chemical equation, are converted into the products at *constant temperature and pressure*. Since most of the reactions occur at constant pressure, it is usual to express the heat of the reaction as **enthalpy change* accompanying the reaction.

When 1 mole of gaseous methane react with 2 moles of oxygen at 298 K and 1 atmospheric pressure to form 1 mole of gaseous CO_2 and 2 moles of liquid water at the same temperature and pressure, 890.35 kJ of energy is given out to the surroundings. It can be represented by thermochemical equation as:

$$CH_{4(g)} + 2O_{2(g)} \rightarrow CO_{2(g)} + 2H_2O_{(l)} \quad \Delta H = -890.35 \text{kJ}$$

A balanced chemical equation which indicates the amount of heat absorbed or evolved in the reaction process is called *thermochemical equation*.

* Enthalpy is the thermodynamic term which denotes the heat change accompanying a chemical reaction at *constant temperature and pressure*. If the reaction is carried out at constant volume then the heat-change accompanying a chemical reaction is denoted by *internal energy*. The symbol for enthalpy change is ΔH.

Heat of reaction depends upon the following factors

(i) Physical state of the reactants and products. Physical state of the reactants and products affects the heat of reaction. For example, the heat of reaction when water vapour is formed from gaseous hydrogen and gaseous oxygen, is different from heat of reaction in the formation of liquid water from the same constituents at same temperature and pressure.

$$H_{2(g)} + \frac{1}{2}O_{2(g)} \rightarrow H_2O_{(l)} \qquad \Delta H = -\ 285.8\ kJ$$

$$H_{2(g)} + \frac{1}{2}O_{2(g)} \rightarrow H_2O_{(g)} \qquad \Delta H = -\ 248.8\ kJ$$

This is because the absorption of latent heat of vaporisation is required to convert liquid water to gaseous state.

(ii) Amounts of reactants involved. If the amount of the reactants is doubled. The heat of reaction is also doubled. For example, if the quantities of hydrogen and oxygen are doubled in the above reaction, the heat of reaction will also be doubled

$$2H_{2(g)} + O_{2(g)} \rightarrow 2H_2O_{(l)} \qquad \Delta H = -571.6 kJ$$

Heat of reaction is generally reported for 1 mole of the reactant.

(iii) Whether the reaction is taking place at constant pressure or constant volume.

(iv) Temperature. The enthalpy change accompanying a chemical reaction varies with temperature. The values of heat of reaction are usually reported at 298 K.

(v) Concentration of the solution. If the solutions are involved in the reaction, their concentration may affect the heat of reaction.

DIFFERENT TYPES OF ENTHALPY OF REACTIONS

Heat of the reaction is a general term used for change or enthalpy change accompanying any reaction. However depending upon the nature of reaction-whether it is neutralization or combustion etc., the enthalpy of the reaction is named accordingly. Some of the important heats of the reaction are discussed below.

(a) Heat of Formation The heat of formation of a substance is defined as *the heat change when 1 mole of the substance is formed from its elements under given conditions of temperature and pressure.* It is usually represented as $\Delta H f$.

(b) Heat of Combustion The heat of combustion of a substance is defined as *the heat change when one mole of the substance is completely burnt in excess of air or oxygen.* For example, the heat of combustion of methane is -890.4 kJ as shown below:

$$CH_{4(g)} + 2O_{2(g)} \rightarrow CO_{2(g)} + 2H_2O_{(l)} \qquad \Delta H = -\ 890.\ 4kJ$$

It should be noted that in determining heat of combustion of a substance, the substance should be completely oxidized.

(c) Heat of Neutralization. The heat of neutralization of an acid is defined as *the heat change when one gram equivalent of an acid is neutralized by a base in dilute solutions.* For example the heat of neutralization of hydrochloric acid by sodium hydroxide in dilute solutions is 57.32 kJ and represented as:

$$HCl_{(aq)} + NaOH_{(aq)} \rightarrow NaCl_{(aq)} + H_2O \qquad \Delta H^o = -57.32kJ$$

The symbol (aq.) indicates that the reaction occurs in such a dilute solution that no more heat changes are observed on further dilution.

Similarly heat of neutralization of a base is defined as *the heat change when one gram equivalent of a base is neutralized by an acid in dilute solutions.* The heat of neutralization of NaOH by HCl is also-57.32kJ and is represented as

$$HCl_{(aq)} + NaOH_{(aq)} \rightarrow NaCl_{(aq)} + H_2O \qquad \Delta H = -57.32kJ$$

Heat of neutralization of strong acid by strong bases

The heat of neutralization of strong acid and strong bases are given in Table 7.1.

Table. 7.1 Enthalpy of Neutralization of strong acids by strong bases

Chemical Equation	*Δ H(kJ/equiv)*
$HCl_{(aq)} + NaOH_{(aq)} \rightarrow NaCl_{(aq)} + H_2O_{(l)}$	– 57.32
$HNO_{3(aq)} + NaOH_{(aq)} \rightarrow NaNO_{3(aq)} + H_2O_{(l)}$	– 57.28
$HCl_{(aq)} + KOH_{(aq)} \rightarrow KCl_{(aq)} + H_2O_{(l)}$	– 57.45
$HCl_{(aq)} + LiOH_{(aq)} \rightarrow LiCl_{(aq)} + H_2O_{(l)}$	– 57.38

From the table, it is evident that enthalpy of neutralization of strong acid with a strong base is identical in all cases. This can be explained on the basis of *theory of ionization.* Strong acid and strong bases are assumed to be completely ionized in dilute aqueous solution. Moreover, the salts they form on mixing are also completely ionized. The reaction between a strong acid like HCl and a strong base like NaOH can be written as:

$$H^+_{(aq)} + Cl^-_{(aq)} + Na^+_{(aq)} + \overline{O}H_{(aq)} \rightarrow Na^+_{(aq)} + Cl^-_{(aq)} + H_2O_{(l)} \Delta H = -57.3 kJ$$

$$H^+_{(aq)} + \overline{O}H_{(aq)} \rightarrow H_2O_{(l)} \qquad \Delta H = -57.3 kJ$$

Thus neutralization is simply a reaction between the H^+ ions given by the acid and $\overline{O}H$ ions given by the base. In case of strong acids and strong bases the number of H^+ ions and $\overline{O}H$ ions produced by one gm equivalent of any strong acid or strong base is always the same. Hence the heat of neutralization between a strong acid and strong base is always same.

Heat of neutralization if either the acid or base is weak

The heat of neutralization of strong acid and weak base or vice versa is always less than -57.32kJ, and is different in different cases as shown in Table 7.2

Table 7.2

Chemical Equation	*ΔH (kJ/equiv.)*
$CH_3COOH + NaOH \rightarrow CH_3COONa + H_2O$	– 55.23
$HCN + NaOH \rightarrow NaCN + H_2O$	– 12.13
$H_2S + 2NaOH \rightarrow Na_2S + 2H_2O$	– 15.90
$HCl + NH_4OH \rightarrow NH_4Cl + H_2O$	– 51.46

In such cases the neutralization process not only involves the combination of hydrogen and hydroxy ions, but also the dissociation of weak acid or base. Consider the neutralization of a weak acid like CH_3COOH with a strong base like NaOH. Acetic acid ionizes to a small extent whereas NaOH ionizes completely.

$$CH_3COOH \rightleftharpoons CH_3COO^- + H^+ \qquad ...(i)$$

$$NaOH \longrightarrow Na^+ + \overline{O}H \qquad ...(ii)$$

$$H^+ + \overline{O}H \longrightarrow H_2O \qquad ...(iii)$$

As H^+ ions given by the acid combine with $\overline{O}H$ ions given by the base, the equilibrium shifts towards the right resulting in dissociation of more of acetic acid. Both the reaction, (i) and (ii) proceed side by side till acetic acid is completely neutralized. *Thus a part of heat produced during the combination of H^+ and $\overline{O}H$ ions is used up for the complete dissociation of acetic acid.* This heat required for dissociation of acetic acid is called heat of dissociation of acetic acid.

The heat of neutralization of acetic acid by NaOH has been found to be – 55.23 kJ which is less than the average value of – 57.23 kJ for the combination of H^+ and $\overline{O}H$ ions. The difference between the two i.e. +2.09 kJ is the heat required for dissociation of acetic acid and hence is called heat of dissociation of acetic acid.

Similarly, the neutralization of a weak base like NH_4OH with a strong acid like HCl can be represented as:

$$NH_4OH \rightleftharpoons NH^+_4 + \overline{O}H \quad ...(i)$$

$$HCl \rightarrow H^+ + Cl^- \quad ...(ii)$$

$$H^+ + \overline{O}H \rightarrow H_2O \quad ...(iii)$$

The heat of neutralization comes out to be – 51.34 kJ which is less than the usual value and hence the enthalpy of dissociation of NH_4OH is 57.32 – 51.34 = 5.98 kJ.

(d) Heat of Solution. The heat of solution of a solute in a particular solvent depends upon the amount of solvent and accordingly it can be defined in two ways:

(i) Integral Heat of solution. The heat of solution of a solute in a particular solvent is defined *as the heat change when one mole of the solute is dissolved in the pure solvent to give a solution of specified concentration.* Thus for integral heat of solution, it is necessary to specify the number of moles of the solvent used. For example if one mole of HCl is dissolved in 25 moles of water, 72.266kJ of heat is evolved. The thermochemical equation for the above dissolution is written as :

$$HCl_{(g)} + 25H_2O_{(l)} \rightarrow HCl(25H_2O) \qquad \Delta H = -72.266 kJ$$

(ii) Differential heat of solution : The differential heat of solution is defined as the heat change *when one mole of solute is dissolved in such a large quantity of the solution that there occurs no appreciable change in concentration of the solution.*

Mathematically differential enthalpy of solution can be represented as:

$$\text{Differential enthalpy of solution} = \frac{d(\Delta H)}{dm}$$

where $d(\Delta H)$ is the small change in value of ΔH when a small amount dm of solute is added to a solution of definite composition.

To illustrate the differential heat of solution, consider a solution containing one mole of HCl dissolved in 20 moles of water. Now if we take large volume of this solution and dissolve 1 mole more of HCl, the heat change produced will be called the differential heat of solution of HCl. This heat change is produced without changing the concentration of the solution appreciably.

It should be mentioned that

- salts like $CuSO_4 \cdot 5H_2O$, $CaCl_2.6H_2O$ etc. or in general *hydrated salts* dissolve with the absorption of heat. For example-

$$CuSO_4.5H_2O + aq \rightarrow CuSO_{4(aq)} \qquad \Delta H = +11.7 \text{ kJ}$$

- Salts which do not form hydrates also dissolve with the absorption of heat.

$$KCl_{(s)} + aq \rightarrow KCl_{(aq)} \qquad \Delta H = +18.6 \text{ kJ}$$

- The anhydrated salts which are capable of forming hydrates e.g $CaCl_2$, $CuSO_4$ etc. are accompanied by evolution of heat and as result the solution heats up.

$$CuSO_{4(s)} + aq \rightarrow CuSO_{4(aq)} \qquad \Delta H = -66.5 \text{ kJ}$$

(e) Heat of Hydration. The heat of hydration is defined as *the heat change when 1 mole of anhydrous substance combines with the requisite amount of water to form a hydrated stable salt.* Hydration of $CuSO_4$ is represented by the following thermochemical equation.

$$CuSO_{4(s)} + 5H_2O(l) \rightarrow CuSO_45H_2O_{(s)} \qquad \Delta H = -78.868 \text{ kJ}$$

Heat of hydration is almost always negative i.e. heat is always evolved in such reactions. Heat of hydration is always calculated from heat of solutions of hydrated and anhydrous substances. For example, heat of hydration of $CuSO_4$ can be calculated from the following thermochemical equations:

$$\text{(i) } CuSO_{4(s)} + 800H_2O_{(l)} \rightarrow CuSO_4(800H_2O) \qquad \Delta H_1 = -68.74\text{kJ}$$

$$\text{(ii) } CuSO_4.5H_2O + 795H_2O_{(l)} \rightarrow CuSO_4(800H_2O) \qquad \Delta H_2 = 10.12\text{kJ}$$

$$\therefore \text{ Heat of hydration of } CuSO_4 = \Delta H_1 - \Delta H_2 = -78.86\text{kJ}$$

(f) Heat of Dilution. Heat of dilution is defined as *the change of enthalpy when a solution containing 1 mole of a solute is diluted from one concentration to another.*

It has been mentioned earlier that heat of solution per mole of a solute varies with the concentration of the solution. This implies that when a given solution is diluted on addition of solvent there must be a change in enthalpy which is called heat of dilution.

HEAT CAPACITY

The heat capacity of a substance is defined as the amount of heat required to raise the temperature of the system from lower to higher temperature divided by the temperature difference, between any two temperatures.

If q calories is the heat absorbed by a substance having mass m, in raising its temperature from T_1 to T_2, the heat capacity will be given by the expression

$$C = \frac{q}{m \times (T_2 - T_1)} \quad ...(i)$$

If m = 1gm, $T_2 - T_1$ = 1 K, then C is defined as *the amount of heat required to raise the temperature of 1gm of a substance through one degree centigrade. This is sometimes also referred to as specific heat of the substance.* Its units are cal $deg^{-1}g^{-1}$ whereas *molar heat capacity* is defined as *the amount of heat required to raise the temperature of 1 mole of a substance through one degree centigrade.* Its units are cal deg^{-1} mol^{-1} or J $deg^{-1}mol^{-1}$.

The relationship between the heat capacity and specific heat of a substance is

$$C = ms \quad ...(ii)$$

where m is the mass and s is the specific heat of the substance.

If we know the specific heat and the amount of a substance, then change in the temperature of the sample after the reaction can tell us the amount of heat (q) that has been absorbed or evolved in particular process. The heat absorbed or evolved will be given by,

$$q = ms\ \Delta t \quad ...(iii)$$

$$= ms\ (t_2 - t_1)$$

where t_2 is the final temperature and t_1 is the initial temperature of the sample. Equation (iii) can also be written as

$$q = C\ \Delta t \quad ...(iv)$$

q is positive for endothermic processes and negative for exothermic processes.

In the laboratory the heat changes in physical and chemical processes are measured with a ***calorimeter*** and the technique is known as ***calorimetry***.

In thermochemistry experiments, the calorimeter used is a Dewar flask or a thermos flask. A simple calorimeter can also be constructed from two thermocol coffee cups, one inserted into the other. The outer cup helps to insulate the reacting mixture from the surroundings.

HEAT CAPACITY OF THE CALORIMETER

In all the thermochemical experiments taking place in calorimeters, the first step will be determination of heat capacity of the calorimeter which is sometimes also referred to as water equivalent of a calorimeter. The heat capacity of a calorimeter is defined as *the heat required to raise the temperature of the calorimeter by 1°C.*

During a reaction taking place in the calorimeter, the calorimeter will also take up some of the heat evolved. This heat should also be taken into account by determining the heat capacity of the calorimeter.

Heat capacity of calorimeter can be obtained by weighing the calorimeter and its components and multiplying their weights by respective specific heats. But when thermos flask is used as calorimeter, the above method is not feasible and hence the heat capacity of the calorimeter is determined by the *method of mixtures* which is discussed in the following experiments.

EXPERIMENT NO. 7.1

AIM **To determine the heat capacity of a calorimeter.**

THEORY The heat capacity or the water equivalent of the calorimeter is determined by the method of mixtures or commonly known as Heat Exchange Method.

In this method a definite amount of cold water is taken in the calorimeter and the same amount of hot water is mixed to this. The temperature of the cold water (T_c), hot water (T_h) and the mixture (T_m) is noted. According to the principle of calorimetry,

Heat gained = Heat lost

∴ Heat gained by (cold water + calorimeter) = Heat lost by hot water

or $[C_p \text{ (Cold water)} + C_p \text{ (Calorimeter)} [T_m - T_c]$

$$= C_p \text{ (hot water)} [T_h - T_m] \quad ...(i)$$

It should be noted that temperature change for cold water and calorimeter is the same.

But, Heat capacity of a substance = mass × specific heat.

Thus, the equation (i) can be written as,

$$[m \times s \text{ (cold water)} + Cp \text{ (calorimeter)}] [T_m - T_c]$$

$$= [m \times s \text{ (hot water)}] [T_h - T_m] \quad ...(ii)$$

Assuming, s (cold water) = s (hot water), equation (ii) takes the form

$$[ms + Cp \text{ (Calorimeter)}] [T_m - T_c] = ms [T_h — T_m] \quad ...(iii)$$

Substituting the value of T_c, T_m, T_h, *m* and *s* is equation (*iii*), Cp (calorimeter) can be determined.

APPARATUS Thermos flask, 6 glass stirrers, thermometer, stop watch, beaker, measuring cylinder.

CHEMICALS Water

PROCEDURE

1. Take 100 ml of distilled water into the calorimeter fitted with the rubber stopper having two holes. Through one hole insert the thermometer and through other hole, a stirrer.
2. Record the temperature of water after every half a minute for about four minutes.
3. Take some water in a beaker and heat it to a temperature of about 25–30° higher than the room temperature Remove the burner and pipette out 100 ml of water in a beaker. Note down the temperature of hot water after every half a minute for about four minutes.
4. Now quickly pour this hot water into the water contained in the thermos flask. Stir and note its temperature for every half a minute for four minutes.

OBSERVATIONS

Observation Table

Time (Sec.)	Temperature °C		Time	Temperature °C
	Cold Water	Hot Water		Mixture

Plot a graph between temperature and time as shown in Fig. 7.1. From the graph find out the temperature of cold water (Tc), hot water (Th) and mixture (Tm) at the time of mixing. This is done by drawing a vertical line at the time of mixing and extrapolating the temperature time curves upto the line. The points of intersection give the respective temperatures.

Note. If the calorimeter used is insulated properly and efficiently then there is no need for the temperatures to be noted from the graph. Simply, temperatures of cold water, hot water and mixture are noted at the time of mixing.

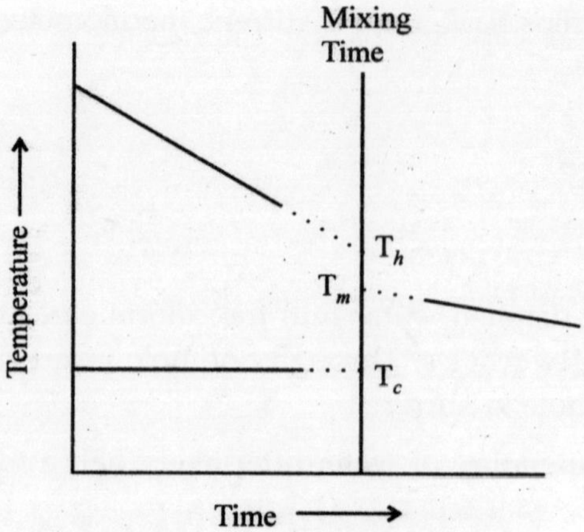

Fig. 7.1 *Temperature time curve*

Volume of cold water taken = 100 ml

Volume of hot water mixed = 100 ml

Temperature of cold water T_c = °C

Temperature of hot water T_h = °C

Temperature of mixture T_m = °C

Density of water is 1g/ml, therefore,

weight of cold water taken = 100 g

weight of hot water mixed = 100 g.

CALCULATIONS

From equation (iii)

$$[ms + Cp\ (calorimeter)]\ [T_m - T_c] = ms\ [T_h - T_m]$$

If specific heat of water is taken as 1 cal/g then the above equation becomes,

$$[m + Cp_{(calorimeter)}\ [T_m - T_c] = m\ [T_h - T_m]$$

$$m + Cp_{\ (Calorimeter)} = \frac{m\,[T_h - T_m]}{[T_m - T_c]}$$

$$Cp_{(Calorimeter)} = \left[\frac{100\,(T_h - T_m)}{(T_m - T_c)} - 100\right] \text{ cal/deg.}$$

The Cp (Calorimeter) can also be determined in S.I. units by multiplying the above equation by 4.185 J as 1 cal = 4.185 J. (specific heat of water in S.I units is 4.185 J/g/°C .

RESULTS The heat capacity of the calorimeter is .... cal/deg or .... Joules/deg.

PRECAUTIONS

1. The temperatures should be noted accurately.
2. The liquids should not spurt while mixing.
3. After mixing, the mixture should be stirred properly.

EXPERIMENT NO. 7.2

AIM **Determination of heat of neutralization of hydrochloric acid with sodium hydroxide.**

THEORY The heat of neutralization of an acid at a given temperature is defined as the amount of heat evolved when one gram equivalent of an acid is neutralized by one gram equivalent of a base in dilute solution at that temperature.

It is found that heat of neutralization for all the strong acids and strong bases is practically a constant quantity. This can be explained on the basis of theory of ionization. Strong acids and strong bases are assumed to be completely ionized in dilute solutions. Moreover the salts they form on mixing are also completely ionized in dilute aqueous solution. The reaction between a strong acid like HCl and a strong base like NaOH can be written as:

$$H^+_{(aq)} + Cl^-_{(aq)} + Na^+_{(aq)} + \overline{O}H_{(aq)} \rightarrow Na^+_{(aq)} + Cl^-_{(aq)} + H_2O_{(l)} \qquad \Delta H = -57.3 \text{ kJ}$$

or

$$H^+_{(aq)} + OH^-_{(aq)} \rightarrow H_2O_{(l)} \qquad \Delta H = -57.3 \text{ kJ}$$

Thus Neutralization is simply a reaction between the H^+ ions given by the acid and $\overline{O}H$ ions given by the base. In case of strong acids and strong bases the number of H^+ ions and $\overline{O}H$ ions produced by one gram of any strong acid or strong base is always the same. Hence the enthalpy of neutralization of a strong acid with a strong base is always the same.

A known volume of HCl solution of known concentration is allowed to react completely with a strong alkali in dilute solution. The rise in temperature is noted. Knowing the heat capacity of the calorimeter, masses of acid and base and their specific heats in dilute solutions, the heat of neutralization can be calculated.

ΔH = – [Heat gained by the calorimeter + Heat gained by the solution]

APPARATUS Thermos flask, glass stirrer, thermometer, stop watch, beaker, measuring cylinder ($50 cm^3$)

CHEMICALS Sodium hydroxide = (0.5 M), Hydrochloric acid = (0.5 M)

PROCEDURE

1. Take a thermos flask and determine its heat capacity as mentioned in experiment 7.1.
2. Now take $50 cm^3$ of HCl in this thermos flask and note down the temperature reading after every half minute for five minutes.
3. Similarly take $50 cm^3$ of NaOH in the thermos flask after removing the acid in a beaker. Note down the temperature of NaOH every half minute for five minutes.
4. Pour the HCl solution in the NaOH already placed in the thermos flask and note down the time of mixing accurately.
5. Note the temperature of the mixture after every half a minute for five minutes.
6. Record the observations as given below.

OBSERVATIONS

Molarity of the acid = 0.5M

Molarity of the base = 0.5 M

Volume of the acid taken = $50 cm^3$

Volume of the base taken = $50 cm^3$

Observation Table

Time (Sec.)	Temperature °C		Time	Temperature °C
	Acid	Base	Sec.	Mixture

Time of mixing = ...

CALCULATIONS

Plot a graph between temperature and time (as shown in the fig. 7.2) From the Fig., calculate , T_a, T_b, T_m, at the time of mixing, where

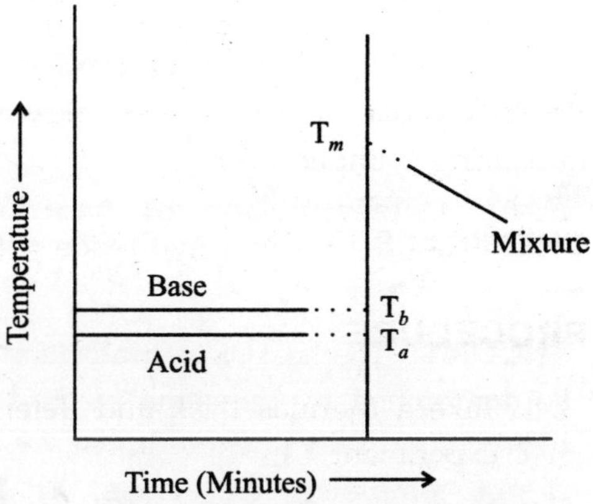

Fig. 7.2 *Temperature Time graph*

T_a is Temperature of the acid at the time of mixing

T_b is Temperature of the base at the time of mixing

T_m is Temperature of the mixture at the time of mixing

This can be done by drawing a vertical line at the time of mixing. Extend the curves to this vertical line. The points of intersection will give the respective temperatures.

The heat evolved during neutralization will raise the temperature of the solution and that of thermos flask, i.e.

$$\begin{aligned}\Delta H &= -\,[\text{Heat gained by the calorimeter} + \text{Heat gained by the solution}]\\ &= -\,[Cp_{(Calorimeter)}\,(T_m - T_b) + [Cp_{(Base)}\,(T_m - T_b)] + [Cp_{(acid)}\,(T_m - T_a)]\\ &= -\,[Cp_{(Calorimeter)}\,(T_m - T_b)] + [V_{base} \times \text{specific heat of the base}\\ &\qquad \times (T_m - T_b)] + [V_{acid} \times \text{specific heat of the acid} \times (T_m - T_a)]\end{aligned}$$

In the case of dilute solutions , the specific heat of the acid and base can be taken as that of water = 4.185 JK^{-1}. Hence,

Specific heat of water = specific heat of acid = specific heat of base = *s*

$$\text{Thus, } \Delta H = -\,[Cp_{(calorimeter)}\,(T_m - T_b) + V_{base}\,.s.\,(T_m - T_b) + V_{acid}\,.s.\,(T_m - T_a)]$$

Substituting the values of $Cp_{calorimeter}$, specific heat of water, V_{base} ,V_{acid} T_m, T_a and T_b calculate ΔH. (S = 4.185 J/g/°C)

This is the enthalpy when 50 cm^3 of 0.5 M HCl is neutralized by 50cm^3 of 0.5 NaOH.

$$\text{Number of moles in } 50cm^3 \text{ of } 0.5M\ HCl = \frac{0.5}{1000} \times 50 = 0.025 \text{ moles}$$

$$\therefore \quad \Delta H \text{ for the neutralization of 1 mole of HCl} = \frac{-\Delta H}{0.025} = \text{...Jmole}^{-1}$$

RESULT The heat of neutralization of hydrochloric acid with sodium hydroxide is ....$Jmol^{-1}$.

EXPERIMENT NO. 7.3

AIM **Determination of heat of neutralization and ionization of acetic acid with sodium hydroxide.**

THEORY The heat of neutralization of acetic acid with sodium hydroxide can be determined by the same method as is discussed for the neutralization of hydrochloric acid in expteriment 7.2. However, acetic acid is a weak acid and is not completely dissociated in dilute aqueous solutions. Thus heat of neutralization is always less than – 57.32 kJ. The reason is that the neutralization process not only involves the combination of hydrogen and hydroxyl ions, but also the dissociation of weak acid or weak base. During the neutralization of a weak acid like CH_3COOH with strong base like NaOH, acetic acid dissociates to a small extent whereas NaOH ionizes completely.

$$CH_3COOH \rightleftharpoons CH_3COO^- + H^+ \quad ...(i)$$

$$NaOH \rightarrow Na^+OH^- \quad ...(ii)$$

$$H^+ + \overline{O}H \rightarrow H_2O \quad ...(iii)$$

As H^+ ions given by the acid combine with $\overline{O}H$ ions given by the base, the equilibrium of the (i) reaction shifts to the right till acetic acid is completely neutralized, thus a part of heat produced by the combination of H^+ and $\overline{O}H$ is used up for the complete dissociation of acetic acid. Thus the enthalpy of neutralization is always less than – 57.32 kJ. The heat required to ionize acetic acid is called heat of ionization of acetic acid and is always positive.

A known volume of CH_3COOH solution of known concentration is allowed to react completely with NaOH in dilute solution. The rise in temperature is noted. Knowing the heat capacity of the calorimeter, masses of acid and base and their specific heats in dilute solutions, the heat of neutralization is calculated.

or $$\Delta H_{ions} + \Delta H_{neut} = \Delta H_{observed}$$

$$\Delta H_{ions} = \Delta H_{obs} - \Delta H_{neut}$$

$$= \Delta H_{obs} - (-57.3 \text{kJ mol}^{-1})$$

$$\therefore \quad \Delta H_{ions} = \Delta H_{obs} + 57.3 \text{kJ mol}^{-1}$$

since ΔH_{obs} is negative and smaller than 57.3 kJ, $\therefore$ ΔH_{ions} is positive i.e. ionization is an endothermic process.

APPARATUS Thermos flask, glass stirrer, thermometer, stop watch, beaker, measuring cylinder.

CHEMICALS Acetic acid (0.5M), Sodium hydroxide (0.5M).

PROCEDURE

1. Take a thermos flask whose heat capacity has been determined as explained in experiment 7.1.
2. Take 50 cm^3 of CH_3COOH in this thermos flask and note down the temperature reading after every half minute for five minutes.
3. Similarly take 50cm^3 of NaOH in the thermos flask after removing the acid in a beaker. Note down the temperature of NaOH every half minute for five minutes.
4. Pour the CH_3COOH solution in the NaOH already placed in the thermos flask and note down the time of mixing accurately
5. Note down the temperature of the mixture after every half a minute for five minutes. Record the observations.

OBSERVATIONS

Molarity of the acid = 0.5M

Molarity of the base = 0.5M

Volume of the acid taken = 50cm^3

Volume of the base taken = 50 cm^3

Observation Table

Time (Sec.)	Temperature °C		Time	Temperature °C
	Acid	Base		Mixture

CALCULATIONS Plot a graph between temperature and time (as shown in the Fig 7.3). From the graph, note down the temperature of the acid, base and mixture.

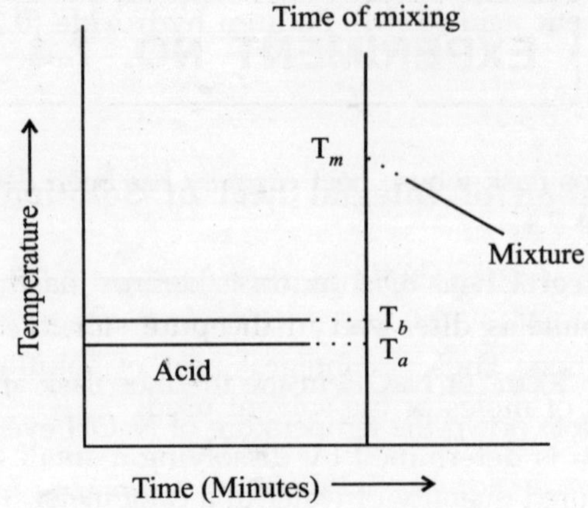

Fig. 7.3 *Temperature time graph*

Temperature of the acid = T_a

Temperature of the base = T_b

Temperature of the mixture = T_m

The heat evolved during neutralization will raise the temperature of the solution and that of thermos flask i.e.

$$\Delta H = -[Cp_{(calorimeter)}(T_m - T_b) + V_{base}.s.(T_m - T_b) + V_{acid}.s.(T_m - T_a)]$$

(As the solution is dilute,

$\therefore$ specific heat of water = specific heat of base

= specific heat of acid = s = 4.185 $JK^{-1} Kg^{-1}$)

Substituting the values of $Cp_{(calorimeter)}$, Specific heat of water,

V_{base}, V_{acid}, T_m, T_a, T_b calculate ΔH.

ΔH gives the enthalpy when $50cm^3$ of 0.5 M CH_3COOH is neutralized by $50cm^3$ of 0.5 M NaOH.

$$\therefore \quad \text{Number of moles neutralized} = \frac{0.5}{1000} \times 50 = 0.025 \text{ moles}$$

$$\therefore \quad \Delta H \text{ for the neutralization for 1 mole} = \frac{-\Delta H}{0.025} = ...\text{J mol}^{-1}$$

Enthalpy of Ionization

$$\Delta H_{ions} = \Delta H_{obs} - (-57.3 \text{ kJ mol}^{-1})$$
$$= \Delta H_{obs} + 57.3 \text{ kJ mol}^{-1}$$

ΔH_{obs} is negative and smaller than 57.3 kJ

$\therefore$ ΔH_{ions} is positive

RESULT The heat of neutralization of acetic acid with sodium hydroxide is .... J mol^{-1} and heat of ionization of acetic acid is ...J mol^{-1}

EXPERIMENT NO. 7.4

AIM Determination of Integral heat of Solution of $CuSO_4 . 5H_2O$.

THEORY The integral heat of solution is defined as the heat change when one mole of the solute is dissolved in the pure solvent to give a solution of specified concentration. Thus for integral heat of solution, it is necessary to specify the number of moles of the solvent used.

Heat of solution is determined by dissolving a small known weight of the given salt in a measured quantity of water in a calorimeter. The initial temperature (T_1) of water and the final temperature (T_2), when a known mass of the solute is completely dissolved in it, is noted. Then ΔH is calculated as follows.

$$q_p = (\text{Heat capacity of the calorimeter + Heat capacity of the products}) \times (\text{Temperature change})$$

$$q_p = [(Cp_{(Calorimeter)} + Cp_{(products)}] (T_2 - T_1)$$

The heat capacity of the calorimeter is determined as in experiment 7.1. The heat capacity of the products is taken as heat capacity of water taken because the solution to be prepared is quite dilute.

$$\therefore \quad Cp_{(products)} = \text{Mass of water (m)} \times \text{specific heat of water(s)}$$

$$= ms = m \times 4.185 \text{ JK}^{-1}$$

Thus for 1 mole of solute, the enthalpy of solution will be given by,

$$\Delta H_{sol} = q_p / \text{No. of moles of solute added.}$$

or

$$\Delta H = q/n = -\frac{[m_1 s + C_{P(calorimeter)}][T_2 - T_1]}{n}$$

$$= -[m_1 s + C_{p\,(calorimetre)}][T_2 - T_1]\frac{M}{m_2}$$

where M = Molecular mass of the solute

m_2= mass of the solute dissolved.

ΔH value is the integral heat of solution of a solute for a specific mole ratio of solute to the solvent. *

---

* **Note.** For example, we want to find out the integral enthalpy of solution of $CuSO_4. 5H_2O$ in water in specific solute solvent ratio of 1 : 100 moles then the weight of $CuSO_4 . 5H_2O$ to be dissolved in $100cm^3$ (100g) of water can be calculated as follows:

100 moles of water will require 1 mole of $CuSO_4. 5H_2O$

(100 × 18)g of water will require 249.68g of $CuSO_4. 5H_2O$

1800g of water will require 249.68 g of $CuSO_4 \cdot 5H_2O$

100g of water will require $= \frac{249.68}{1800} \times 100\text{g of } CuSO_4. 5H_2O$

$= 13.87 \text{ g of } CuSO_4 \cdot 5H_2O$

APPARATUS Thermos flask, glass stirrer, thermometer, stop watch, beaker measuring cylinder

CHEMICALS Solid $CuSO_4 \cdot 5H_2O$

PROCEDURE

1. Determine the heat capacity of the thermos flask as explained in the experiment-7.1.
2. Now weigh appropriate mass of $CuSO_4$. $5H_2O$ on a glazed paper. (For 50 cm^3 of water, you can weigh 6.925 g for 1 : 100, solute : solvent ratio).
3. Take 50 cm^3 of water in the thermos flask and note down its temperature for about 4 minutes at an interval of half a minute.
4. Add the weighed $CuSO_4.5H_2O$ to the water. Stir the solution and note down the temperature of the solution after every half a minute for another 4 minutes. Plot temperature-time curve on a graph paper and note down the initial (T_1) and final temperature (T_2) at the time of mixing from the graph.
5. Record the observations as given below

OBSERVATIONS

Volume of water taken = $50 cm^3$

Mass of water taken m_1 = 50g = 50×10^{-3} kg

Weight of $CuSO_4.5H_2O$ taken = 6.925g

Observation Table

Time (Sec.)	Temperature °C Cold Water	Time (sec)	Temperature °C Mixture

Time of mixing...

CALCULATION

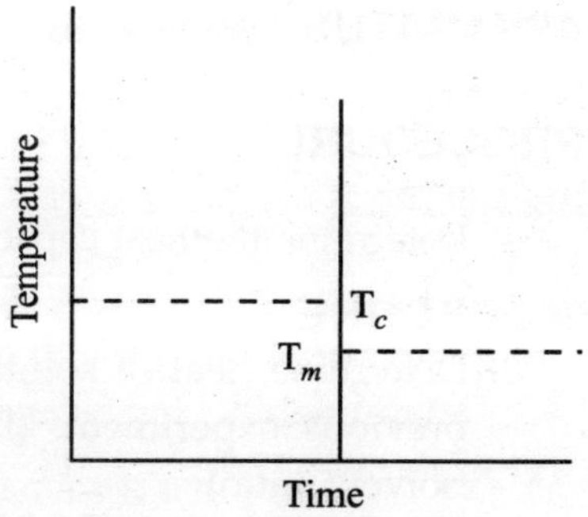

Fig. 7.4

The heat capacity of the thermos flask as calculated in experiment 7.1 = Cp(calorimeter)

Temperature of cold water = T_c

Temperature of solution = T_m

Calculate the heat change for dissolving the specific mass of $CuSO_4.5H_2O$ as follows:

$$\Delta H = \frac{qp}{n} = -\frac{[m_1 s + C_{p\,(calorimeter)}][T_m - T_c]}{n}$$

$$\Delta H = -[m_1 s + C_{p\,(calorimeter)}][T_2 - T_1] \times \frac{\text{Molecular mass of the solute}}{\text{Amount of solute}}$$

RESULT The integral enthalpy of solution of copper sulphate was found to be ...J mol^{-1}.

EXPERIMENT NO. 7.5

AIM Determination of heat of hydration of $CuSO_4.5H_2O$

THEORY The heat change when one mole of an anhydrous substance combines with the requisite number of moles of water to form a hydrated stable salt is known as heat of hydration. Hydration of $CuSO_4$ is represented by the following thermochemical equation.

$$CuSO_{4(s)} + 5H_2O_{(l)} \rightarrow CuSO_4.5H_2O_{(s)} \qquad \Delta H = -78.868 \text{ kJ}$$

Heat of hydration is almost always negative i.e. heat is always evolved in such reaction. Heat of hydration is always calculated from heat of solutions of hydrated and anhydrous substances.

Heat of hydration of $CuSO_4$ can be calculated by determining the heat of solution of anhyhdrous and hydrated salt separately. The difference of the two gives the heat of hydration of $CuSO_4$.

$$CuSO_4.5H_2O_{(s)} + aq \rightarrow CuSO_{4(aq)} \qquad \Delta H_1$$

$$CuSO_{4(s)} + aq \rightarrow CuSO_{4(aq)} \qquad \Delta H_2$$

$\therefore$ Heat of hydration of $CuSO_4$ = $\Delta H_2 - \Delta H_1$

APPARATUS Same as in experiment number 7.4

PROCEDURE

1. Determine the heat capacity of the thermos flask as explained in experiment 7.1.
2. Determine heat of solution of hydrated $CuSO_4.5H_2O$ as explained in the previous experiment. (Dissolve 6.925g of $CuSO_4.5H_2O$ for 1:100, solute :solvent ratio)
3. Determine heat of solution of anhydrated copper sulphate in a similar way, at the same concentration. (Dissolve 4.42of $CuSO_4$ for 1 : 100, solvent : solvent ratio).

OBSERVATIONS

1. For hydrated $CuSO_4.5H_2O$

Volume of water taken = 50 cm^3

Mass of water = 50g = 50×10^{-3} kg

Weight of $CuSO_4$ .5 H_2O taken = 6.925g.

Observation Table

Time (Sec.)	Temperature of water °C	Time (Sec.)	Temperature °C Mixture

Plot temp-time curve on a graph paper. From the graph paper, note down the initial and final temperature at the time of mixing of $CuSO_4$. $5H_2O$.

Let the initial temp of water = T_1

The final temp of solution = T_2

2. For hydrated copper sulphate

Volume of water taken = 50cm^3

Weight of water taken = m_1' = 50g = 50×10^{-3} kg

Weight of $CuSO_4$ taken = 4.42 g

Observation Table

Time (Sec.)	Temperature of water °C	Time (Sec)	Temperature °C Mixture

Plot temp.-time curve on a graph paper and note down the initial and final temperature at the time of mixing

Let the initial temperature of water = T'_1

the final temperature of solution = T'_2

CALCULATIONS

1. The heat change for dissolving specific amount of hydrated copper sulphate is given by

$$\Delta H_1 = -[m_1 s + Cp_{(cal)}][T_2 - T_1] \times \frac{\text{Molar mass of } CuSO_4\ 5H_2O}{\text{Amount of } CuSO_4 \cdot 5H_2O}$$

2. The heat change for dissolving specific amount of anydrated copper sulphate is given by

$$\Delta H_2 = -[m_1' + C_{p(cal)}][T_2 - T_1] \times \frac{\text{Molar mass } CuSO_4}{\text{Amount of } CuSO_4}$$

3. Heat of hydration of $CuSO_4$ =

Heat of solution of $CuSO_4$ – Heat of solution of $CuSO_4$ $.5H_2O$

$$= \Delta H_2 - \Delta H_1$$

RESULT The heat of hydration of $CuSO_4$ is...

EXERCISE

1. *What is the significance of thermochemistry?*

Ans. Thermochemistry is the branch of physical chemistry which deals with the heat changes accompanying chemical reactions or change of phase. It is based on the first law of thermodynamics.

2. *Define heat of reaction or enthalpy of reaction?*

Ans. Heat of reaction is defined as the amount of heat absorbed or evolved in a reaction when the number of moles of reactants as represented by the balanced chemical equation, are converted into the products at *constant temperature and pressure*. Since most of the reactions occur at constant pressure, it is usual to express the heat of the reaction as *enthalpy change* accompanying the reaction.

3. *What are the factors on which enthalpy depends?*

Ans. Heat of reaction depends upon the following factors:

- Physical rate of reactants and products.
- Amounts of reactants involved.
- Whether the reaction is taking place at constant pressure or constant volume.
- Concentration of the solution.

4. *Define heat of neutralization?*

Ans. **Heat of Neutralization.** The heat of neutralization of an acid is defined as *the heat change when one gram equivalent of an acid is neutralized by a base in dilute solutions*. For example the heat of neutralization of hydrochloric acid by sodium hydroxide in dilute solutions is 57.32 kJ and represented as:

$$HCl_{(aq)} + NaOH_{(aq)} \rightarrow NaCl_{(aq)} + H_2O \qquad \Delta H^\circ = -57.32kJ$$

The symbol (aq.) indicates that the reaction occurs in such a dilute solution that no more heat changes are observed on further dilution.

Similarly heat of neutralization of a base is defined as *the heat change when one gram equivalent of a base is neutralized by an acid in dilute solutions*. The heat of neutralization of NaOH by HCl is also-57.32kJ and is represented as

$$HCl_{(aq)} + NaOH_{(aq)} \rightarrow NaCl_{(aq)} + H_2O \qquad \Delta H = -57.32kJ$$

5. *Why heat of neutralization of strong acid by strong bases are same?*

Ans. This can be explained on the basis of *theory of ionization*. Strong acid and strong bases are assumed to be completely ionized in dilute aqueous solution. Moreover, the salts they form on mixing are also completely ionized. The reaction between a strong acid like HCl and a strong base like NaOH can be written as:

$$H^+_{(aq)} + Cl^-_{(aq)} + Na^+_{(aq)} + \bar{O}H_{(aq)} \rightarrow Na^+_{(aq)} + Cl^-_{(aq)} + H_2O_{(l)} \quad \Delta H = -57.3kJ$$

$$H^+_{(aq)} + \bar{O}H_{(aq)} \rightarrow H_2O_{(l)} \qquad \Delta H = -57.3kJ$$

Thus neutralization is simply a reaction between the H^+ ions given by the acid and $\bar{O}H$ ions given by the base. In case of strong acids and strong bases the number of H^+ ions and $\bar{O}H$ ions produced by one gm equivalent of any strong acid or strong base is always the same. Hence the heat of neutralization between a strong acid and strong base is always same.

6. *Heat of neutralization is less than 57.3 kJ/mol, if either the acid or base is weak?*

Ans. In such cases the neutralization process not only involves the combination of hydrogen and hydroxy ions, but also the dissociation of weak acid or base. Consider the neutralization of a weak acid like CH_3COOH with a strong base like NaOH. Acetic acid ionizes to a small extent whereas NaOH ionizes completely.

$$CH_3COOH \rightleftharpoons CH_3COO^- + H^+ \quad ...(i)$$

$$NaOH \longrightarrow Na^+ + \bar{O}H \quad ...(ii)$$

$$H^+ + \bar{O}H \longrightarrow H_2O \quad ...(iii)$$

As H^+ ions given by the acid combine with $\bar{O}H$ ions given by the base, the equilibrium shifts towards the right resulting in dissociation of more of acetic acid. Both the reaction, (i) and (ii) proceed side by side till acetic acid is completely neutralized. *Thus a part of heat produced during the combination of H^+ and $\bar{O}H$ ions is used up for the complete dissociation of acetic acid.* This heat required for dissociation of acetic acid is called heat of dissociation of acetic acid.

7. *Distinguish between integral and differential heat of solution?*

Ans. The heat of solution of a solute in a particular solvent depends upon the amount of solvent and accordingly it can be defined in two ways:

(i) Integral Heat of solution. The heat of solution of a solute in a particular solvent is defined *as the heat change when one mole of the solute is dissolved in the pure solvent to give a solution of specified concentration.* Thus for integral heat of solution, it is necessary to specify the number of moles of the solvent used.

$$HCl_{(g)} + 25H_2O_{(l)} \rightarrow HCl(25H_2O) \qquad \Delta H = -72.266 kJ$$

(ii) Differential heat of solution : The differential heat of solution is defined as the heat change *when one mole of solute is dissolved in such a large quantity of the solution that there occurs no appreciable change in concentration of the solution.*

Mathematically differential enthalpy of solution can be represented as:

$$\text{Differential enthalpy of solution} = \frac{d(\Delta H)}{dm}$$

where $d(\Delta H)$ is the small change in value of ΔH when a small amount dm of solute is added to a solution of definite composition.

8. *Define heat of hydration, heat of dilution?*

Ans. Heat of Hydration. The heat of hydration is defined as *the heat change when 1 mole of anhydrous substance combines with the requisite amount of water to form a hydrated stable salt.* Hydration of $CuSO_4$ is represented by the following thermochemical equation.

$$CuSO_{4(s)} + 5H_2O(l) \rightarrow CuSO_45H_2O_{(s)} \qquad \Delta H = -78.868 \text{ kJ}$$

Heat of Dilution. Heat of dilution is defined as *the change of enthalpy when a solution containing 1 mole of a solute is diluted from one concentration to another.*

9. *Define heat capacity?*

Ans. The heat capacity of a substance is defined as the amount of heat required to raise the temperature of the system from lower to higher temperature divided by the temperature difference, between any two temperatures.

If q calories is the heat absorbed by a substance having mass m, in raising its temperature from T_1 to T_2, the heat capacity will be given by the expression

$$C = \frac{q}{m \times (T_2 - T_1)} \quad ...(i)$$

If m = 1gm, $T_2 - T_1$ = 1 K, then C is defined as *the amount of heat required to raise the temperature of 1gm of a substance through one degree centigrade. This is sometimes also referred to as specific heat of the substance.* Its units are cal $deg^{-1}g^{-1}$ whereas *molar heat capacity* is defined as *the amount of heat required to raise the temperature of 1 mole of a substance through one degree centigrade.* Its units are cal deg^{-1} mol^{-1} or J $deg^{-1}mol^{-1}$.

10. *What is the relation between heat capacity and specific heat of a substance?*

Ans. The relationship between the heat capacity and specific heat of a substance is

$$C = ms \quad \text{...(ii)}$$

where m is the mass and s is the specific heat of the substance.

❑❑❑

8

Chemical Kinetics

CHEMICAL KINETICS

Chemical kinetics is that branch of chemistry which deals with the study of the rates of chemical reactions, the factors affecting the rates of the reactions and the mechanism by which the reaction proceeds.

RATE OF REACTION.

The rate of a chemical reaction is the speed with which the reaction takes place. It is measured either in terms of decrease in the concentration of a reactant per unit time or increase in the concentration of a product per unit time. Hence it can be defined as the change in concentration of any one of the reactants or products per unit time.

For a reaction,

A → B, the rate of reaction is given by,

$$-\frac{d[A]}{dt} = +\frac{d[B]}{dt}$$

The minus sign in case of $-d[A]/dt$ implies that the concentration of A is decreasing with time and positive sign given to B indicates that the concentration of B is increasing with time. However, rate of reaction is always a positive quantity.

For the reaction of the type,

$aA + bB \rightarrow lL + mM$, the reaction rate can be expressed as,

$$r = \frac{1}{a}\frac{d[A]}{dt} = -\frac{1}{b}\frac{d[B]}{dt} = +\frac{1}{l}\frac{d[L]}{dt} = +\frac{1}{m}\frac{d[M]}{dt}$$

where a, b, l and m etc. are the stoichiometric coefficients.

Units of Rate. Rate has the units of concentration divided by time.

If the concentration is expressed in mol/L and time in sec, then rate of the reaction will have the units, mol L^{-1} sec^{-1}.

RATE CONSTANT OR VELOCITY CONSTANT

Let us again take a reaction,

$$A \rightarrow B$$

The rate of the reaction is given by,

$$r = -\frac{d[A]}{dt} = k[A]$$

where k is called the rate constant or velocity constant of the reaction at the given temperature. It can be defined as the fraction of A that reacts per unit time. If concentration of A is unity then

$$r = k$$

Thus, at a given temperature, *the rate constant or velocity constant of a reaction is equal to the rate of reaction when the concentration of the reactant is unity.*

For a general reaction,

$$aA + bB \rightarrow cC + dD$$

$$r = k[A][B]$$

If [A] = [B] = 1, then r = k. Thus, at a given temperature, the rate constant or velocity constant of a reaction is numerically equal to the rate of reaction when the concentration of each reactants is taken as unity.

ORDER OF REACTION

The order of a reaction is defined as the total number of reacting species whose concentration actually alters during the course of chemical reaction. It has been found out that the rate of reaction is not influenced by all the molecules as represented by a chemical equation but is influenced by those molecules whose concentration actually alters during the course of chemical reaction.

It may be emphasized here that order of reaction can be determined *only experimentally* and a chemical equation do not supply the information about the order of the reaction. Let us take a hypothetical reaction.

$$A + B \rightarrow C$$

Now, if the rate of reaction, as determined experimentally, depends upon the concentration of A and do not alters on changing the concentration of B, then we can say that rate of reaction is proportional to the concentration of A only.

$$r \propto [A]$$

$$r = k[A]$$

where k is the velocity constant. Thus this would be called *first order reaction.*

If on the other hand rate of reaction depends upon the concentration of A and B both then

$$r = k\,[A]\,[B]$$

and the reaction would be *second order reaction.*

If, however, experimental observations show that the rate of reaction varies as the concentration of A and as a square of concentration of B i.e.

$$r = k\,[A]\,[B]^2$$

then the order of the reaction would be 1 + 2 = 3.

If the reaction rate is independent of the concentration of the reacting species, the reaction is said to be a *zero order reaction.*

The expression which gives the dependence of rate upon concentration of reacting substance in known as the ***rate law.***

Alternatively, *order of reaction can be defined as the sum of the powers to which the concentration terms are raised in the rate law equation.*

The order of a reaction may not always be a whole number. It can be a fractional quantity also. It is particularly so in heterogeneous reactions.

Units of Rate Constant. Units of rate constant vary according to the order of reaction. As is evident from the above discussion,

$$\text{rate} = \frac{dx}{dt} = k\,[\text{Conc}]^n$$

$$k = \frac{dx}{dt}\cdot\frac{1}{[\text{Conc}]^n} = \frac{\text{Conc.}}{\text{Time}}\times\frac{1}{[\text{Conc}]^n} = \frac{1}{\text{Time}}\times\frac{1}{[\text{Conc.}]^{n-1}}$$

For zero order reaction, $n = 0$

$$k = \frac{1}{\text{Time}}\times\frac{1}{[\text{Conc.}]^{-1}} = \textbf{moles L}^{-1}\ \textbf{sec}^{-1}$$

For 1st order reaction, $n = 1$

$$k = \frac{1}{\text{Time}}\times\frac{1}{[\text{Conc.}]^{0}} = \textbf{sec}^{-1}$$

For 2nd order reaction, $n = 2$

$$k = \frac{1}{\text{Time}}\times\frac{1}{[\text{Conc}]^{1}} = \frac{1}{\text{Time}}\times\frac{1}{\text{Moles.L}^{-1}} = \textbf{L mol}^{-1}\ \textbf{sec}^{-1}$$

MOLECULARITY OF A REACTION

The molecularity of a reaction is defined as the number of reactant molecules which take part in an elementary reaction. If only one molecule is involved, the reaction is said to be unimolecular. For example,

$$N_2O_{5\,(g)} \rightarrow N_2O_{4\,(g)} + \frac{1}{2}O_{2\,(g)}$$

If two molecules are involved, the reaction is termed as bimolecular.

$$H_2(g) + I_{2(g)} \rightarrow 2HI_{(g)}$$

Order and Molecularity The order and molecularity of a reaction are generally different and the two should not be confused with one another. Consider the following simple reactions,

$$N_2O_{5(g)} \rightarrow 2NO_{2(g)} + \frac{1}{2}O_{2(g)}$$

Molecularity = 1

Order = 1

$$CH_3COOC_2H_5 + NaOH \rightarrow CH_3COONa + C_2H_5OH$$

Molecularity = 2

Order = 2

$$2FeCl_3 + SnCl_2 \rightarrow 2FeCl_2 + SnCl_4$$

Molecularity = 3

Order = 3

In the above examples, the order is identical with the molecularity. But this is not always the case. In several reactions, the order is different from molecularity. This is particularly so when one of the reactants is present in large excess.

For example,

$$CH_3COOC_2H_5 + H_2O \xrightarrow{H^+} CH_3COOH + C_2H_5OH$$

$$C_{12}H_{22}O_{11} + H_2O \xrightarrow{H^+} C_6H_{12}O_6 + C_6H_{12}O_6$$

In the above reactions, the molecularity of each reaction is 2. But the order of each reaction is 1. The reason being that water is present in such a large excess that its concentration remains almost constant in the course of reaction in each case. From the experiment we know that rate of reaction varies directly as the concentration of ethyl acetate in the first case and that of sucrose in second case. Such reactions are termed as *pseudounimolecular reactions*.

Order and Molecularity of Complex Reactions A number of reactions occur in two or more steps. Such reactions are termed as complex reactions. Each step has its molecularity depending upon the number of molecules of the reactants taking part in the reaction. It is not desirable to express molecularity for the overall reaction as different steps of a reaction may involve different number of molecules.

However, the various elementary reactions that constitute the overall chemical reaction, do not proceed at the same rate. The rate of the reaction is determined by the rate of the slowest step. This is because rate of the overall reaction, cannot be faster then the rate of the slowest reaction.

The *order* of the complex reaction is given by the order of the slowest step. For example, the decomposition of nitrogen peroxide represented as

$$2N_2O_5 \rightarrow 4NO_2 + O_2$$

This is a bimolecular reaction and order of the reaction is 1. The various elementary steps are:

(i) $N_2O_5 \rightarrow NO_2 + NO_3$ (slow)

(ii) $NO_2 + NO_3 \rightarrow NO_2 + O_2 + NO$

(iii) $NO_2 + NO_3 \rightarrow N_2O_5$

(iv) $NO + N_2O_5 \rightarrow 3NO_2$

The molecularity of the reactions i, ii, iii and iv is 1, 2, 2, and 2 respectively. The order of the overall reaction is 1.

It must be noted that order of the reaction may be integral, zero, negative or fractional . But the molecularity must necessarily be an integer, usually 1 or 2, but can never be zero, fractional or negative.

KINETIC EQUATION FOR FIRST ORDER REACTION

Consider a reaction

$$A \rightarrow \text{Products}$$

Suppose we start with C_0 moles per litre of A. After time t, suppose x moles per litre of it have decomposed. Therefore after time t, the concentration of A would be $(C_0 - x)$. According to law of mass action,

Rate of reaction $\alpha\ (C_0 - x)$

$$\frac{dx}{dt}\ \alpha\ (C_0 - x)$$

$$\frac{dx}{dt} = k\ (C_0 - x) \qquad \text{...(i)}$$

where k is rate constant.

Equation (i) can be written as,

$$\frac{dx}{(C_0 - x)} = k\ dt \qquad \text{...(ii)}$$

Integrating equation (ii) we get,

$$\int \frac{dx}{(C_0 - x)} = \int k\,dt$$

$$-\ \text{In}\ (C_0 - x) = \text{kt} + \text{I} \qquad \text{...(iii)}$$

where I is constant of integration

In the beginning when $t = 0$, $x = 0$ because no substance has decomposed at the start of the reaction.

Putting these values in equation (iii), we get

$$-\ \text{In}\ (C_0 - 0) = \text{k} \times 0 + \text{I}$$

$$-\ \text{In}\ C_0 = \text{I}$$

Substituting this value of I in equation (iii), we get

$$-\ \text{In}\ (C_0 - x) = \text{kt} + (-\ \text{In}\ \ C_0)$$

$$kt = \text{In Co} - \text{In}\ (C_0 - x) = \text{In}\ \frac{C_0}{C_0 - x}$$

$$k = \frac{1}{t}\,\text{In}\,\frac{C_0}{C_0 - x} \qquad \text{...(v)}$$

$$k = \frac{2.303}{t}\log\frac{C_0}{C_0 - x} \qquad \text{...(vi)}$$

EXPERIMENT NO. 8.1

AIM Determination of reaction rate constant of acid catalyzed hydrolysis of ester

THEORY The hydrolysis of ethyl acetate catalyzed by dil. HCl can be represented as:

$$CH_3COOC_2H_{5(aq.)} + H_2O_{(l)} \rightarrow CH_3COOH_{(aq.)} + C_2H_5OH_{(aq.)}$$

In the above reaction the concentration of water being in excess, is very high and practically remains constant during the reaction. Concentration of H^+ which catalyze the reaction also remains constant. Thus the rate of reaction only depends upon the concentration of the ester i.e. it is a *first order reaction.*

$$\text{rate} = k\,[CH_3COOC_2H_5]$$

Reactions of the above type in which one of the reactant is present in large excess, are though bimolecular but follow the kinetics of first order are called *pseudounimolecular reactions.*

The value of rate constant is calculated from the equation of first order reaction.

$$k = \frac{2.303}{t} \log \frac{C_o}{C_o - x} \quad ...(1)$$

Where C_o = Conc. of the ester

$C_o - x$ = Conc. of the ester at a time t.

Since acetic acid is formed as a result of hydrolysis, the kinetics of the reaction can be followed by determining the amount of acetic acid produced at definite intervals of time. For this, known amounts of the ester and the mineral acid are mixed and equal volumes of the reaction mixture are withdrawn at definite intervals. This reaction mixture so withdrawn is titrated against standardized alkali (NaOH) using phenolphthalein. The volume of the alkali used for each titration corresponds to the amount of acid used as catalyst and the amount of acetic acid produced at that time. The final reading is taken after keeping the remaining reaction mixture in water bath for 24hrs. or keeping it in hot water at 50–60° C for about 30 minutes.

Let V_o = Volume of NaOH used at time $t = 0$

V_t = Volume of NaOH used at time t

V_∞ = Volume of NaOH used at time $t = \infty$

Now, V_o α The amount of HCl present in the reaction mixture.

V_t α The amount of HCl present in the reaction mixture + acetic acid produced at time t

V_∞ α The amount of HCl present in the reaction mixture + acetic acid produced after the completion of reaction.

Therefore,

Amount of acetic acid produced when the hydrolysis is complete

$= V_\infty - V_o$ ...(2)

Amount of acetic acid produced at time t $= V_t - V_o$ ...(3)

Further,

$V_\infty - V_o$ α The initial conc. of the ester, C_0. ...(4)

$(V_\infty - V_o) - (V_t - V_o) = (V_\infty - V_t)$ α The conc. of ester at time t, $C_0 - x$ ...(5)

Using equations (4) and (5) in equation (1) we get,

$$k = \frac{2.303}{t} \log \frac{V_\infty - V_o}{V_\infty - V_t} \quad ...(6)$$

The value of k is determined using equation (6) for the different sets of values corresponding to different intervals of time.

From equation (6)

$$\log (V_\infty - V_t) = \log (V_\infty - V_o) - \frac{kt}{2.303} \quad ...(7)$$

Graphically, k can be calculated by plotting log $(V_\infty - V_t)$ against t. As per equation (7) this should give a straight line where

$$\text{slope} = \frac{-k}{2.303}$$

or $k = -2.303 \times \text{slope}$.

APPARATUS 50 cm^3 burettes, 100cm^3 conical flasks, 250cm^3 conical flask, graduated pipette, stopwatch, thermometer, water bath or thermostat.

CHEMICALS Ethyl acetate, hydrochloric acid (0.5N), Phenolphthalein indicator, sodium hydroxide (0.1 N), ice cubes (prepared from CO_2 free water.)

PROCEDURE

1. Take 50 cm^3 of 0.5 N HCl in 250 cm^3 conical flask and 10ml of ethyl acetate in a test tube.

2. Cork both of them and place them in thermostat maintained at room temperature. Let the two solutions attain the temperature of the bath.
3. In the mean time, rinse and fill a burette with 0.1 N NaOH solution. Fix it on a burette stand. Also arrange five conical flasks of 50 cm^3 capacity and put 25ml of ice cold water in one of them.
4. Now pipette out 5ml of the ester kept in the thermostat and transfer it to the conical flask containing hydrochloric acid. Start the stop watch when nearly half the volume of ester has been added.
5. Mix the contents quickly and at once pipette out 5ml of the reaction mixture into the conical flask containing 25 cm^3 of ice cold water. This freezes or arrests the reaction (due to lowering of temperature)
6. Now titrate it quickly against 0.1 N NaOH using phenolphthalein as indicator. Appearance of light pink colour, marks the end point. The titre value represents V_o.

   (The volume of sodium hydroxide used in the above titration nearly corresponds to the concentration of hydrochloric acid in the reaction mixture before the hydrolysis starts.)
7. Similarly, after a time interval of 10 minutes, again pipette out 5ml of the reaction mixture into a conical flask in which 25cm^3 of ice cold water has already been added. Titrate it against 0.1 N NaOH as before using phenolphthalein as indicator.
8. Repeat the process of pipetting 5cm^3 of reaction mixture into a conical flask containing phenolphthalein and 25 ml of ice cold water and titrating against 0.1 N NaOH after every 10 minutes for 1 hour. Each of the titre values from second onwards constitute V_t

   (The titre value required for each titration after *t* minutes corresponds to the volume of alkali used to neutralize hydrochloric acid and acetic acid formed after time *t*).
9. To obtain the volume of NaOH required when the hydrolysis is complete, the corked flask containing the reaction mixture is kept in a water bath at a temperature of 50–60° C for about 30 minutes to 1 hour. [To save time, this can be done just after first reading by withdrawing 20 ml of the reaction mixture in a separate flask and keeping it in water bath at 50°–60°C for 30 minutes to 1 hour] Cool the conical flask to room temperature and titrate 5 cm^3 of this mixture against 0.1 N NaOH as above. In this titration, there is no need to add ice cold water. The last titre value represents the infinite reading V_∞.

OBSERVATION

Room temperature = ... °C

Observation Table

Time (Minutes)	Burette Reading		Titre Value	$\log (V_\infty - V_t)$
	Initial	Final		

CALCULATIONS

The value of k is now calculated according to the first rate expression,

$$k = \frac{2.303}{t} \log \frac{V_\infty - V_o}{V_\infty - V_t}$$

1. The values of k are calculated at different intervals of time t

   (i) $k = \ ...\text{sec}^{-1}$

   (ii) $k = ...\text{sec}^{-1}$

   (iii) $k = ...\text{sec}^{-1}$

2. A graph between t versus $\log (V_\infty - V_t)$ is plotted. A straight line is obtained. Value of k is also calculated from the graph

   $k = -\ 2.303 \times \text{slope} = ...\ \text{sec}^{-1}$

RESULT k (by calculation) = ... sec^{-1}

k (by graphical method) = ...sec^{-1}

Instructions for the Lab. Assistant

1. **Preparation of 0.5 N HCl solution.** Transfer 45cm^3 of conc. HCl (11N) in a measuring flask containing approx. 500 cm^3 of distilled water. Make up the volume to 1dm^3 (1 litre).
2. **0.1 N NaOH.** Dissolve 8g of NaOH pellets in 2 litres of distilled water.

   Standardize it with N/10 oxalic acid solution
3. **Phenolphthalein indicator.** Dissolve 1g of phenolphthalein in 100 ml of ethanol. Add 100 ml of water stirring it continuously.

EXERCISE

1. *What do you mean by rate of reaction?*

Ans. The rate of reaction is defined as the change in concentration of reactant or product per unit time.

2. *What are the factors which affect rate of reaction?*

Ans. Rate of reaction is influenced by:

(i) Nature of reactants (ii) Concentration of reactants (iii) Temperature (iv) Presence of catalyst.

3. *What do you mean by rate constant?*

Ans. It is defined as the rate of reaction when the concentration of each reactant is taken as unity. Rate constant is sometimes also referred to as specific reaction rate.

4. *Define order of reaction.*

Ans. The sum of the concentration terms on which the rate of a reaction actually depends as observed experimentally is called the order of the reaction.

5. *Distinguish between order and molecularity of a reaction.*

Ans. The number of atoms, ions or molecules that must collide with one another simultaneously so as to result into chemical reaction is called molecularity of the reaction.

Order of reaction is experimentally determined quantity whereas molecularity can be calculated by simply adding the molecules of the rate determining step (slowest step)

Order of reaction need not be a whole number whereas molecularity is always a whole number.

6. *What are the units of rate constant of (i) zero order (ii) first order (iii) second order and third order*

Ans. (i) $L\,mol^{-1}\,s^{-1}$ (ii) s^{-1} (iii) $mol\,L^{-1}\,s^{-1}$ (iv) $mol^2\,L^{-2}\,s^{-1}$

7. *What are pseudo-unimolecular reactions?*

Ans. The reactions which are not truly first order but under certain conditions become reactions of the first order are called pseudounimolecular reactions. These are bimolecular reactions which follow the kinetics of the first order reaction.

8. *Give examples of pseudounimolecular reactions.*

Ans. (i) Hydrolysis of ethyl acetate

(ii) Inversion of cane sugar

9. *Why the reaction is speeded up with increase in temperature?*

Ans. At temperature increase causes greater fraction of the molecules to have energy equal to or greater than activation energy.

10. *Which of the following will react: (i) fastest, (ii) at highest rate?*

(a) 1 mol A + 1 mol B in a 1L vessel, (b) 2 mol A + 2 mol B in 2L vessel, (c) 0.2 mol A + 0.2 mol B in a 0.1 L vessel.

Ans. (*i*) (*b*), since there is more reactant present so more products are produced per unit time.

(ii) (c), since the vessel with greatest concentration experiences greatest change in concentration per unit time.

11. *How does half-life period of a first order reaction vary with temperature? Explain.*

Ans. Half-life *decreases* with *rise* of temperature, because $t_{1/2} \propto 1/k$ and k increases with rise of temperature.

12. *How is rate constant related to concentration of reactants?*

Ans. $k = \dfrac{1}{C^{n-1}}$, where n = order of reaction.

13. *For the reaction in acid solution:* $HCOOCH_3 + H_2O \rightarrow HCOOH + CH_3OH$*, the rate* $= k\,[HCOOCH_3]\,[H^+]$*. Why does* $[H^+]$ *appear in the rate law, when it does not appear in the balance equation?*

Ans. H^+ (a *catalyst* for the reaction) actually acts as a reactant in an early intermediate stage of the reaction and is then released back into the solution at a later stage.

14. *Why is reaction speeded up in presence of a catalyst?*

Ans. A catalyst *lowers* the activation energy, therefore more molecules in the reactants possess sufficient energy to react.

□□□

9

Surface Tension

The existence of strong intermolecular forces of attraction in liquids gives rise to a very important property known as surface tension. In the interior of a liquid a molecule is completely surrounded by the other molecules and hence experiences the force of attraction equally in all directions. (Fig. 9.1)

Fig. 9.1

On the other hand, a molecule at the surface is surrounded by liquid molecules on one side and that of vapour on the other. Thus the molecule at the surface experiences the attractive forces or inward pull perpendicular to the surface. (This is because of the difference of the number molecules per unit volume in the liquid and the vapour phase). Thus, under the influence of molecular forces of attraction from within, a liquid develops a tendency to contract its surface such that it acquires the smallest possible area for a given volume. This is due to the above reason that drops of a liquid acquire a spherical shape. As a result of the tendency to contract, a liquid surface behaves as if it were in a state of tension. This effect is known as *surface tension*

Definition. Surface tension is defined in SI units as the force in newtons (N) acting at right angle to any line of one meter length along the surface of the liquid. The surface tension is represented by a greek symbol γ.

Units. In SI system — Newton m^{-1}

In C.G.S system — Dynes cm^{-1}

Surface Energy: The surface energy is a characteristic property of the surface and arises due to surface tension. The effect of surface tension is to reduce the area of the surface to minimum. It we wish to increase the area of the surface of a liquid, we have to work against the force of surface tension.

The work required to increase the surface area by one unit is called surface energy of the liquid

Units. Surface energy has units of work per unit area

In S.I System — joule m^{-2}

C.G.S system — erg cm^{-2}

It can easily be shown that,

$$\text{erg cm}^{-2} = \text{dyne} \times \text{cm cm}^{-2} = \text{dyne cm}^{-1}$$

In other words, the surface tension and surface energy are dimensionally equal.

Factors Affecting Surface Tension.

1. **Effect of Temperature:** As temperature increases, the kinetic energy of the molecules increases. Thus, the intermolecular forces decrease and hence surface tension of the liquid decreases. The surface tension of a liquid is expected to disappear at its critical temperature but in most of cases it vanishes at $T_c - 6$.
2. **Effect of surfactants:** Surfactants are surface active materials and consist of molecules having polar and nonpolar parts. Presence of surfactants in aqueous solutions lowers the surface tension.

Some effects of surface tension

1. Drops of liquid or bubbles of a gas are spherical in shape because of surface tension.
2. Water in the capillary tube is pulled upwards due to the force of surface tension and is pulled downwards due to force of gravity. Water continues to rise in the capillary till a balance between the two forces is reached.
3. When water rises in a glass, the shape of the meniscus is concave. This indicates that the adhesive forces between water and glass are stronger than the cohesive forces between water molecules. In some cases, e.g in case of mercury, the shape of the meniscus is convex . This is due to stronger cohesive forces as compared to adhesive forces.
4. In metallurgy, the froth floatation process for the concentration of ore is based on surface tension.

MEASUREMENT OF SURFACE TENSION

There are various methods for determining the surface tension of a liquid. Here, we will be dealing with stalagmometer method only

EXPERIMENT NO. 9.1

AIM **Determine the surface tension of a given liquid by drop number method.**

THEORY When a liquid is allowed to flow through a capillary tube, a drop is formed at its lower end. It increases to a certain size and falls off. The size of the drop depends on the radius of the capillary and the surface tension of the liquid. The surface tension acting along the circumference of the capillary tube supports the drop in the upward direction. (Fig. 9.2)

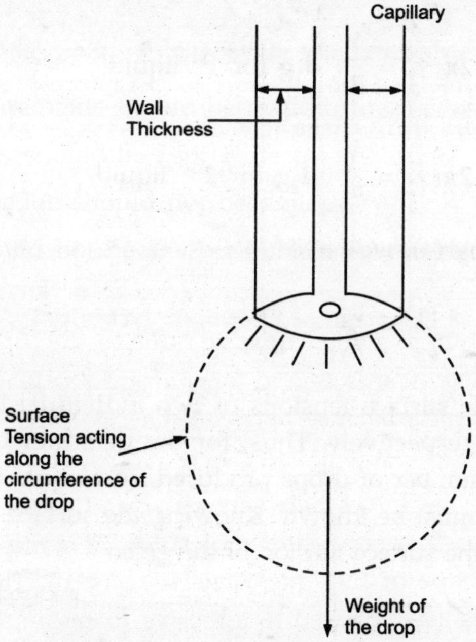

Fig. 9.2 *Forces acting on the drop formed in stalagmometer*

The measurement of surface tension of liquid is based on the fact that the drop of the liquid at the lower end of capillary falls down when weight of drop becomes just equal to the surface tension.

1. Now the force of gravity exerted on the drop, is given by v.d.g.

   Where v is the volume of the drop, d its density and g its gravity. The force of gravity will pull down the drop .

2. The force tending to uphold the drop is given by $2\pi r\gamma$.

   where 2πr is circumference of the capillary of radius *r*.

When the two forces are balanced.

$$2\pi r\gamma = v.d.g \quad ...(1)$$

If n is the number of drops in volume V of the liquid then volume of each drop, will be

$$v = \frac{V}{n},$$

From eq. (1) we have

$$\therefore \quad 2\pi r\gamma = \frac{V}{n}\, dg. \quad ...(2)$$

Consider two liquids of densities d_1 and d_2 having the surface tension γ_1 and γ_2. Let the number of drops counted for the same volume V of the two liquids be n_1 and n_2 respectively. Then,

$$\therefore \quad 2\pi r\gamma_1 = \frac{V}{n_1} d_1 g \text{ for } 1^{st} \text{ liquid} \quad ...(3)$$

$$\therefore \quad 2\pi r\gamma_2 = \frac{V}{n_2} d_2 g \text{ for } 2^{nd} \text{ liquid} \quad ...(4)$$

Dividing equation (3) by (4), we have

$$\frac{\gamma_1}{\gamma_2} = \frac{n_2}{n_1} \times \frac{d_1}{d_2}$$

Where γ_1 and γ_2 are the surface tensions of two individual liquids and d_1 and d_2 are their densities, respectively. Thus, for the determination of surface tension of any liquid , the number of drops produced from equal volume of two liquids and their densities must be known. Knowing the surface tension of the reference liquid (e.g water) the surface tension of the given liquid can be determined

APPARATUS Stalagmometer, beakers, rubber tubing, pinch cock, relative density bottle, thermometer.

Stalagmometer. Stalagmometer is essentially a pipette with a capillary at the lower end and a bulb in the middle. The end is flattened and grounded carefully so that there is a large dropping surface . There are two marks A and B above and below the bulb respectively (Fig. 9.3). These marks define the volume to be delivered for counting the drops. In case the rate of flow of the liquid is rapid leading to rapid drop formation, a piece of rubber tubing with a pinch cock is attached to the top of the stalagmometer and

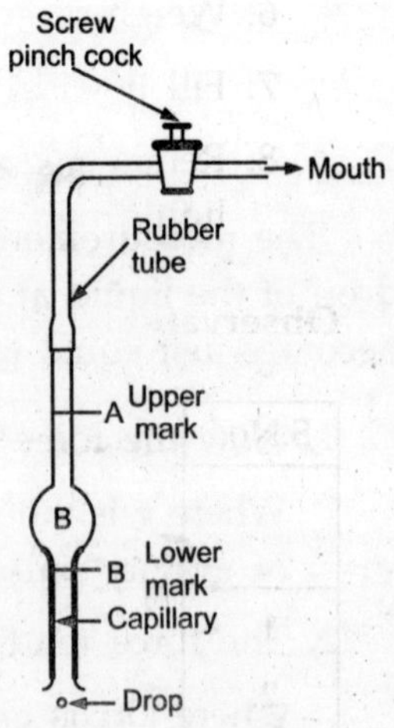

Fig. 9.3 *Stalagmometer*

the rate of flow is adjusted by altering the pressure of the screw so that the drops are formed at a rate not greater than 15 drops per minute.

(ii) *Relative Density Bottle.* The density of liquid is conveniently measured by means of a relative density bottle or specific gravity bottle. It is a round bottomed glass vessel, fitted with a glass cork containing a fine capillary

PROCEDURE

1. Clean the stalagmometer first with NaOH, then with chromic acid solution to remove oil, grease etc. sticking on the inner capillary surface. Wash it with distilled water and finally with acetone and dry by passing air. (In case water is used as reference liquid there is no need to dry the apparatus)
2. Immerse the lower end of the stalagmometer in a beaker containing distilled water. *Suck the water until it rises above the mark A (Control the rate of flow of water with the help of screw pinch cock so that the number of drops per minute is about 15-20)*
3. Now let the water flow out and start counting the number of drops when the water meniscus just reaches the mark A, and counting is stopped when the water meniscus passes the lower mark B. Let the number of drops be n_w.
4. Repeat to get three readings.
5. Clean the stalagmometer and dry it. Fill it with liquid until it rises above the mark A and count the number of drops when a fixed volume of liquid flows between the mark A and B, as before. Let the number of drops be n_l.
6. Weigh the relative density bottle and note down its weight.
7. Fill it with water and weigh it again.
8. Remove water and dry the relative density bottle. Now fill it with the given liquid and weigh.

Observation Table

S.No.	Water		Given Liquid (C_2H_2OH)	
	No. of Drops	Mean	No. of Drops	Mean
1.				
2.				
3.		n_w	n_l	

CALCULATIONS

Weight of empty specific gravity bottle = ω_1g

Weight of empty specific gravity bottle with water = ω_2g

Weight of empty specific gravity bottle with liquid = ω_3g

$\therefore$ Weight of water = $(\omega_2 - \omega_1)$g

Weight of liquid = $(\omega_3 - \omega_1)$g

CALCULATIONS

$$\text{Relative density of the liquid} = \frac{d_l}{d_\omega} = \frac{\omega_3 - \omega_1}{\omega_2 - \omega_1}$$

$$\frac{\gamma_l}{\gamma_w} = \frac{d_l}{d_w} \times \frac{n_w}{n_l}$$

where γ_l, d_l, n_1 are the surface tension, density and number of drops of liquids.

γ_w, d_w, n_w are the surface tension, density and number of drops of water.

By putting the respective values, the surface tension of the liquid relative to water is found out.

Absolute surface tension of the liquid.

$$\gamma_1 = \frac{n_\omega}{n_l} \times \frac{d_1}{d_\omega} \times \gamma_\omega = \ldots \text{Nm}^{-1}.$$

The value of γ_w is obtained from the table. 9.1

Table 9.1 Surface tension of water at different temperatures.

Temp	5	10	15	20	25	30	35	40
Surface tension (dynes/cm)	75.09	74.36	73.62	72.88	72.14	71.70	70.66	69.92

RESULT The surface tension of the given liquid is. . . dynes cm^{-1}

PRECAUTIONS

1. Surface tension is highly sensitive to contamination of even minute traces of grease and impurities. Stalagmometer therefore should be thoroughly cleaned with chromic acid.
2. Stalagmometer should be in a vertical position throughout the counting process.
3. Drop formation should be adjusted at a slower rate and should not exceed fifteen drops per minute.

EXERCISE

1. *Define surface tension.*

Ans. The surface tension γ is the magnitude F of the force exerted parallel to the surface of a liquid divided by the length L of the line over which the force acts:

$$\gamma = \frac{F}{L}$$

2. *What are the units of surface tension?*

Ans. **The common units for surface tension are dynes/cm or N/m. These units are equivalent.** Units of surface tension ($N\text{-}m^{-1}$) are equivalent to joules per square metre ($J\text{-}m^{-2}$). This means that surface tension can also be considered as **surface energy**. If a surface with surface tension γ is expanded by a unit area, then the increase in the surface's stored energy is also equal to γ.

3. *What are the forces responsible for surface tension?*

Ans. The **cohesive** forces between liquid molecules are responsible for the phenomenon known as surface tension.

4. *What is the surface tension of water?*

Ans. The **surface tension** of water is 72 dynes/cm at 25°C. It would take force of 72 dynes to break a surface film of water 1 cm long

5. *Why is Surface Tension important?*

Ans. If any formulation changes at the molecular level then the surface tension will change. If the formulation changes due to the addition of another chemical, the addition of a surfactant, or if anything contaminates the fluid in question, then the surface tension will change. Measuring surface tension is a direct indicator of the quality of any chemical and any formulation.

6. *Why hot water is considered better cleaning agent?*

Ans. Hot water is a better cleaning agent because the lower surface tension makes it a better "wetting agent" to get into pores and fissures rather than bridging them with surface tension. Soaps and detergents further lower the surface tension.

7. *What is the difference between adhesive and cohesive forces? On the basis of these forces discuss capillary action.*

Ans. Molecules in liquid state experience strong intermolecular attractive forces. When those forces are between like molecules, they are referred to as *cohesive forces*. For example, the molecules of a water droplet are held together by cohesive forces, and the especially strong cohesive forces at the surface constitute **surface tension**.

When the attractive forces are between unlike molecules, they are said to the *adhesive forces*. The adhesive forces between water molecules and the walls of a glass tube are stronger then the cohesive forces, then this leads to an upward turning meniscus at the wall of the vessel and contribute to capillary action.

8. *How small insects can walk on water?*

Ans. Small insects such as the water strider can walk on water because their weight is not enough to penetrate the surface.

9. *Why it is advised, not to touch the tent during rainy season?*

Ans. Common tent materials are somewhat rainproof because the surface tension of water will bridge the pores in the finely woven material. But if we touch the tent material with our finger, we break the surface tension and the rain will drip through.

10. *How the property of surface tension can be used as a test for jaundice?*

Ans. Normal urine has a surface tension of about 66 dynes/cm but if bile is present (a test for jaundice), it drops to about 55. In the Hay test, powdered sulphur is sprinkled on the urine surface. It will float on normal urine, but sink if the surface tension is lowered by the bile.

11. *Discuss capillary action.*

Ans. Capillary action or capillarity is the ability of a narrow tube to draw a liquid upwards against the force of **gravity**.

Capillary action is the result of **adhesion** and **surface tension**. Adhesion of water to the walls of a vessel will cause an upward force on the liquid at the edges and result in a meniscus which turns upward. The surface tension acts to hold the surface intact, so instead of just the edges moving upward, the whole liquid surface is dragged upward.

12. *Which phenomena is responsible for the shape of liquid droplets?*

Ans. **Surface tension** is responsible for the shape of liquid droplets. Although easily deformed, droplets of water tend to be pulled into a spherical shape by the **cohesive** forces of the surface layer.

13. *Why the surface of water has a concave shape when it is confined in a tube?*

Ans. When liquid water is confined in a tube, its surface (meniscus) has a concave shape because water wets the surface and creeps up the side.

14. *Why mercury does not wet the glass?*

Ans. Mercury does not wet glass because the cohesive forces within the drops are stronger than the adhesive forces between the drops and glass. When liquid mercury is confined in a tube, its surface (meniscus) has a convex shape because the cohesive forces in liquid mercury tend to draw it into a drop.

15. *What are surfactants?*

Ans. Surfactants, also known as wetting agents, lower the **surface tension** of a liquid, allowing easier spreading, and the **interfacial tension** between two liquids. The term surfactant is a contraction of "Surface active agent". Surfactants are usually **organic** compounds that are **amphipathic,** meaning they contain both **hydrophobic** groups (their "tails") and **hydrophilic** groups (their "heads"). Therefore, they are typically sparingly soluble in both organic solvents and water. Surfactants reduce the surface tension of water by **adsorbing** at the air-water interface.

16. *Why does the surface tension of water decrease on adding detergents? What is the phenomenon behind decrease in surface tension of water?*

Ans. In the solid and liquid phase, water molecules are bonded to each other by virtue "hydrogen bonding." The result is a kind of "skin" on liquid water's surface. Soaps and detergents have a polar head to which water is attracted and a non-polar tail that is hydrophobic — water repelling. When these substances are present, they weaken the strength of the skin by interfering with hydrogen bonding between water molecules because the polar end of the soap or detergent is also attracted to the water.

17. *What is the effect of temperature on surface tension?*

Ans. Surface tension arises due to cohesive forces within the molecules. With increase in temperature, kinetic energy of the molecules of liquid increases and hence force of attraction between them decreases. Hence surface tension of a liquid decreases with rise in temperature.

EXPERIMENT NO. 9.2

AIM **To determine the composition of a liquid mixture A and B (acetic acid and water) by surface tension method.**

APPARATUS Stalagmometer, thermometer, beaker etc.

CHEMICALS Pure A (acetic acid) pure B (water) and the given solution.

THEORY When the surface tension of two liquids constituting the mixture do not differ greatly, the surface tension of the mixture varies with composition.

A number of known solutions of A (acetic acid) and B (water) are prepared and the number of drops for each one of these is calculated as in the previous experiment. The number of drops of unknown solution is also determined. Let it be x. A graph is plotted between the number of drops against the percent composition. From the graph the concentration of the unknown solution can be determined.

PROCEDURE

1. Prepare a number of solutions of A and B containing 90%, 80%, 70%, 60%, 50%, 40% etc. of the component A.
2. Determine the number of drops of each of the solution by stalagmometer as described in the previous experiment. Also determine the number of drops of unknown mixture.

OBSERVATIONS

Composition of the mixture		Number of drops
A	B	
90%	10%	
80%	20%	
70%	30%	
60%	40%	
50%	50%	
unknown	—	

Note The composition can also be found out by determining the surface tension of each of the solution and then plotting a graph between surface tension versus composition. (The density of each solution has to be measured in this case).

CALCULATION Plot a graph between the number of drops against the percentage composition. The point corresponding to the number of drops of the unknown mixture is found. A perpendicular is drawn on the composition axis from that point and the value gives the composition of the unknown mixture.

RESULT The composition of the given mixture is .....

PRECAUTIONS Same as in experiment 9.1

EXPERIMENT NO. 9.3

AIM **Compare the cleansing powers of two samples of a detergent.**

THEORY The cleansing action of soaps and detergents is due to their property of lowering the surface tension of water. *Greater the reduction in the surface tension, more will be the cleansing power.*

To compare the cleaning powers of different samples, the surface tension of their respective solutions containing equal weights of each detergent is determined by the procedure adopted in experiment 9.1.

APPARATUS Stalagmometer

CHEMICALS Two different samples of detergents

PROCEDURE

1. Weigh equal amounts (0.2 to 0.5 g) of the two samples of detergents.
2. Prepare the solutions of the two detergents separately by adding the weighed amounts in 100 ml of preheated water. Stir the solutions very gently till the detergents completely dissolve.
3. Bring the solutions to room temperature by allowing the solutions to cool down.
4. Remove the insoluble impurities through filteration
5. Determine the number of drops formed from a fixed volume of both the solutions separately using stalagmometer as explained in experiment number 9.1.
6. Determine the number of drops of the same volume of distilled water

OBSERVATIONS

No. of drops of solution 1 (n_1)	No. of drops of solution 2 (n_2)	No. of drops of water (n)
1.	1	1
2.	2	2
3.	3	3

CALCULATIONS

For solution - 1

$$\frac{\gamma_1}{\gamma} = \frac{n d_1}{n_1 d} \qquad \text{...(i)}$$

where γ_1 is the surface tension of solution 1 and

γ is the surface tension of water

d_1 is the density of solution

d is the density of water

n_1 is the number of drops of solution

n is the number of drops of water from the same volume.

For solution 2

$$\frac{\gamma_2}{\gamma} = \frac{n d_2}{n_2 d} \qquad \text{...(ii)}$$

where γ_2, d_2 and n_2 are the surface tension, density and number of drops formed from the same volume of solution 2.

Since the solutions are very dilute, their densities will almost be the same as that of water, Hence,

$$\gamma_1 = \frac{n}{n_1} \cdot \gamma \qquad \text{...(iii)}$$

$$\gamma - \gamma_1 = \gamma - \left(\frac{n}{n_1} \cdot \gamma\right) = \left(1 - \frac{n}{n_1}\right)\gamma \qquad \text{...(iv)}$$

Similarly

$$\gamma_2 = \frac{n}{n_2} \cdot \gamma$$

$$\gamma - \gamma_2 = \gamma - \left(\frac{n}{n_2} \cdot \gamma\right) = \left(1 - \frac{n}{n_2}\right)\gamma \qquad \text{...(v)}$$

Ratio of cleansing powers will be,

$$\frac{\gamma-\gamma_1}{\gamma-\gamma_2} = \frac{1-n/n_1}{1-n/n_2} \quad \text{...(vi)}$$

Substitute the values of n, n_1 and n_2 in equation (vi) and find the ratio of cleansing powers.

RESULT The cleaning power of solution_______ is more than solution_______.

PRECAUTIONS

1. The contents should be stirred very gently so that no foam is formed.
2. The stalagmometer must be filled with solutions by gentle suction so that foam formation does not take place.

EXERCISE

1. How can you define a soap ?

Ans. Soaps are the Na or K salt of higher fatty acids having cleansing properties.

2. How the cleansing action is related to the surface tension ?

□□□

10

Viscosity

Different liquids flow with different speeds. For example, glycerine, castor oil, honey etc. flow slowly whereas water, ether alcohol etc. flow rapidly. There is some sort of internal friction operating in between the layers of the liquid which checks the flow of liquids and varies from one liquid to other. In case of liquids, the internal friction is primarily due to the forces of attraction between the molecules. This resistance to the flow of liquid is known as *viscosity.*

The phenomenon of viscosity is further elucidated by considering a laminar (stream lined) flow of the liquid through a narrow pipe. (The flow in which a liquid moves in layers is known as Laminar flow). In such a case, all parts of the liquid do not move with the same velocity. The layer immediately in contact with the surface of the inner wall of the pipe is stationary. The velocity of each successive layer increases as we move away from the surface towards the centre of the tube.

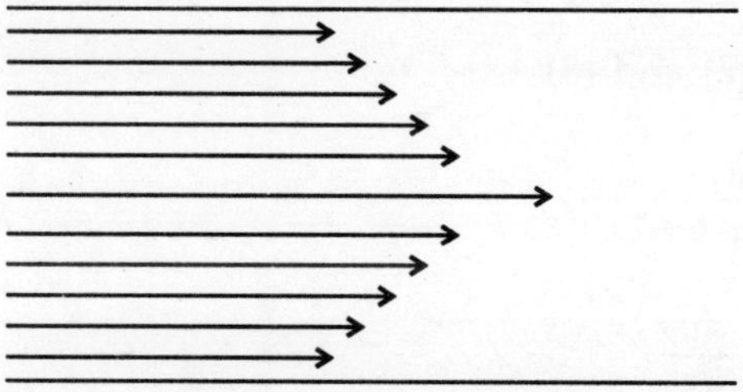

Fig. 10.1 Flow of liquid

As a result, a velocity gradient is set up between different layers of the liquid. Moreover, the molecules moving in any one layer will tend to retard the movement of the molecules in the adjacent faster moving layer due to intermolecular attraction. Thus, there arises a frictional force or drag between the two layers which gives rise to viscosity which can be defined as *the resistance that one*

Note. A streamline flow is that in which all the particles of the liquid move in an orderly manner along fixed paths. The velocity of particles of the liquid passing one after the other through a given point on a stream line remains unchanged in magnitude as well as in direction at that point.

The flow remains streamlined as long as the velocity is not large and remains below a certain velocity called critical velocity beyond which the flow becomes turbulent and highly irregular. In the turbulent flow the velocity of a water particle passing through a point is not the same as that of the particle which precede it.

part of a liquid flowing with one velocity offers to another part of a liquid flowing with different velocity.

The internal friction which resists the flow of a liquid can be measured in terms of the tangential force which is needed to keep the speeds of different layers constant. If A is the area of contact of two adjacent layers, dx the distance between two layers and du the velocity difference between two adjacent layers, then the tangential force required to maintain a constant velocity difference du will depend upon the following factors:

(i) It is directly proportional to A, the area of contact. Larger is A, greater will be the effect of intermolecular forces and hence larger would be the decrease in velocity. Therefore, larger force is required to maintain the velocity constant.

$$F \propto A \qquad ...(1)$$

(ii) It is also directly proportional to du, the velocity difference between two adjacent Layers.

$$F \propto du \qquad ...(2)$$

(iii) It is inversely proportional to dx. Larger is the distance, lesser would be intermolecular attractions and hence lesser force is required to maintain the velocity.

$$F \propto \frac{1}{dx} \qquad ...(3)$$

Combining (1), (2) and (3)

$$F \propto A \frac{du}{dx}$$

or

$$F = \eta A \frac{du}{dx} \qquad ...(4)$$

where η is known as *coefficient of viscosity* of a liquid or simply viscosity of the liquid. Rearranging (4)

$$\eta = \frac{F}{A(du/dx)} \qquad ...(5)$$

Thus η can be defined *as force per unit area required to maintain a velocity difference of unity between two parallel layers of liquid which are unit distance apart.*

Equation (4) is also known as *Newton's law of viscosity*. The fluid flow to which this equation is applicable is called laminar (streamlined) flow. The liquids which do not obey the above equation are called non-Newtonian liquids.

Units of Viscosity: From equation (5)

(i) In C.G.S system, $\eta = \dfrac{F}{A\,(du/dx)} = \dfrac{F}{A}\cdot\dfrac{dx}{du}$

$$\text{Units of } \eta = \frac{\text{dyne cm}}{\text{cm}^2\ \text{cm s}^{-1}} = \text{dyne cm}^{-2}\ \text{s} = \text{poise}$$

Thus, in the C.G.S system the unit of viscosity is **dyne cm^{-2} s** and is known as poise.

$$\text{poise} = \text{dyne cm}^{-2}\ \text{s}$$
$$= \text{gcm s}^{-2}.\ \text{cm}^{-2}\ \text{s} = \text{g cm}^{-1}\ \text{s}^{-1}$$

In other words 1 poise = 1g cm^{-1} s^{-1}

(ii) In S.I system

$$\text{Units of } \eta = \frac{\text{Newton } m}{m^2\ ms^{-1}}$$
$$= \text{Nm}^{-2}\ \text{s}$$

Since Nm^{-2} = pascal (Pa) (unit of pressure)

$\therefore$ Units of η = Pas

Thus in S.I system the unit of viscosity is Pascal second (Pas).

Note. 1 poise = 1g cm^{-1} s^{-1} = 1 (10^{-3}kg) (10^{-2} m)$^{-1}$ s^{-1}

$$= 10^{-1}\ \text{kg m}^{-1}\ \text{s}^{-1}$$
$$= 10^{-1}\ \text{Pas}$$

Hence 1 poise = 10^{-1} Pas.

Since the viscosities of most of the liquids are in small magnitude, therefore these are usually expressed in the units of centipoise and millipoise.

1 Centipoise = 10^{-2} poise

1 millipoise = 10^{-3} poise.

The reciprocal of viscosity is called fluidity and is represented by the symbol ϕ so that,

$$\phi = \frac{1}{\eta}$$

Fluidity is the measurement of the ease with which the liquid flows. Its units are *Poise*$^{-1}$.

FACTOR ON WHICH THE VISOSITY OF THE LIQUID DEPENDS

1. Temperature. The viscosity of a liquid decreases with increase of temperature. In liquids, as the temperature rises, the kinetic energy of the molecules increases and the intermolecular forces of attraction become weak, resulting in the subsequent decrease in the viscosity. It has been found that the decrease is about 2% per degree rise of temperature.

The molecule taking part in liquid flow should have sufficient energy to overcome the forces of intermolecular attraction due to surrounding molecules. This energy is the activation energy for viscous flow and is designated as E_a. The variation of viscosity with temperature is best expressed by empirical relation given by Arrhenius

$$\eta \ \alpha \ e^{E_a/RT}$$

$$\eta = Ae^{E_a/RT} \qquad ...(6)$$

where A is a constant of proportionality, T is the temperature and R is gas constant.

It is clear from equation (6) that as the temperature is increased the resistance to flow or the viscosity may be expected to decrease in a reciprocal manner.

Taking log on both sides of equations (6)

$$\ln \eta = \ln A + \frac{E_a}{RT}$$

$$\log \eta = \log A + \frac{E_a}{2.303R} \times \frac{1}{T} \qquad ...(7)$$

The plot of log η against $\frac{1}{T}$, therefore, gives a straight line demonstrating the fact that the viscosity of a liquid rapidly decreases with rise in temperature.

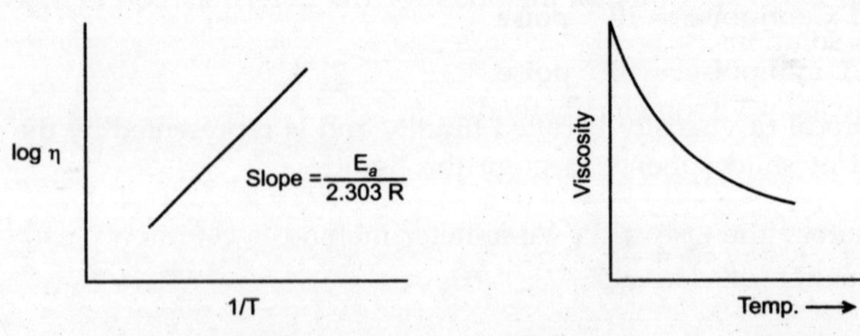

Fig. 10.2

In contrast to liquids, η increases with increasing temperature in the case of gases.

2. Effect of Pressure. The viscosity of the liquids increases with increase in pressure . As the pressure is increased the molecules of the liquid come closer and it becomes difficult for liquid molecules to move around and thus it becomes more difficult for them to flow.

3. Viscosity and Chemical Constitution. Though there is no significant relationship between viscosity and chemical constitution, however some general results have been observed.

(i) The viscosity increases as the size of the molecules increases. For example, in a homologous series, as we go up, the viscosity increases and increase per CH_2 group is almost constant.

(ii) In the compounds having chain isomerism, the n-isomers have generally greater viscosity than the branched chain. This is due to the fact that area of contact between the two layers in n-isomer is more and hence the molecules experience greater intermolecular attraction thereby increasing the viscosity.

However, molecules with flexible chains have very high viscosity due to entanglement of side chains.

(iii) Non-polar molecules have lower viscosities than polar molecules.

(iv) Presence of intermolecular hydrogen bonding increases the viscosity of the liquid. Glycerol and ethylene glycol have comparatively larger values of coefficients of viscosity since they have more than one hydroxyl group which results into hydrogen bonding.

(v) Presence of impurities increases the viscosity. Viscosity is greatly enhanced due to the presence of lyophillic colloids.

MEASUREMENT OF VISCOSITY

There are various experimental methods for the determination of viscosities of liquids and solutions.

(i) Ostwald's Viscometer method.

(ii) Fall of solid spheres through the liquids.

In laboratory the Ostwald's Viscometer method is commonly employed for the determination of viscosity.

EXPERIMENT NO. 10.1

AIM **Determination of coefficient of viscosity of the given liquid by Ostwald's Viscometer method.**

THEORY The Ostwald's Viscometer method is based on Poiseuille's equation. This relates the rate of flow of a liquid through a capillary tube with the coefficient of viscosity and is expressed by the equation,

$$\eta = \frac{\pi r^4 tP}{8Vl} \quad ...(7)$$

where V = Volume of the liquid of viscosity η flowing in time t, through a capillary tube of radius r and length l

P = Hydrostatic pressure of the liquid.

Thus the determination of the absolute viscosity by means by Poiseuille's expression involves the determination of V, r , t, l and P.

The method is however tedious and laborius one. Hence, a simpler method is used wherein we compare the viscosities of the two liquids. If the coefficient of viscosity of one liquid is known, then that of the other can be calculated.

If t_1 and t_2 are the flow times required to flow for equal volumes of two liquids through the same length of a capillary tube then from equation (7) we have

$$\eta_1 = \pi P_1 r^4 t_1/8Vl$$

$$\eta_2 = \pi P_2 r^4 t_2/8Vl$$

or
$$\frac{\eta_1}{\eta_2} = \frac{P_1 t_1}{P_2 t_2} \quad ...(8)$$

Now $P = hdg$ ...(9)

where h is the height of the liquid column (It is constant for all the liquids if taken for identical points and liquids are taken in equal volumes for a particular set of observations.)

g = acceleration due to gravity

d = density of the liquid

Since in this case for two liquids h and g are same hence,

$$\frac{\eta_1}{\eta_2} = \frac{d_1}{d_2}\frac{t_1}{t_2}$$

This will give us the relative viscosity of the given liquid. Relative viscosity has no units. The absolute viscosity $\eta = \frac{d_1}{d_2} \cdot \frac{t_1}{t_2} \cdot \eta_2$

APPARATUS Ostwald's viscometer, stop watch, specific gravity bottle, 10cm^3 pipette, distilled water, rubber tubing.

Description. (i) *Ostwald's Viscometer:* It consists of a capillary tube connected at its upper end with a bulb A and its lower end with a U-tube provided with bulb B. Bulb B is of larger size than A (Fig. 10.3) Marks X and Y are etched on the capillary tube above and below the bulb A. The bulbs are necessary to maintain the hydrostatic pressure during flow of liquid. Through the capillary, the liquid flows with a measurable speed.

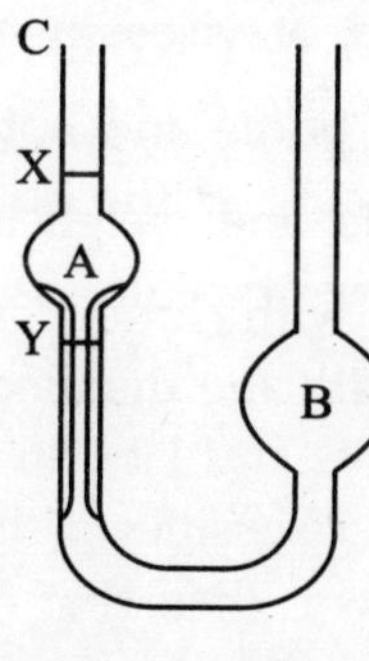

Fig. 10.3 *Viscometer*

(ii) *Relative Density Bottle.* The density of liquid is conveniently measured by means of a relative density bottle or specific gravity bottle. It is a round bottomed glass vessel, fitted with a glass cork containing a fine capillary Fig. 10.4

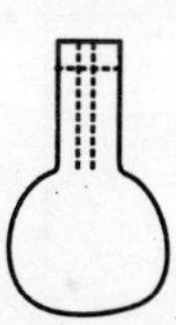

Fig. 10.4 *Relative Density Bottle*

PROCEDURE

1. Clean the viscometer with chromic acid ($K_2Cr_2O_7$ + Conc. H_2SO_4) and then wash it several times with distilled water. It is finally washed with alcohol and ether and then dried.
2. Attach a piece of clean rubber tube to the end C and clamp the viscometer vertically in air.

3. Now introduce a sufficient volume of the given liquid with the help of a pipette in bulb B so that the bend portion of the U-tube and more than half of bulb B are filled up.
4. Through the rubber tube, suck up the liquid until it rises above the mark X. Make sure that there is no air bubble inside the liquid.
5. Now, allow the liquid to fall freely through the capillary upto the mark X. Start the stop watch and note the time t_1 for the flow of the liquid from mark X to mark Y.
6. Repeat the experiment thrice. The values should be concordant.
7. Remove the liquid and clean and dry the viscometer again.
8. Repeat the experiment by taking the same volume of the distilled water and note down the time taken t_2 for flow of water from mark X to Y. Repeat thrice.
9. Weigh the relative density bottle and note down its weight.
10. Fill it with the given liquid and weigh it again.
11. Remove the liquid. Wash it with chromic acid and then distilled water. Dry in the oven. Now fill it with the distilled water and weigh it

OBSERVATIONS

Room temperature = ... °C.

Observation Table

S	Given liquid		Water	
No.	Time of Flow	mean (t_1)	Time of flow	Mean (t_2)
1.				
2.				
3.				

Weight of empty specific gravity bottle = ... ω_1g

Weight of empty specific gravity bottle + given liquid = ... ω_2g

Weight of empty specific gravity bottle + distilled water = ... ω_3g

CALCULATIONS

$$\frac{\text{Density of solution}}{\text{Density of water}} = \frac{d_1}{d_2} = \frac{\omega_2 - \omega_1}{\omega_3 - \omega_1}$$

Relative viscosity of the given liquid $= \dfrac{\eta_1}{\eta_2} = \dfrac{d_1}{d_2} \times \dfrac{t_1}{t_2}$

Absolute Viscosity $= \dfrac{d_1}{d_2} \times \dfrac{t_1}{t_2} \times \eta_2$

RESULT The relative viscosity of the given solution is ... at ... °C.

The absolute viscosity of the given solution is ... Pas at ... °C.

EXERCISE

1. *Define Viscosity*

Ans. Viscosity is the resistance that one part of a liquid flowing with one velocity offers to another part of liquid flowing with different velocity.

2. *What are the units of viscosity?*

Ans. Units of Viscosity:

(i) In the C.G.S system the unit of viscosity is dyne cm^{-2} s and is known as poise.

$$\text{poise} = \text{dyne cm}^{-2}\,\text{s}$$
$$= \text{gcm s}^{-2}.\ \text{cm}^{-2}\,\text{s} = \text{g cm}^{-1}\,\text{s}^{-1}$$

In other word 1 poise $= 1\text{g cm}^{-1}\,\text{s}^{-1}$

(ii) In S.I system the unit of viscosity is Pascal second (Pas).

Note. $1 \text{ poise} = 1\text{g cm}^{-1}\,\text{s}^{-1} = 1\ (10^{-3}\text{ kg})\ (10^{-2}\text{ m})^{-1}\,\text{s}^{-1}$
$$= 10^{-1}\text{ kg m}^{-1}\,\text{s}^{-1}$$
$$= 10^{-1}\text{ Pas}$$

Hence 1 poise = 10^{-1} Pas.

3. *What do you mean by fluidity?*

Ans. The reciprocal of viscosity is called fluidity and is represented by ϕ so that,

$$\phi = \frac{1}{\eta}$$

Fluidity is the measurement of the ease with which the liquid flows. Its units are *Poise*$^{-1}$.

4. *How viscosity of a liquid varies with temperature?*

Ans. The viscosity of a liquid decreases with increase of temperature. As the temperature rises, the kinetic energy of the molecules increases and intermolecular forces of attraction become weak, resulting in subsequent decrease in the viscosity.

The variation of viscosity with temperature is best expressed by empirical relation given by Arrhenius

$$\eta \propto e^{Ea/RT}$$

$$\eta = Ae^{Ea/RT} \quad ....(6)$$

where A is a constant of proportionality, T is the temperature and R is gas constant.

In contrast to liquids, η increases with increasing temperature in the case of gases.

5. *What is the effect of pressure on viscosity of a liquid?*

Ans. The viscosity of the liquids increases with increase in pressure. As the pressure is increased the molecules of the liquid come closer and it becomes difficult for liquid molecules to move around and thus it becomes more difficult for them to flow.

6. *How viscosity is related to chemical constitution?*

Ans. Though there is no significant relationship between viscosity and chemical constitution, however some general results have been observed.

(i) The viscosity increases as the size of the molecules increases.

(ii) In the compounds having chain isomerism, the n-isomers have generally greater viscosity than the branched chain.

However, molecules with flexible chains have very high viscosity due to entanglement of side chains.

(iii) Non-polar molecules have lower viscosities than polar molecules.

(iv) Presence of intermolecular hydrogen bonding increases the viscosity of the liquid.

(v) Presence of impurities increases the viscosity. Viscosity is greatly enhanced due to the presence of lyophilic colloids.

7. *What are the applications of viscosity measurements?*

Ans.

- Viscosity Measurements are useful in predicting the behaviour of lubricants and oils in hot and cold conditions.
- Viscosity is also used for determination of molar masses of polymers.
- For distinguishing between branched and chain isomers. Chain isomers generally have high viscosities.

❑❑❑

11

Cryoscopy

DETERMINATION OF MOLAR MASS OF A NON VOLATILE SUBSTANCE-RAST METHOD

It is a well known fact that when a non-volatile solute is added to a solvent, it lowers the freezing point or raises the boiling point of the solvent. Lowering of freezing point and elevation in boiling point are colligative properties (*The property which depends only on the number of solute particles and not on their nature in called colligative property*). Thus, the measurement of any of these properties may be used to determine the molar mass of a non-volatile substance.

MEASUREMENT OF DEPRESSION IN FREEZING POINT

The freezing point of a liquid is defined as the temperature at which the solid phase begins to separate out from the liquid. At this temperature solid and liquid phases are in equilibrium.

The addition of a non-volatile solute to a solvent decreases the tendency of the solvent molecules to escape into the gaseous phase i.e. it lowers the vapour pressure of the solvent. Thus the vapour pressure of a solution containing a non-volatile solute is less than that of pure solvent and hence it will boil at higher temperature as compared to the temperature of the pure solvent.

Similarly, it can also be concluded that the solution will freeze at a lower temperature (depression in freezing point) as compared to the pure solvent.

The difference in the temperatures between the freezing point of the pure solvent and that of the solution is known as *depression in freezing point*. (ΔT).

Relation between ΔT and molecular weight of solute (m_1). From Raoult's law for dilute solutions, we have,

$$\frac{P - P'}{P} = \frac{n_1}{n_2} \qquad ...(i)$$

where P and P′ are the vapour pressure of the pure solvent and the solution respectively. n_1 and n_2 are the number of moles of solute and solvent respectively. Now,

$$n_1 = \frac{w_1}{M_1} \text{ and } n_2 = \frac{w_2}{M_2} \qquad ...(ii)$$

where w_1 = weight of solute

w_2 = weight of solvent

M_1 = molar mass of solute

M_2 = molar mass of solvent.

From (i) and (ii)

$$\frac{P - P'}{P} = \frac{w_1}{M_1} \cdot \frac{M_2}{w_2}$$

$$\Delta P = P \frac{w_1}{M_1} \cdot \frac{M_2}{w_2} \qquad ...(iii)$$

For a given solvent at its freezing point, P and M_2 are constants, hence from (iii), we have

$$\Delta P = \text{constant} \times \frac{w_1}{w_2 M_1}$$

or

$$\Delta P \ \alpha \ \frac{w_1}{w_2 M_1} \qquad ...(iv)$$

Since lowering of vapour pressure is directly proportional to depression in freezing point, i.e.

$$\Delta P \ \alpha \ \Delta T \qquad ...(v)$$

where ΔT is lowering of vapour pressure, from equation (iv) and (v) we have

$$\Delta T \ \alpha \ \frac{w_1}{w_2 M_1}$$

or

$$\Delta T = k \frac{w_1}{w_2 M_1} \qquad ...(vi)$$

where k is a constant.

If $\frac{w_1}{M_1} = 1$ and $w_2 = 1000g$, then equation (vi) reduces to

$$\Delta T = k/1000 = Kf \qquad ...(vii)$$

Note. A solute is said to be normal if its molecules neither dissociate nor associate nor interact with the solvent molecules.

where K_f is known as ***molal depression constant*** *or cryoscopic constant.*

Molal depression constant or the cryoscopic constant, K_f, of a solvent is defined as *the depression in freezing point which would be produced by dissolving 1g mole of a solute in 1000g of the solvent.* In other words the molal depression constant is the depression in freezing point as observed for one *molal* solution of any normal solute in the solvent.

From equation (vi) and (vii) we have,

$$\Delta T = \frac{1000\, K_f\, w_1}{w_2\, M_1} \qquad \text{...(viii)}$$

or $$\Delta T = K_f\, m' \qquad \text{...(ix)}$$

where m' is the molality of the solution.

(Molality is defined as the number of moles of the solute dissolved in 1000g of the solvent)

From equation (viii) we have,

$$M_1 = \frac{K_f . 1000 \cdot w_1}{\Delta T \cdot w_2} \qquad \text{...(x)}$$

The K_f value for a number of solvents is known and hence determining w_1, w_2 and ΔT experimentally, molar mass of the given solute can be calculated.

EXPERIMENT NO. 11.1

AIM Determination of molar mass of a substance by Rast Method.

THEORY This method was devised by Rast in 1922. Since in this method, only a few milligrams of substance is required, it is called micro-method. The basis of this method for determining molar mass of a substance is the use of camphor as a solvent. Camphor is used as a solvent due to the following reasons.

- It has a high value for freezing point depression constant K_f i.e. 40°C for 1 mole of solute in 1000g of solvent. Thus a very small quantity of solute is sufficient to depress the freezing point of camphor to a considerable extent such that the difference could be easily read by an ordinary thermometer.

- Camphor is a good solvent for most of the substances.

A known weight of the given substance is melted with about 10 times the weight of camphor and then cooled to form a solid. The melting point of this solid solution is determined. From the depression from the melting point of pure camphor (174°), the molecular weight of the solute is calculated using the formula–

$$M = \frac{K_f \cdot 1000\ w_1}{\Delta T \cdot w_2}$$

where M = Molar mass of the substance

K_f = Molal depression in freezing point constant (for camphor, it is $40 mol^{-1}$)

ΔT = Observed depression in freezing point.

= Melting point of pure camphor – Melting point of the mixture.

w_1 = Weight of the substance

w_2 = Weight of camphor

It may be noted that melting point of solid phase and the freezing point of liquid phase of any substance are the same. Therefore, whether we measure depression in the melting point or depression in the freezing point, it is one and the same thing.

APPARATUS Beaker 100cm^3, capillary tube, thermometer (0.1° C), test tube, apparatus for determining melting point.

CHEMICALS Camphor, experimental substance (acetanilide, urea or thiourea), liquid paraffin

PROCEDURE

1. Take a clean, thin-walled pyrex glass test tube and weigh it accurately.
2. Introduce about 10mg of the given powdered substance into the tube and weigh it again accurately.
3. Now add about 100mg of pure powdered camphor into the tube and weigh it again.
4. Draw the open end of this tube to a long capillary and then seal it, so that the vaporization of camphor is prevented. (or you can close the mouth of the tube with the help of a cork).

5. Place liquid paraffin in a beaker and heat it 5–10° C above the melting point of pure camphor. Now immerse the bottom of the tube containing camphor and the substance in it. The mixture will melt. Keep on rotating the tube to allow proper mixing of the two components.
6. Cool the tube quickly in a water bath. This is done to avoid the loss of camphor due to sublimation.
7. Break open the test after wiping off its outer surface with filter paper. Take out the solid and powder it with pestle and mortar.
8. Introduce the powdered solid in a capillary tube sealed at one end.
9. Attach the capillary with the bulb of thermometer and then immerse it in a melting point appartus as shown in the figure. 11.1 (a) or (b).

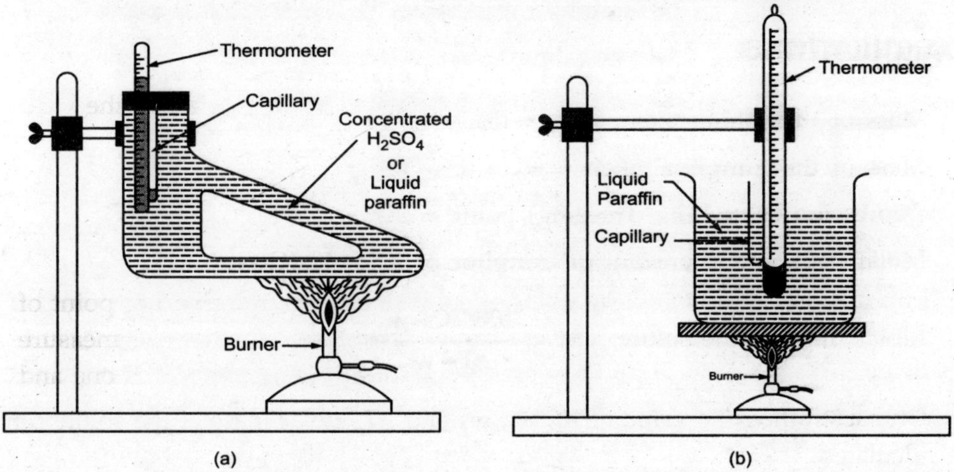

Fig. 11.1 *Melting Point apparatus*

10. Heat gently and determine the melting point of the mixture when the last crystal has just melted.
11. Allow the bath liquid to cool so that the liquid in the capillary freezes. Find the melting point of the mixture again. In this way, take atleast three readings and take mean of these readings (t_1).
12. Take pure camphor, powder it and fill it up in the capillary sealed at one end. Determine its melting point in the same way as described above. Take three readings and take their mean (t_2).

OBSERVATIONS

(a) Mass of the empty tube = ...g

(b) Mass of the tube + substance = ...g

(c) Mass of the tube + substance + camphor = ...g

Melting Point

S.no.	Melting point of Mixture °C	Melting point of pure camphor °C
	Mean (t_1) =	Mean (t_2) =

Melting point of mixture t_1 = ... °C

Melting point of camphor t_2 = ... °C.

CALCULATIONS

Mass of the solute taken = w_1 = (b) – (a) g

Mass of the camphor taken = w_2 = (c) – (*b*) g

Depression in melting (freezing) point = $\Delta Tf = t_2 - t_1$

Molal depression constant of camphor or Kf = 40 (Given)

Molar mass of the solute, $M = \dfrac{1000\,Kf \cdot w_1}{\Delta Tf \cdot w_2}$

On substituting the value of Kf, w_1, w_2 and ΔT, molar mass of the solute can be calculated.

RESULT The molar mass of the given substance was found to be ... g mol^{-1}.

Limitations of Rast Method

1. The method is applicable only to those substances which are soluble in camphor, do not react with it chemically or do not decompose below the melting point of camphor (about 177° C)
2. Although the method is very convenient but gives only approximate results because of the various assumptions made during the derivation of the theory.
3. Commercial samples of camphor vary in their character and hence it becomes necessary to determine the depression constant for the given sample of camphor using a solute of known molar mass such as naphthalene etc.

PRECAUTIONS

1. During the preparation of homogeneous mixture of the substance with camphor, the contents of the tube should be shaken vigorously while heating.
2. Heating should be done slowly.
3. Melting point should be noted when the last trace of the crystal just melts.
4. Camphor is easily powdered by moisturising it with a little ether.
5. Evaporation of camphor should be avoided.
6. Camphor to be used should be pure.
7. Amount of camphor taken should be approximately ten times the weight of the substance

EXERCISE

1. What do you mean by colligative property?
2. What is freezing point constant? How is it related to molecular weight?
3. Define Raoult's Law
4. What is the principle of Rast's method for determining the molar mass?
5. Rast's method is applicable to which type of substances?

   [Hint: The substances which are soluble in camphor and do not react chemically with it or do not decompose below melting point of camphor (177°C)]
6. Define molal depression constant?
7. What are the units of molal depression constant *Kf*?

   [Hint. In SI units, the units are K kg mol^{-1}]
8. Why camphor is preferred for determining the molar mass of a non-volatile solute?
9. Which is the other method employed for determining the molar mass of a substance cryoscopically?
10. What is the *Kf* value for water, acetic acid, benzene, naphthalene and camphor

    [Hint:

Solvent	***Kf***	**Solvent**	***Kf***
Water	= 1.86°	Naphthalene	= 6.8°
Acetic acid	= 3.9°	Camphor	= 40°
Benzene	= 1.86°		

11. Name some compounds whose melting point can be determined by Rast's method?

    [Hint: Naphthalene, acetanilide]
12. What are the advantages of Rast method.

    [Hint: Can be performed with common apparatus such as ordinary thermometer.]

12

Polymers

POLYMER

The word 'polymer' is derived from the classical Greek words, *Poly* meaning 'many' and *meres* meaning 'parts'. Simply stated, a polymer is a molecular compound with high molecular mass, ranging into thousands and millions of gram. It is composed of a large number of repeating units of identical structure called *monomers.* For example, polythene is a polymer formed by linking together of a large number of ethene molecules.

$$\underset{\text{(Monomer)}}{\underset{\text{Ethene}}{n\,CH_2 = CH_2}} \longrightarrow \underset{\text{(Polymer)}}{\underset{\text{Polyethene}}{(CH_2 - CH_2)}}$$

Both natural and synthetic polymers impinge on every aspect of our existence. The idea is often expressed as that we live in "a plastic age". The range of polymeric materials is so wide that they have found applications in practically every branch of industry. About 80% of today's chemists, including biochemists, work with polymers.

The number of repeating units in a chain formed in a polymer, is known as the "**degree of polymerization**". Polymers with high degree of polymerization are termed as "high polymers" and those with low degree of polymerization are called oligopolymers.

CLASSIFICATION OF POLYMERS

Synthetic polymers have been classified into various groups, in various manners.

(i) On the basis of the chemical structure of their backbone, polymers are of two types.

Homopolymer. If a polymer consists of identical monomer, the polymer is termed as homopolymer.

$$\sim A - A - A - A - A \sim \qquad \sim A - B - A - B - A - B \sim$$

Homopolymer — Copolymer

Copolymer. If the polymer is a mixture of more than one type of monomer it is termed as copolymer.

∽ A — A — A — A ∽ Linear Homopolymer

∽ A — B — A — B — A — B ∽ Linear Copolymer

(ii) On the basis of their polymeric structures, the polymers can be classified as linear, branched or crosslinked.

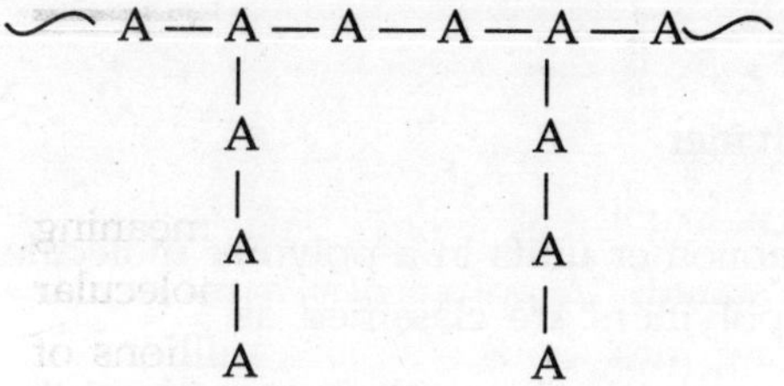

Branched Chain Homopolymer

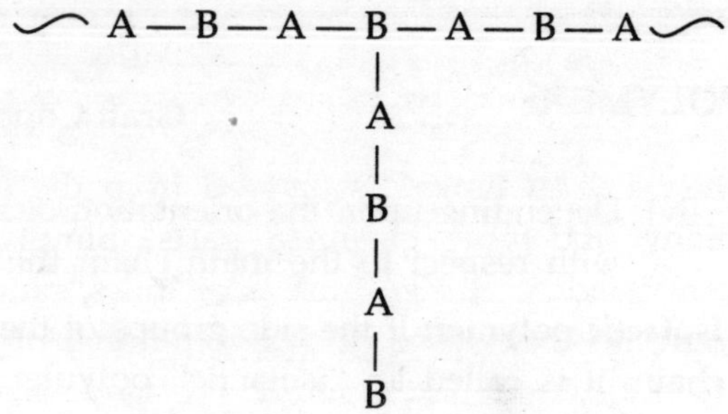

Branched Chain Copolymer

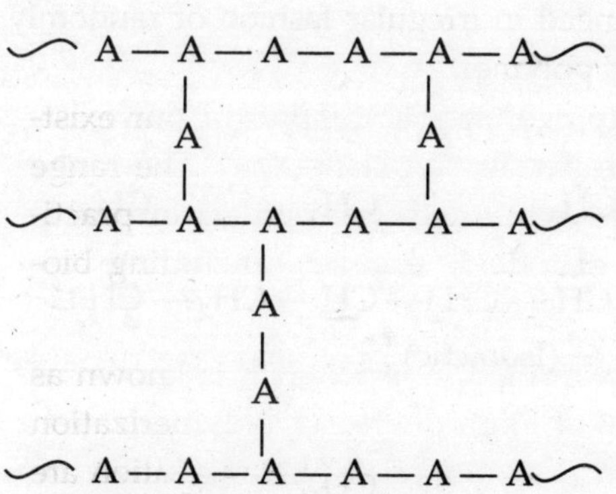

Cross Linked Homopolymer

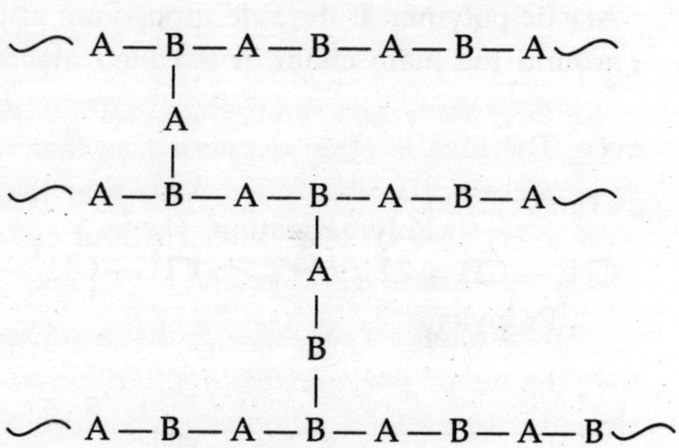

Cross Linked Copolymer

(iii) Depending upon the arrangement of the monomers in the case of copolymers, the polymers are divided into block copolymers and graft copolymers.

Block Copolymers. Linear polymers in which the identical monomeric units occur in relatively long sequences are called block co-polymers.

— A — A — A — B — B — B — A — A — A — B — B — B —

Block Copolymer

Graft Copolymers. Graft copolymers are branched copolymers in which the backbone is formed from one type of monomer and branches are formed of the other, i.e. the monomer segments on the branches and backbone are not the same.

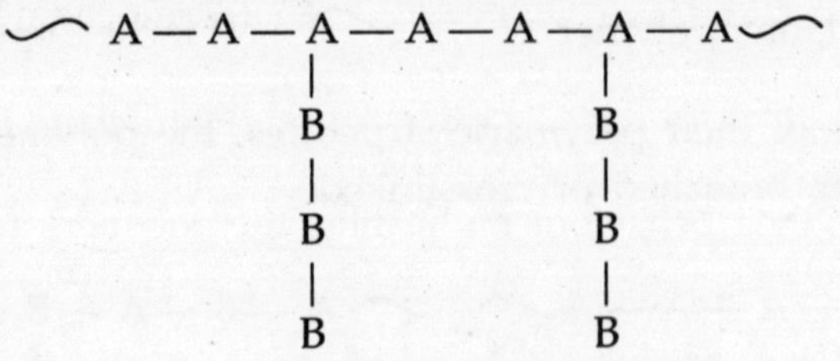

Graft Copolymer

(iv) Depending upon the orientation of monomer units in a polymer molecule with respect to the main chain the polymers are classified as

Isotactic polymer: If the side groups of the monomers lie on the same side of the chain, it is called an "isotactic" polymer.

Syndiotactic polymer. If the side groups are arranged in an alternate fashion, it is called syndiotactic polymer.

Atactic polymer: If the side groups are arranged in irregular fashion or randomly around the main chain, it is called atactic polymer.

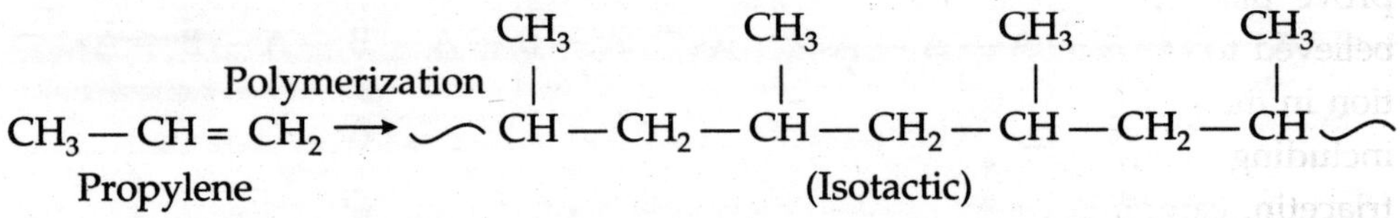

(Isotactic)

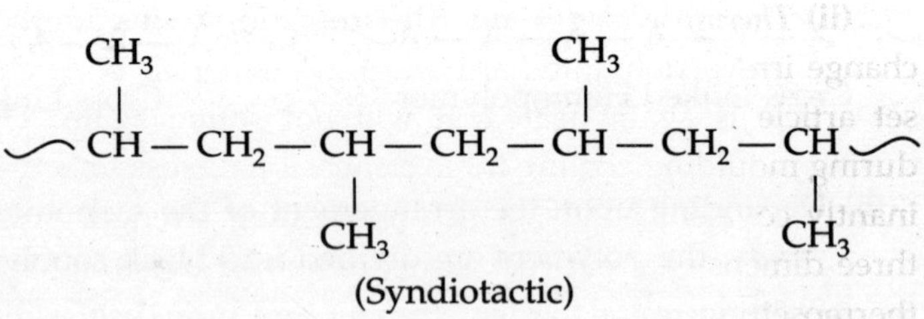

(Syndiotactic)

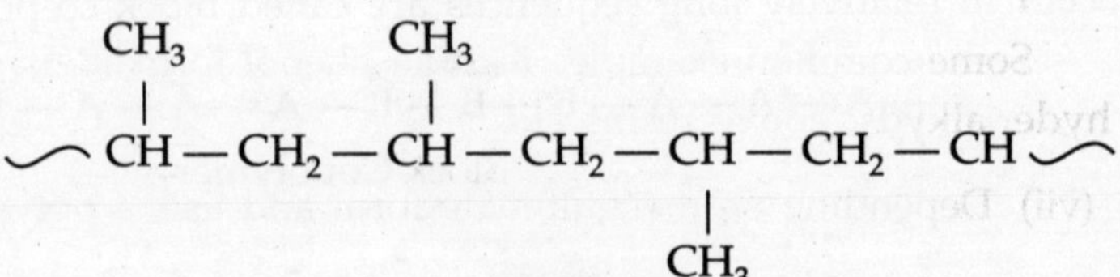

(Atactic)

(v) If the main chain is composed of the atoms of the same specie, the polymer is called homochain and if the main chain is made up of different atoms then they are called heterochain polymers.

$$\sim C-C-C-C-C-\sim \quad \text{(Homochain Polymer)}$$

$$\sim C-C-O-C-O-C\sim \quad \text{(Hetrochain Polymer)}$$

(vi) According to their behaviour on heating, the polymers are divided into two groups.

(i) *Thermoplastics.* Thermoplastics or resins soften on heating and become plastic so that they can be converted to any desired shape by moulding. On cooling, they become hard and rigid but their hardness is a temporary property subject to change with increase or decrease in temperature. In thermoplastic resins, the chemical structure or the molecular weight are not changed during the heating or moulding operations, because the changes involved are of physical nature. Only the secondary bonds between the individual molecular chains are broken on heating which results in their softening. On cooling, these secondary bonds are again formed and hence they become hard again.

Some common examples of thermoplastics are polythene, polypropylene, polystyrene, polyvinyl chloride, teflon, polyacrylonitrile etc.

Plasticizer. Sometimes, the substances called plasticizers are added to improve plasticity and flexibility of the plastics. The plasticizing properties are believed to be mainly due to the neutralization of intermolecular forces of attraction in the resin molecules. A variety of organic materials are used as plasticizers including non-drying vegetable oils, tributyl phosphate, triphenyl phosphate, triacetin, camphor, esters of oleic or stearic acid.

(ii) *Thermosetting resins.* Thermosetting resins are those which upon heating, change irreversibly into hard and rigid materials on heating. After cooling, if the set article is again heated, it will not soften again. The thermosetting resins during moulding acquire three dimensional cross linked structures with predominantly covalent bonds, which retain their strength even on heating. Thus, the three dimensional network structure joined by strong covalent bonds, make the thermosetting resins harder, stronger and more brittle than thermoplastic resins. If heating of a thermoset resin is prolonged, these undergo change in chemical composition to give a infusible and insoluble mass.

Some common examples include phenol formaldehyde, melamine formaldehyde, alkyds, epoxy resins etc.

(vii) Depending on their ultimate form and use, a polymer can be classified as:

Plastics. All the synthetic polymers (other than elastomers) are usually referred to as plastics. The polymers which are shaped into hard and tough utility articles by the application of heat pressure are called plastics. At some stage of manufacture, these are in plastic condition (Plasticity is the property by virtue of which a material undergoes permanent deformation under stress).

Elastomers. The polymers which are capable of being stretched rapidly at least 150 percent of their original length without breaking and return to their original shape on release of stress are known as elastomers. Synthetic rubbers constitute important examples of this group.

FUNCTIONALITY

The functionality of a molecule is the number of reactive sites it has. For a substance to act as a monomer, it must have two reactive sites. In other words, they should have a functionality of atleast two. A compound assumes functionality because of the presence of reactive functional groups like – OH, – COOH, – NH_2, – SH etc. The number of functional group present in a compound defines its functionality.

TYPES OF POLYMERIZATION

There are two types of polymerization processes:

1. Addition Polymerization
2. Condensation polymerization.

ADDITION POLYMERIZATION

Addition polymerization is characterized by self-addition of the monomer molecules to each other, without the loss of any material. The product is an exact multiple of the original monomeric molecule. Examples:

(i) Addition by breaking up of multiple bonds:

$$\underset{\text{(ethene)}}{nCH_2 = CH_2} \longrightarrow \underset{\text{(Polyethene)}}{(CH_2 - CH_2)_n}$$

(ii) Addition by opening up of ring structures:

$$CH_3OH + (n+1)\ \underset{\diagdown\ O\ \diagup}{CH_2 - CH_2} \longrightarrow CH_3O + CH_2 - CH_2 - O +_n CH_2CH_2OH$$

Vinyl compound (CH_2 = CHX), allyl compounds (CH_2 = CH · CH_2X) olefins (CH_2 = CHR) and dienes (CH_2 = CR – CH = CH_2) typically undergo addition polymerization. Since a majority of these fall under the 'vinyl' category, addition polymerization is also customarily termed as vinyl polymerization.

CONDENSATION POLYMERIZATION

Condensation polymerization is brought about by monomers containing two or more reactive functional groups with the elimination of small molecules like H_2O, HCl etc. For example.

$$n\,HO-\overset{\overset{\displaystyle O}{\|}}{C}-(CH_2)_4-\overset{\overset{\displaystyle O}{\|}}{C}-OH + n\,H_2N-(CH_2)_6-NH_2$$

(adipic acid) (hexamethylene diamine)

$$\downarrow$$

$$HO-\!\left[\overset{\overset{\displaystyle O}{\|}}{C}-(CH_2)_4-\overset{\overset{\displaystyle O}{\|}}{C}-NH-(CH_2)_6-NH\right]_n\!-H + (2n-1)\,H_2O$$

nylon 6, 6

It is seen from the above reaction that the polymer formed still contains both the reactive functional groups at its chain ends and hence, is 'active' and not 'dead'.

In the above example, the different type of reactive functional groups, viz.., $-NH_2$ and $-COOH$ are present on two different monomers. In some cases, both the reactive groups may be present on the same monomer and hence lead to self condensation reaction. For example, polycondensation of amino acids.

$$n\,H_2N-\!\!-\!(CH_2)_6\!-COOH \longrightarrow H\!-\!\left[HN(CH_2)_6-CO\right]\!-OH + (n-1)\,H_2O$$

PREPARATION OF POLYMERS

The common techniques employed for the production of polymers are:

1. *Bulk Polymerization.* The initiator or the free radical catalyst is dissolved in the monomer from which the polymer has to be prepared. The mixture is then heated and stirred in a vessel. As the polymerization takes place the viscosity of the system increases. The product obtained is of high purity.

2. *Solution Polymerization.* Here an inert solvent is required. The monomer and the initiator are dissolved in a suitable inert solvent. The resulting solution is heated with stirring. The technique is useful when the polymer is to be used in solution form such as in the case of adhesives and surface coatings.

3. *Suspension Polymerization.* Here, a solution of the catalyst in the monomer is dispersed as fine droplets in an inert solvent (water). The suspension is stabilized by the addition of water-soluble protective colloids such as polyvinyl alcohol or starch. The mixture is stirred continuously. The product is in the form of separable beads which are washed by water to remove the protective colloids.

4. *Emulsion Polymerization.* In this technique, the particle size of the monomer is greatly reduced. This is achieved by vigorous stirring and by the use of synthetic surfactants. The catalyst used are persulphate or hydrogen peroxide, both of which are water soluble. The product so obtained is in the form of fine particles dispersed in water. It can be used as such (in case of adhesive or emulsion paints) or can be separated.

5. *Melt Polycondensation.* The monomers involved in condensation are heated together in a definite stoichiometric ratio above the melting point of the product. The temperature should be such that the starting monomers and the product should be stable. To prevent oxidation at high temperature, oxygen has to be excluded from the reaction chamber.

6. *Interfacial Polycondensation.* In this technique the polymerization is made to take place at the interface of an organic and an aqueous medium. The solutions of the monomers are prepared separately in two immiscible solvents (aqueous and organic) and the reaction is allowed to take place at interface. The technique is good for the reactants with quite reactive functional groups.

EXPERIMENT NO. 12.1

AIM Preparation of Nylon 66

Polyamides are a group of polymers which contain the amide (– CONH –) linkage in the main polymer chain. Polymers of this type are the synthetic linear aliphatic polymides which are capable of fibre formation.

Nylon is the generic name for synthetic polyamides capable of forming fibres. Polyamides (nylons) are prepared by the polycondensation between dicarboxylic acids and diamines. Nylons are designated by two numbers, the first rep resenting the diamine. Polyamides of commercial importance are Nylon 6, 6; Nylon 6, 10, Nylon 6 etc.

Nylon 6,6: Nylon 6, 6 is prepared by the polycondensation of hexamethylene diamine with adipic acid.

$$\underset{\text{Adipic acid}}{HOOC-(CH_2)_4-COOH} + \underset{\text{Hexamethylene diamine}}{H_2N-(CH_2)_6-NH_2} \longrightarrow$$

$$-\overset{\overset{O}{\|}}{C}-(CH_2)_4-\overset{\overset{O}{\|}}{C}-NH-(CH_2)_6-NH-$$

THEORY For the prepation of Nylon 66 the method of *interfacial condensation* is used. The diamine is dissolved in water and diacid chloride is dissolved in an organic solvent like CCl_4 or $CHCl_3$. The aqueous layer is poured on the top of organic layer carefully. Diamine is soluble in both the solvents and hence diffuses from aqueous layer to organic layer and react with the reactive diacid chloride at the interface to form the product. The polymer is removed with the help of a forcep and the process continues.

$$\underset{\text{Hexamethylene diamine}}{H_2N-(CH_2)_6-NH_2} + \underset{\text{Adipoyl Chloride}}{n\,Cl-\overset{\overset{O}{\|}}{C}-(CH_2)_4-\overset{\overset{O}{\|}}{C}-Cl} \longrightarrow$$

$$H\!\left[(CH_2)_6-NH-\overset{\overset{O}{\|}}{C}-(CH_2)_4-\overset{\overset{O}{\|}}{C}\right]\!Cl + (2n-1)\,HCl$$

Nylon 66, Poly [(hexamethylene adipamide)]

APPARATUS 100 cm^3 Beaker, glass rod, forcep.

CHEMICALS Hexamethylene diamine solution (~ 1% in water) * Adipoyl chloride solution (~ 2% in CCl_4)

PROCEDURE

1. Take about 50 ml of the solution of adipoyl chloride (in CCl_4) in 100 ml beaker.
2. Take 25 ml of the aqueous solution of hexamethylene diamine in a separate beaker.
3. Carefully pour about 25 ml of the aqueous solution of hexamethylene diamine along the sides of the beaker containing solution of adipoyl chloride, so that a separate aqueous layer, over heavier CCl_4 solution is formed.
4. The formation of a film of nylon 6, 6 at the interface of the two liquids will take place.
5. Carefully insert a glass rod or a copper wire into the solution and pull out the polymer. A pair of forceps can also be used.
6. Wrap the polymer around a clean test tube as it comes out in the from of thread (Do not handle the reaction mixture with hands as the reactants are harmful)
7. Rotate the test tube to pull more and more of nylon.
8. Wash the polymer with water by placing the test tube under tap water.

RESULT = ... g of nylon was obtained from .... g of adipic acid

PRECAUTIONS

1. Hexamethylene diamine is corrosive and hence must be handled carefully.
2. The diamine solution should be added very slowly along the sides of the beaker so that the lower organic layer is not disturbed.
3. Do not pour unreacted polymerization mixture in the sink. Stir it with a glass rod till there is no more polymerization and then discard the solvent.

PROPERTIES OF NYLONS

1. Nylons are characterized by a combination of high strength, elasticity, toughness and abrasion resistance.
2. They have very good moiture resistance.
3. They are insoluble in common solvents.
4. The molecular chains are held together to each other by hydrogen bonds having linear structure, permitting side-by-side alignment.
5. They are very flexible and retain their original shape after use.
6. Nylons provide excellent resistance to wear and abrasion
7. Due to the presence of hydrogen bonding, all nylons are water sensitive. Water acts as plasticizer, which reduces tensile strength and dimensional stability.

USES

1. Nylon 6, 6 is used as a plastic as well as a fibre. Its most important fibre applications include automobile tire cords, ropes, threads, cords having high tenacity and good elasticity.
2. Nylon 6,6 is also used to make textile fibres for use in dresses, undergarments, socks etc.

***Note.** Adipoyl chloride can also be prepared in the lab from adipic acid and thionyl chloride. For the preparation of adipoyl chloride following procedure is adopted:

APPARATUS Round bottomed flask, air condenser, beakers.

CHEMICALS Adipic acid, thionyl chloride, dimethyl formamide, carbon tetrachloride, NaOH,

PROCEDURE In a 50 cm^3 round bottomed flask, fit an air condenser. Take the glass funnel and fix a rubber tube with it. The other end of the rubber tube is fixed with the condenser with the help of a glass tube fitted in a cork. Put the funnel (inverted) in a beaker containing water or NaOH. Now place 1g of adipic acid in 3-4 drops of dimethylformamide in the flask and add 1 cm^3 of thionyl chloride dropwise with stirring. Heat the flask on water bath for 15 minutes till the evolution of gas ceases and the solid disappears. Allow the flask to cool and add 30 cm^3 of CCl_4 to dissolve the product. The product so obtained is the solution of adipoyl chloride in CCl_4.

3. Nylon 6,6 being a tough plastic, is used as substitute for metals in gears and bearings etc. Nylon 6,6 can be moulded in one piece whereas similar item in metal require assembling of several parts. Moreover nylon bearings and gears perform quietly and need little or no lubrication.
4. It is also used in making rollers, slides and door latches.

EXPERIMENT NO. 12.2

AIM **Preparation of urea formaldehyde resin.**

THEORY *Urea formaldehyde resin* is prepared by the condensation reaction between urea and formaldehyde in neutral or alkaline conditions. The first products formed during the formation of resin are monomethylol and dimethylol ureas.

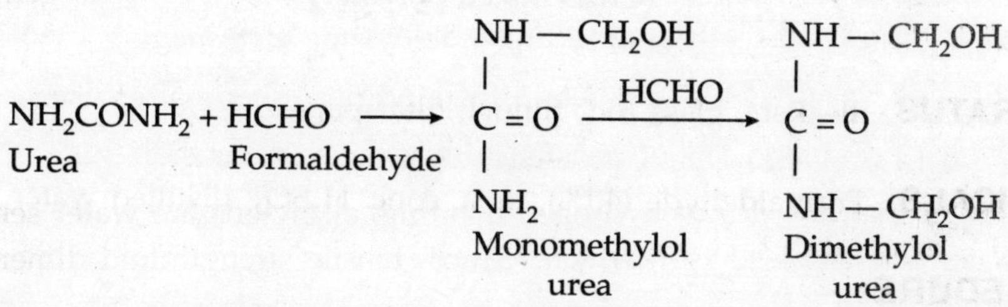

Polymerization can take place from mono or dimethylol urea, or possibly through both, with the formation of long chains.

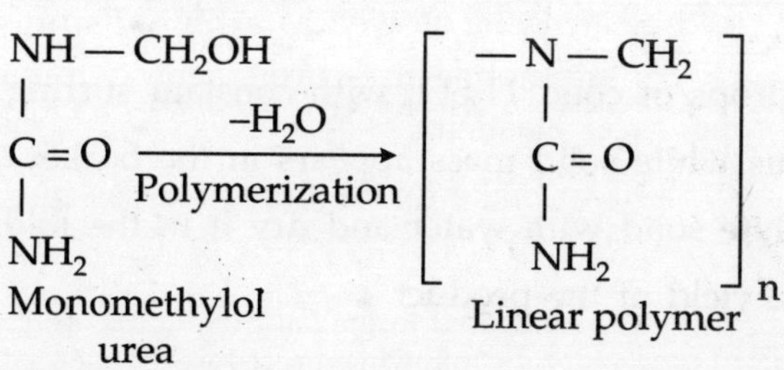

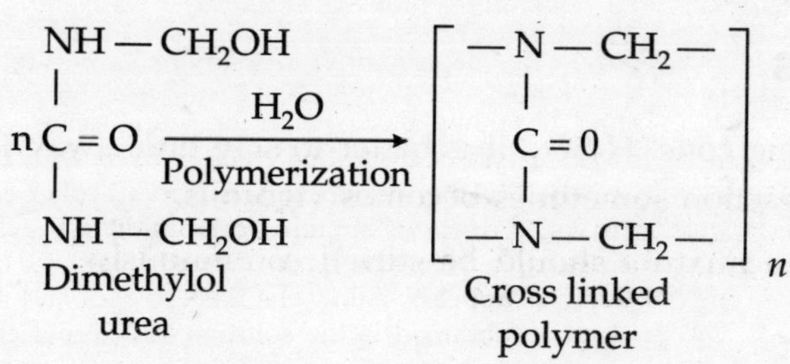

Theromosetting of the resin takes place during the moulding process. For moulding, the methylol derivatives are compounded with fillers, plasticizers, pigments etc. and then are subjected to heat and pressure, resulting into a hard, infusible product. A fully cross linked urea-formaldehyde resin can be represented as

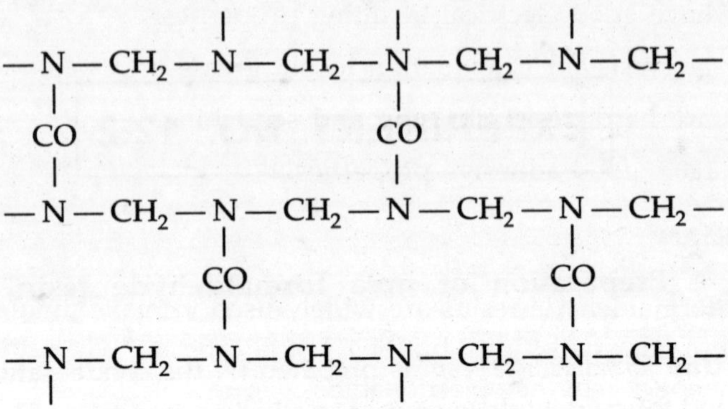

Urea-formaldehyde resin
(Cross linked polymer)

APPARATUS Beakers, glass rod, funnel, filter paper

CHEMICALS Formaldehyde (40%), urea, conc. H_2SO_4, distilled water.

PROCEDURE

1. Place about 20 ml of 40% formaldehyde solution in 100 ml beaker.
2. Add about 10 g of urea with constant stirring till saturated solution is obtained.
3. Add a few drops of conc. H_2SO_4 with constant stirring.
4. A voluminous white solid mass appears in the beaker.
5. Wash the white solid with water and dry it in the folds of filter paper.
6. Calculate the yield of the product.

RESULT The yield of urea formaldehyde resin is ....

PRECAUTIONS

1. While adding conc. H_2SO_4, it is better to stay little away from the beaker since the reaction sometimes becomes vigorous.
2. The reaction mixture should be stirred continuously.

PROPERTIES

1. The urea formaldehyde resins are light in colour and enable the production of light or pastel coloured objects.
2. Their heat and moisture resistance are lower than those of phenolic resins.
3. They have good electrical insulting properties.
4. They are resistant to oxidation to oil, grease and weak acids.
5. They are hard, resist abrasion and scratching.
6. They have good adhesive properties.

USES

1. Urea-formaldehyde resins are widely used to make moulded articles like vacuum flask cups, jugs, buttons, bottle caps etc. where the darker colour of phenolics may be objectionable.
2. They are also used for adhesive applications for the production of plywood and laminating.
3. They are used for the manufacture of coated abrasive paper and binder for foundry cores.
4. They are used for the manufacture of cation exchange resins.
5. These also find use in the manufacture of electrical switches, plugs and insulating foams.
6. Their applications also include the treatment of textile fibers for improving their shrink and crease resistance.

EXPERIMENT NO. 12.3

AIM **Preparation of phenol formaldehyde resin. (Bakelite)**

THEORY

Phenolic Resins

Phenol formaldehyde resins are the most important and extensively used synthetic plastics and perhaps the oldest synthetic plastics. In these resins one of the monomers is always a phenol molecule. One of the most important member of this class is phenol formaldehyde resin.

The phenol formaldehyde reaction is catalyzed by alkalies and acids. The nature of the product depends on several factors, the major one being the nature of the catalyst and the proportion of the reactants (phenol formaldehyde ratio). The formation of phenol formaldehyde resin comprises of the following steps:

(*i*) **Methylolation:** When phenol and formaldehyde react together, the first step is the entry of methylol (CH_2 OH) groups in ortho and para positions to the hydroxy group. The reaction takes place in the presence of acid or alkali. Depending upon phenol formaldehyde ratio, various phenol-alcohols may be formed.

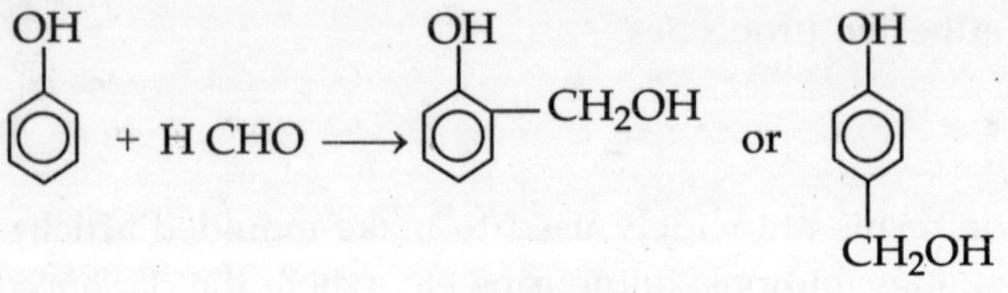

Monomethylol phenols

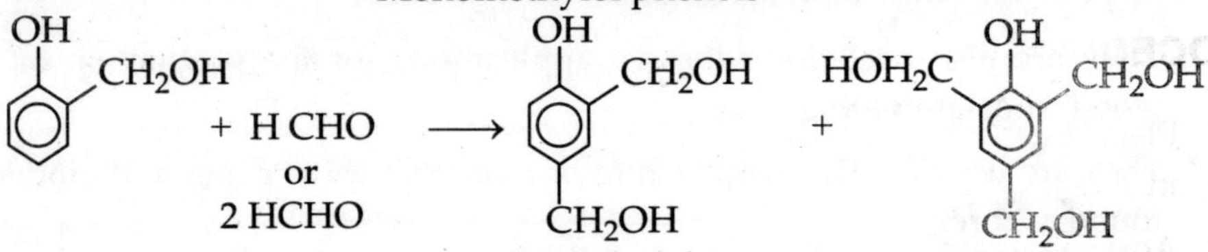

Dimethylol phenols

Trimethylol phenols

(*ii*) **Novolac formation:** Depending upon the ratio of phenol formaldehyde (P/F) different resins, namely novolac and resol resins are obtained. In the presence of acid catalyst, when the P/F ratio is greater than unity, the methylol derivatives condense with phenol to form a linear polymer with little methylol groups. The product is thermoplastic in nature and is known as novolac.

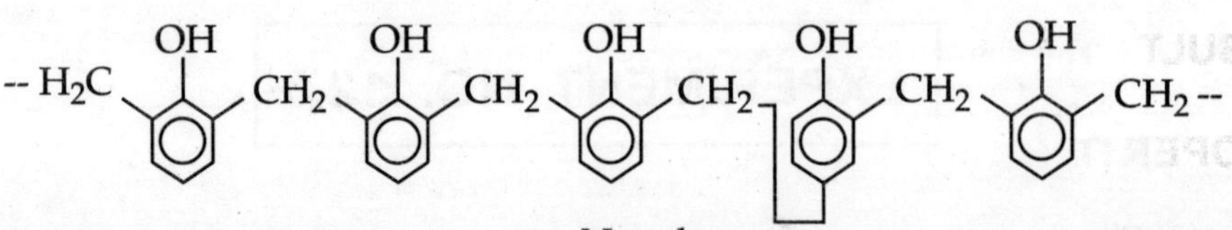

Novolac

(*iii*) **Resole formation:** In the presence of alkaline catalysts with P/F ratio less than 1, the methylol phenols condense to form linear structure called resols. The resols are soluble in the reaction mixture and have excess of methylol groups capable of further reaction during continued heating. Resols, therefore have limited shelf-life. However, the reaction may be stopped at the desired step by cooling.

(*iv*) **Resite formation:** On further heating, resole is transformed into a three dimensional polymer, called resite, due to the presence of active groups and mobile hydrogens. Resite is an infusible and insoluble material.

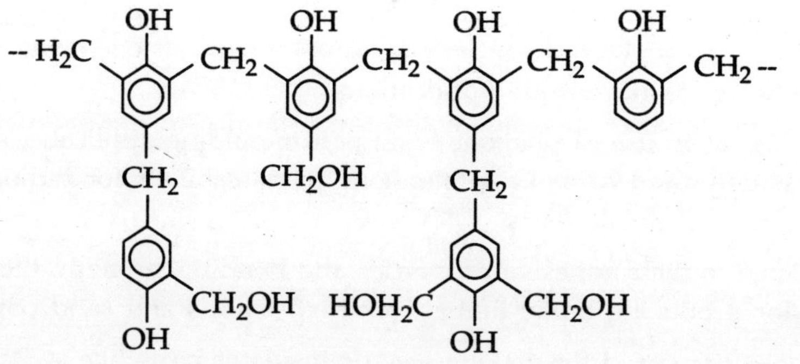

Resite

APPARATUS Glass rod, beakers, funnel, filter paper

CHEMICALS Glacial acetic acid, 40% formaldehyde solution, phenol, conc. HCl.

PROCEDURE

1. Place 5 ml of glacial acetic acid and 2.5 ml of 40% formaldehyde solution in a 500 ml beaker.
2. Add 2 g of phenol to it.
3. Wrap a cloth loosely round the beaker. Add a few ml of conc. HCl into the mixture carefully and heat it slightly.
4. A large mass of plastic, pink in colour is formed
5. The residue is washed with water and is filtered.
6. The product is dried and the yield is calculated

RESULT The yield of the product is ....

PROPERTIES

(*i*) Phenol formaldehyde moulding resins have excellent heat resistance.

(*ii*) These have high dimensional stability.

(*iii*) To improve the properties further, fillers such as wood, flour, mica, asbestos etc. are added to them.

(*iv*) Phenolic resins have good dielectric properties.

(*v*) These have remarkale adhesive properties and bonding strength.

(*vi*) They are however attacked by alkalies, because of the presence of free hydroxy groups.

(*vii*) They are hard, rigid and scratch-resistant.

USES

Phenolic resins have wide applications.

(*i*) The main use of phenolic resin is in moulding applications. They are widely used in making telephone parts, cabinets for radio, television and automobile parts.

(*ii*) Due to their adhesive properties and bonding strength, they are used for producing brake linings, abrasive wheels and sand paper.

(*iii*) They are used for making electric insulator parts like switches, plugs, switch boards, heater-handles etc.

(*iv*) Phenolic resins are used for impregnating paper, wood and other fillers.

(*v*) These are also used in varnishes, paints and protective coatings.

(*vi*) These are used in the production of ion exchange resin for water softening.

MOLECULAR WEIGHT OF POLYMERS

The molecular weight of polymers is related to the chain length and the extent of crosslinking between different chains. The extent of cross-linking depends on the concentration of the monomer during polymerization. The length of the polymer chain depends on the random encounter between the monomer and the reactive site of the chain. Due to the randomness, some polymeric chains grow longer than the others. Thus, a polymer sample is a mixture of different molecules having different chain lengths. Hence molecular weight of a polymer is always expressed as some sort of average of molecular weight. The molecular weights can be expressed as–

(i) Number Average Molecular Weight M_n:

$$M_n = \frac{\text{Total Mass of the polymer sample}}{\text{Number of molecules present in the sample}}.$$

or

$$M_n = \frac{n_1 M_1 + n_2 M_2 + n_3\, M_3 + \ldots}{n_1 + n_2 + n_3 + \ldots}$$

$$= \frac{\Sigma n_i\, M_i}{\Sigma n_i}$$

where n_1, n_2, n_3 etc. are the number of molecular species having molecular mass M_1, M_2, M_3 etc ... respectively. Measurement of an appropriate colligative property affords the Number Average Molecular Weight.

(ii) Weight Average Molecular weight Mw:

$$M_w = \frac{W_1M_1 + W_2M_2 + W_3\,M_3 + \dots}{W_1 + W_2 + W_3 + \dots} = \frac{W_1\,M_1 + W_2M_2 + W_3M_3 + \dots}{W}$$

where W_1 = Total mass of the species having molecular weight M_1.

$W_1 = n_1M_1$

W = Total weight of the polymer sample

$$\therefore \quad M_w = \frac{\Sigma n_i\,M_i^2}{W}$$

(iii) Viscosity Average Molecular Weight, $\overline{M}_v$:

$$\overline{M}_v = \left[\frac{\Sigma Ni\,Mi^{(1+a)}}{\Sigma NiMi}\right]^{1/a}$$

where Ni is the number of molecules of molecular weight Mi a is the exponent in Mark-Kuhn-Houwink equation i.e. $[\eta] = KM^a$, where η is the intrinsic viscosity and K is a constant.

EXPERIMENT NO. 12.4

AIM **To determine the molecular weight of a polymer (polystyrene) by using viscometric method.**

THEORY When a very small amount of a polymer is added to a solvent of low viscosity, the viscosity of the resulting solution increases sharply. This increase in viscosity depends upon the molecular weight of the polymer, concentration, size and shape of the solute molecules.

* The relative viscosity (η_r) of the polymer solution will be given by

$$\eta_r = \frac{\eta_s}{\eta_0} \quad \text{...(i)}$$

where η_s is the coefficient of viscosity of the polymer solution.

η_o is the coefficient of viscosity of pure solvent at the same temperature.

***Note** For the details of viscosity please refer to the chapter on viscosity.

Now $$\frac{\eta_s}{\eta_0} = \frac{P_s\, t_s}{P_0\, t_0} \quad \text{...(ii)}$$

where P_s and P_0 are the densities of polymer solution and the pure solvent, and t_s and t_0 are the corresponding time of flow of specified volume through the capillary in viscometer.

For a dilute solution, it can be assumed that $P_s = P_0$, hence equation (ii) becomes

$$\eta_r = \frac{t_s}{t_0} \quad \text{...(iii)}$$

Further, the specific viscosity (η_{sp}) of the polymer solution is obtained by:

$$\eta_{sp} = \frac{\eta_s - \eta_0}{\eta_0} \quad \text{...(iv)}$$

where $\eta_s - \eta_0$ is the increase in viscosity of the solvent due to the presence of a solute.

or $$\eta_{sp} = \frac{\eta_s}{\eta_0} - 1 \quad \text{...(v)}$$

$$= \eta_r - 1 = \frac{t_s}{t_0} - 1 \quad \text{...(vi)}$$

The ratio of specific viscosity of the solution to its concentration C (expressed in g/100ml) is known as *Viscosity Number* or *Reduced viscosity* η_{red}.

$$\eta_{red} = \frac{\eta_{sp}}{C} \quad \text{...(vii)}$$

The plot of η_{red} versus concentration is a straight line (Fig. 12.1). It is given by the equation,

$$\eta_{red} = \frac{\eta_{sp}}{C} = mC + \text{constant} \quad \text{...(viii)}$$

where m represents the slope of the line.

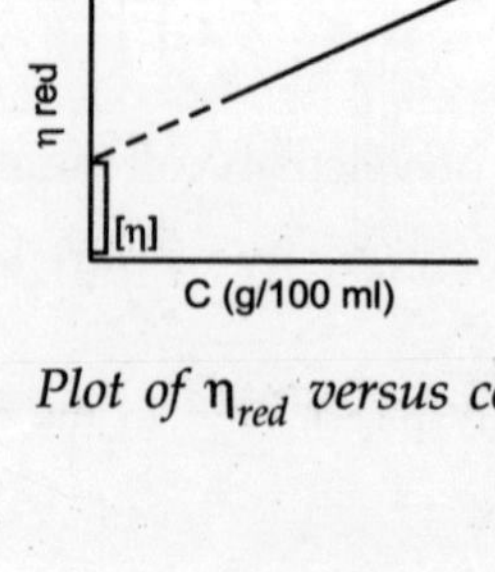

Fig. 12.1 *Plot of η_{red} versus concentration.*

The value of constant in equation (viii) is given by the intercept on the ordinate, which is obtained by extrapolating the graph to zero concentration or infinite dilution. This is termed as *limiting Viscosity* or *Intrinsic Viscosity* $[\eta]$. Actually, intrinsic viscosity is the limiting value of reduced viscosity

$$\text{Thus, } [\eta] = \lim_{C \to 0} \left(\frac{\eta_{sp}}{C} \right) = \lim_{C \to 0} \frac{\eta_s - \eta_0}{\eta_0 \times C} \quad \text{...(ix)}$$

For linear polymers, the intrinsic viscosity $[\eta]$ is related to the molecular weight of the polymer by *Mark-Kuhn-Houwink equation*:

$$[\eta] = KM^a \quad \text{...(x)}$$

where M is the average molecular weight of the polymer; K and a are constants for a given polymer-solvent system.

Thus molecular weight of the polymer can be determined from viscometric measurements using equation (x) . For most systems 'a' lies between 0.6 to 0.8.

APPARATUS Ostwald's viscometer, stop watch.

CHEMICALS Polystyrene, toluene or benzene.

PROCEDURE

1. Weigh accurately 0.5 g of polystyrene and transfer it in a 100 ml measuring flask.
2. Add about 90-95 ml of the solvent (benzene or toluene) and dissolve polystyrene by shaking it. Make up the volume to 100 ml to prepare 0.5% of the solution.
3. Similarly prepare the solutions of different concentrations (0.1%, 0.2%, 0.3% and 0.4%) by diluting the above solution.
4. Wash the Oswald's viscometer with chromic acid, distilled water and dry in an oven. Pipette out 20 ml of the pure solvent (benzene or toluene) in the bulb B of the viscometer which is clamped vertically in a stand, placed in thermostat at 25°C or 30°C Fig. 12.2.

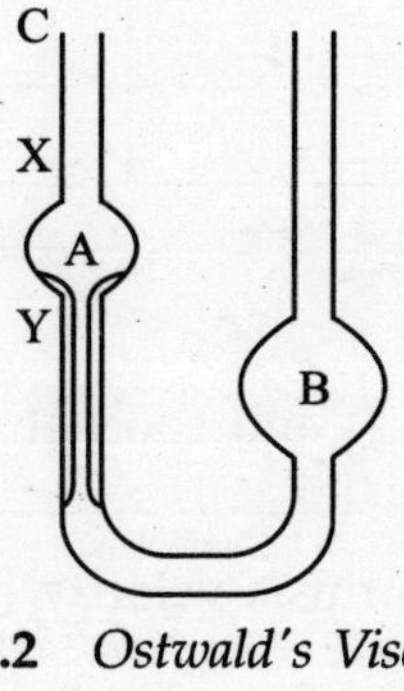

Fig. 12.2 *Ostwald's Viscometer*

5. Through a rubber tube attached to upper arm of bulb A, suck up the solvent untill it rises above the mark *X*.
6. Allow the solvent to fall freely through a capillary upto the mark *X*. Start the stop watch and note the time t_1 for the flow of liquid from mark *X* to mark *Y*.
7. Repeat to get five readings and take the mean as the flow time t_0.
8. Remove the solvent clean the viscometer and dry it.
9. Pipette out 20 ml of one of the solutions prepared above (say 0.5%) and determine the flow time as above. Repeat to get three readings and take their mean.
10. Similarly, determine the flow times for solutions of different concentrations after proper cleaning and drying of the viscometer after each set of readings.

OBSERVATIONS.

Temperature of the experiment =

Solvent used =

Value of the constant '*K*' for styrene/solvent at the above temperature (from the table - 12.1) =

Value of constant '*a*' (from the table 12.1) =

Observation Table.

	Conc. of polymer solution	Flow time (Mean of three readings	Relative viscosity $\eta_r = \frac{\eta_s}{\eta_0} = \frac{t_s}{t_0}$	specific viscosity $\eta_{sp} = \eta_r - 1$	Reduced viscosity $\eta_{red} = \eta_{sp}/C$
1.	Pure solvent	t_0			
2.					
3.					
4.					

CALCULATIONS

1. Plot a graph between η_{red} and concentration and extrapolate the graph to zero concentration.
2. Find out the value of intrinsic velocity $[\eta]$ from the graph.

   $[\eta]$ = Intercept

3. Now, we know, $[\eta] = KM^a$

Substitute the values of [η], K, and a and calculate M, the molecular weight of the polymer.

Using logarithm,

$$\log [\eta] = \log K + a \log M$$

$$\log M = \frac{\log[\eta] - \log K}{a}$$

$$M = \text{Antilog } \frac{[\log[\eta] - \log K]}{a}$$

RESULT The molecular weight of the given polymer (polystyrene) is . . .

PRECAUTIONS

1. All measurements should be made using a constant volume of the liquid
2. Viscometer should be kept perfectly vertical.
3. Mouth suction should not be done.
4. There should not be any air bubbles in the viscometer.

Table 12.1 Values of constants *'K'* and *'a'*

Polymer	Solvent	Temperature	'K × 10⁴'	'a'
Polystyrene	Benzene	25	1.02	0.74
		30	1.10	0.735
Polystyrene	Toluene	25	1.1	0.72
		30	1.1	0.725

EXERCISE

1. What is the significance of determination of molecular weight of a polymer?

Ans. *Many of the important properties (mechanical strength, softening temperature, melt viscosity etc.) depend on the molecular weight.*

2. How are the values of K and a in equation determined?

Ans. *Log form of equation*

$$\log [\eta] = \log K + a \log M$$

Thus, a plot of log [η] versus log M will be a straight line where.

Slope = *a*

Intercept = log *K*

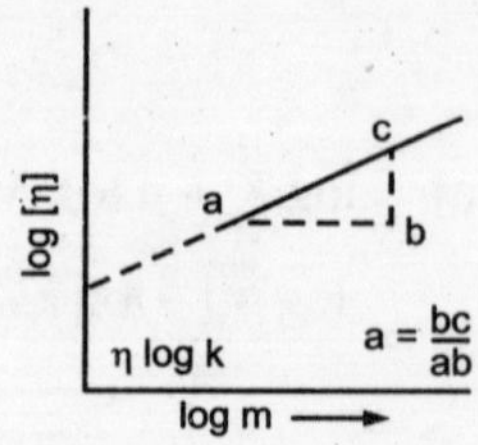

Fig. 12.4 Plot of log [η] versus log M

Thus the values of constants K and *a* can be determined by working with samples of the polymer of known molecular weights in the same solvent.

3. What is intrinsic viscosity? How is it obtained?

4. Define relative viscosity, specific viscosity and reduced viscosity.

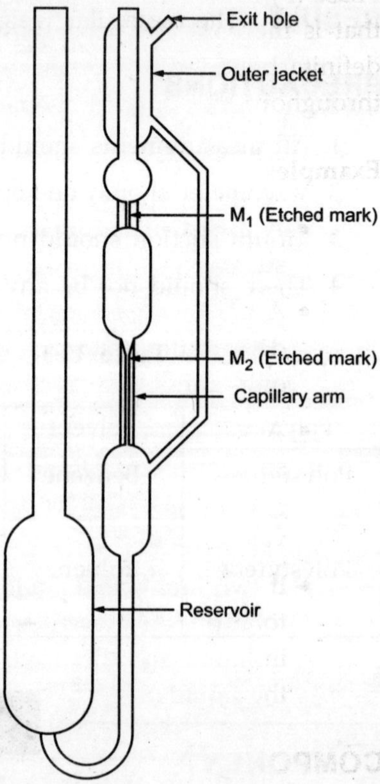

Fig. 12.5 *Viscometer ubbelohde*

Note. For accurate measurement of viscosity another type of viscometer known as ubbelholder viscometer is employed Fig. 12.5. It is an improvement over Ostwald's viscometer. It is less affected by incorrect alignment. Moreover as the bottom of the capillary is above the level of the liquid so the pressure head is independent of the volume of the liquid present in the reservoir.

□□□

13

Phase Equilibria

PHASE

Phase (P) is defined as a physically distinct and homogeneous part of the system that is mechanically separable from other parts of the system by means of a definite boundary. In otherwords, phase is a state of matter that is uniform throughout, not only in chemical composition but also in physical state.

Examples:

- A crystal of a substance is a single phase. But if the crystals of two substances are mixed then there exist two phases.
- A gas or a gaseous mixture consist of a single phase, since there is no demarcation (interface) between one gas and another. Air is in one phase only although it is a mixture of number of gases.
- A solution of two miscible liquids will exist in one phase only as far as solution is concerned. But the system as a whole will have two phases - one phase will be the solution (liquid) and the other would be the vapour above it.
- If two immiscible liquids are mixed together to form a solution, then the total phases present would be three. Two phases are for the substance in liquid and the third one for vapours, containing the vapours of both the liquids.

COMPONENT

The number of components (C) in a system is the smallest number of substances, in terms of which the composition of all the phases in the system can be described separately.

Examples:

- Phase equilibria containing ice, water and its vapour is *one* component system since the composition of each phase can be expressed in terms of only one component i.e. water.

- A mixture of water and ethanol is a *two* component system.
- The sulphur system consists of four phases viz., rhombic sulphur$_{(s)}$, monoclinic sulphur$_{(s)}$, liquid sulphur$_{(l)}$ and sulphur vapour$_{(g)}$. Each phase is chemically made up of sulphur only and hence it is *one* component system.
- Decomposition of $CaCO_3$: The thermal decomposition of calcium carbonate in a closed vessel takes place as follows

$$CaCO_{3(s)} \rightleftharpoons CaO_{(s)} + CO_{2\,(g)}$$

The system consists of three phases - solid $CaCO_3$, solid CaO and gaseous CO_2. Due to the existence of equilibrium, composition of each of the three phases can be expressed in terms of any two constituents $CaCO_3$, CaO and CO_2 as shown below:

For example if CaO and CO_2 are two constituents chosen, then the composition of different phases is represented as:

Phase $CaO = CaO + 0\ CO_2$

Phase $CO_2 = 0\ CaO + CO_2$

Phase $CaCO_3 = CaO + CO_2$

Thus it is a *two* component system because each of the three phases can be expressed by two components. Also, if $CaCO_3$ and CO_2 are chosen, then again the three phases can be expressed by two components.

Phase $CO_2 = 0\ CaCO_3 + CO_2$

Phase $CaO = CaCO_3 - CO_2$

Phase $CaCO_3 = CaCO_3 + 0\ CO_2$

- Dissociation of ammonium chloride in a closed vessel is expressed as,

$$NH_4\,Cl_{(s)} \rightleftharpoons NH_{3\,(g)} + HCl_{(g)}$$

The system consists of two phases, solid NH_4Cl and a gaseous mixture of NH_3, HCl and NH_4Cl. The composition of either of the two phases can be expressed in terms of one component as the proportions of NH_3 and HCl are equivalent i.e. both the phases have formal composition 'NH_4Cl'. However, if NH_3 or HCl is in excess, the system becomes a two component system.

If there are 'S' substances and 'R' relations (equations representing chemical equilibria) among them, then the number of components (C) can be calculated as-

$$C = S - R$$

For example, consider a solution of glucose in water. There are two substances and hence S = 2, but there is no relation between them so R = 0, therefore

$$C = S - R$$

or $$C = 2 - 0 = 2$$

Hence it is a two component system.

Now consider $NH_4Cl_{(s)}$ in equilibrium with its vapours containing NH_3 and HCl. Here S = 3 and R = 2 because there are two relations as shown,

1. $NH_4Cl_{(s)} \rightleftharpoons NH_{3\,(g)} + HCl_{(g)}$

2. Conc. of NH_3 = Conc. of HCl

Therefore C = S – R = 3 – 2 = 1. Thus it is *one* component system.

DEGREE OF FREEDOM

The degree of freedom (F) of a system is defined as the number of independent variables, such as temperature, pressure and composition which are to be specified to describe a system completely. It is also termed as variance.

- For a system consisting of Ice, water and vapours,

$$Ice_{(s)} \rightleftharpoons Water_{(l)} \rightleftharpoons Vapours_{(g)}$$

if all the three phases are present in equilibrium, then no condition has to be specified. This is because the three phases of water can exist together in equilibrium only at a particular temperature and pressure (corresponding to the freezing point) and no factor is necessary to be specified to define the system. Thus it is a non-variant system or invariant system i.e. F = 0.

- For a system consisting of water in contact with its vapour i.e.$_{(l)}$

$$Water_{(l)} \rightleftharpoons water\ vapour_{(g)}$$

Such a system can be completely defined by specifying either pressure or temperature. This is because on fixing one variable, the other becomes automatically fixed. Hence either temperature or pressure is required to specify the system. The system is therefore, univariant i.e. F = 1.

- For a system consisting of water vapours only, both temperature and pressure are to be specified in order to describe the system completely. Thus it is a bivariant system i.e. F = 2.
- For a pure gas, F = 2. This is because PV = RT, if the value of pressure (P) and temperature (T) are fixed, volume automatically gets fixed.
- A system of saturated NaCl solution has F = 1. Because if only one variable i.e. temperature is fixed, composition and vapour pressure get automatically fixed.

PHASE RULE

Willard Gibbs (1876) deduced a simple relationship among the number of phases in equilibrium, the number of components and degrees of freedom. This relationship is known as phase rule. According to phase rule, *in a heterogeneous equilibrium the number of degrees of freedom (F) of the system is defined by the number of components (C) and the number of phases (P) according to the equation,*

$$F = C - P + 2 \quad ...(i)$$

provided the equilibrium is not influenced by gravitational, electrical or magnetic forces but is only influenced by temperature, pressure and concentration.

In the above equation (i), F-stands for degrees of freedom, C stands for the number of components and P stands for number of phases and the numeral 2 stands for the effect of temperature and pressure.

Significance of phase rule. Phase rule predicts quantitatively the equilibrium existing between different phases of a heterogeneous systems i.e. the conditions that must be specified for a system to exhibit equilibrium.

Some important conclusions from the phase rule equation for a ***one component*** *system:*

- When the system consists of only one phase, we have

$$C = 1 \text{ and } P = 1$$

According to phase rule

$$F = C - P + 2 = 1 - 1 + 2 = 2$$

The system is bivarient i.e. only two variables are required in order to define the system completely. These two variables are temperature and pressure.

- When the system consists of two phases in equilibrium, we have

$$C = 1 \text{ and } P = 2$$

Hence $F = C - P + 2 = 1 - 2 + 2 = 1$

The system is univariant and only one variable is required to specify the state of the system. This means that the pressure is fixed automatically for a given temperature.

- When the system has three phases in equilibrium, we have

$$C = 1 \text{ and } P = 3$$

Hence $F = C - P + 2 = 1 - 3 + 2 = 0$

thus the system is non variant or invariant and no variables are required to specify the state of the system. This condition can be attained at a definite temperature and pressure.

APPLICATION OF PHASE RULE TO ONE COMPONENT SYSTEM

As is evident from the above examples, for a system having a specified number of components, greater the number of phases, smaller is the number of degrees of freedom or vice versa.

The minimum number of phases that can be possible in any system is one. According to the phase rule,

$$F = C - P + 2 = 1 - 1 + 2 = 2$$

Thus, a one component system requires maximum of two variables to be fixed in order to define the system completely. The two variables are temperature and pressure. Thus, a one component system can be studied by plotting the conditions of pressure against corresponding temperatures.

PHASE DIAGRAMS

A phase diagram may be defined as a plot showing the conditions of pressure and temperature under which two or more physical states can exist together in a state of dynamic equilibrium. Fig. 13.1 is a typical case for a one-component systems. The diagram consists of

(a) Areas

(b) Curves/Lines

(c) Triple point

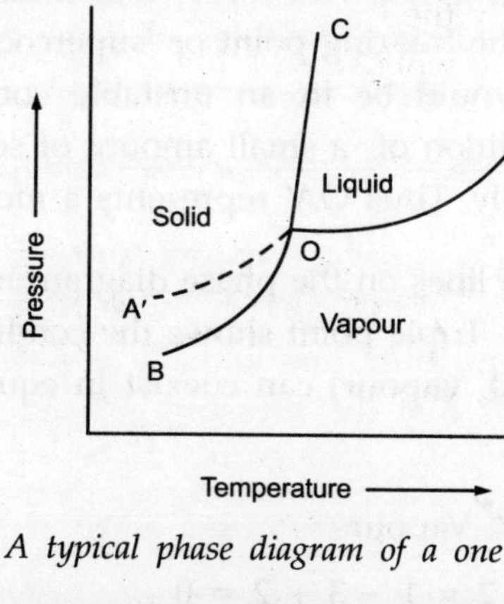

Fig. 13.1 *A typical phase diagram of a one component system.*

(a) **Areas**. The diagram has 3 regions or areas labelled as solid, liquid and vapour i.e. COB, COA and BOA respectively. Each of the three areas shows the conditions of temperature and pressure under which the respective phase can exit. When only one phase is present,

$$F = 1 - 1 + 2 = 2$$

Thus, each phase has two degrees of freedom; both temperature and pressure must be specified to define the state of the system.

(b) **Curves**. There are three lines or curves separating the areas. These show the conditions of equilibrium between any two of the three phases:

- Curve OA represents the equilibrium,

$$\text{Liquid} \rightleftharpoons \text{Vapour}$$

This is referred to as **vapour pressure curve**. Above the curve liquid is stable phase while below it, vapour is the stable phase.

- Curve OC represents the equilibrium

$$\text{Solid} \rightleftharpoons \text{Liquid}$$

This is referred to as **melting curve** or **fusion curve** or **freezing point curve** and indicates the influence of pressure in melting point of solid.

- Curve OB represents the equilibrium

$$\text{Solid} \rightleftharpoons \text{vapour}$$

This is referred to as **sublimation curve**.

- **Curve OA′, Metastable equilibrium:** The vapour pressure curve OA of liquid phase terminates at O, when the liquid is converted into solid. However it is possible to cool the liquid below the point O without solidification as shown by the curve OA′. This means that the liquid can be cooled far below the freezing point or 'supercooled' without solidification. This liquid would be in an unstable condition. On slightest disturbance or by addition of a small amount of seed crystal, the entire liquid solidifies rapidly. Thus OA′ represents a metastable equilibrium.

(c) **Triple Point**. The three lines on the phase diagram intersect at a common point called the **Triple Point**. Triple point shows the condition under which all the three phases (solid, liquid, vapour) can coexist in equilibrium. Thus triple point can be represented as,

$$\text{Solid} \rightleftharpoons \text{Liquid} \rightleftharpoons \text{Vapour}$$

$$F = C - P + 2 = 1 - 3 + 2 = 0$$

which predicts that system has no degrees of freedom. At triple point both pressure and temperature are fixed. If we try to change temperature or pressure the three phases would no longer co exist and one of the three phases will disappear.

APPLICATION OF PHASE RULE TO TWO COMPONENT SYSTEMS

When a single phase is present in two component system, the degree of freedom is three,

$$F = 2 - 1 + 2 = 3$$

Hence three variables must be specified in order to define the system completely. The three variables are pressure, temperature and concentration. We must have a phase diagram in three coordinate axes to understand such a system. For the sake of understanding, it is simplified as a two dimensional plot, keeping one of the three variables constant. In this case, each type of equilibria is studied separately. For example, for a solid/liquid equilibrium, the gas phase is usually absent and the effect of pressure on equilibrium is very small. So the effect of pressure can be discarded. Such a solid/liquid system with the gas phase absent is called **Condensed system.**

PHASE RULE FOR THE CONDENSED SYSTEM

Measurement on solid-liquid equilibria in condensed system is carried out at atmospheric pressure. Since the degree of freedom in such a case is reduced by one (as pressure is constant) the phase rule is modified for such cases as

$$F = C - P + 1 \qquad \text{...(ii)}$$

Equation (ii) is known as *condensed phase rule.* Now for two component systems equation (ii) becomes

$$F = 3 - P$$

when $P = 1$, then $F = 2$

Thus, it has two variables which are temperature and composition. Therefore solid-liquid equilibria are best represented by temperature composition diagram.

THERMAL ANALYSIS

A convenient method to study solid/liquid equilibria is through thermal analysis. In this method, solids of different compositions are separately heated above their melting points. The resulting liquids are cooled slowly and cooling curves are constructed by plotting temperature against time.

From the cooling curve for any mixture of a definite composition, it is possible to determine its

(i) Freezing point

(ii) Eutectic point

EUTECTIC POINT

The term eutectic implies 'easy melting'. A binary system consisting of two substances, which are miscible in all proportions in the liquid phase and do not react chemically, is known as **eutectic system**. The two substances involved have the property of lowering each other's freezing point. Further, solid solution of two or more substances having the **lowest freezing point** of all the possible mixtures of the components is known as **Eutectic mixture**. The minimum freezing point attainable corresponding to the eutectic mixture, is termed as the **eutectic point** (lowest melting point).

A mixture of lead and silver alloy, a mixture of bismuth-cadmium etc. are the few examples of eutectic mixtures. Eutectic mixtures have number of important applications including safety devices where low melting alloys can be used, fire-sprinklers, soldering etc.

ANALYSIS OF THE COOLING CURVES

1. **Pure substance**. When a pure substance is cooled slowly and the temperature is noted at definite intervals, a curve of the type shown in Fig. 13.2 is obtained. Point 'b' denotes the temperature at which the liquid starts solidifying or freezing. After this the temperature remains constant till whole of the liquid is completely solidified (point c). Thereafter, the fall in temperature will again become continuous as is denoted by the curve cd. This curve represents the cooling of the solid mixture.

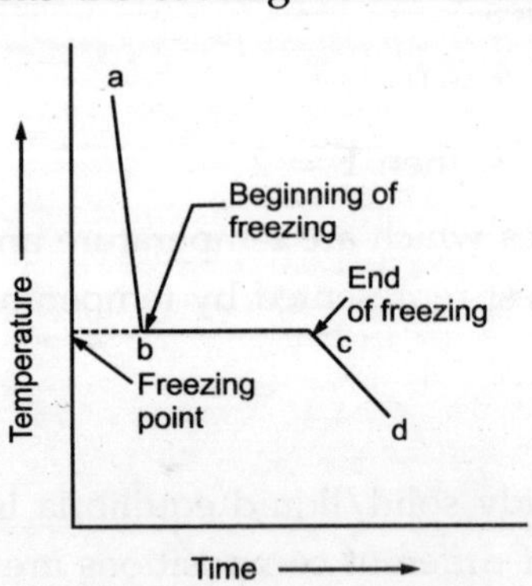

Fig. 13.2 *Cooling curve for pure substances*

2. **Mixture of two solids**. When the mixture of two liquids is cooled slowly and the cooling curves is constructed by plotting temperature against time, a cooling curve shown in Fig. 13.3 is obtained. Point b marks the beginning of freezing or appearance of solid. However the temperature does not remain constant, as in the case of pure substance. The temperature decreases continuously, but at a different rate (compare with cooling curve of a pure substance).

Actually at point b one of the components begins to solidify or crystallize. During this solidification process, heat is liberated and therefore, the temperature falls more slowly. From point b to c as the solidification proceeds, the composition of the mixture changes continuously. At a certain temperature T, known as eutectic temperature, the remaining liquid solidifies as a whole, i.e. both the components crystallize out separately. The temperature remains constant from point *c* to *d*. This is also known as *temperature halt.*

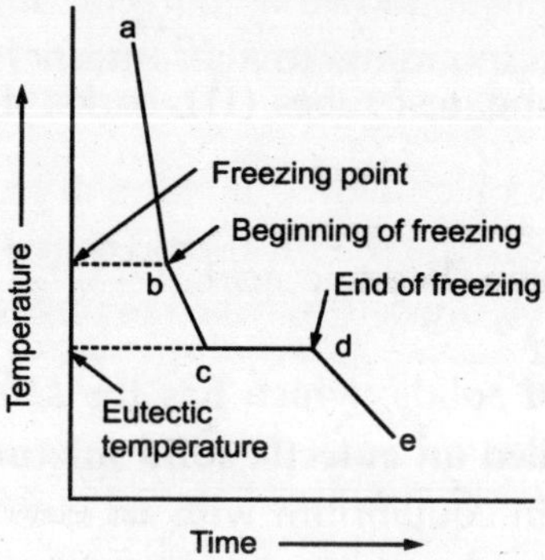

Fig. 13.3 *Cooling curve of a binary miscible mixture*

PHASE DIAGRAM FOR A TWO COMPONENT SYSTEM

In order to obtain a phase diagram a number of mixtures of two components A and B, having different compositions are prepared. Each mixture is heated to a liquid and the cooling curve is obtained for each mixture. From the cooling curve the freezing point (temperature at which the separation of solid phase just starts) and the temperature of complete halt (eutectic temperature) is determined. A graph is plotted between composition of the mixture and its freezing point. This is the phase diagram for this eutectic system. (Fig. 13.4) 'C' represents the eutectic temperature.

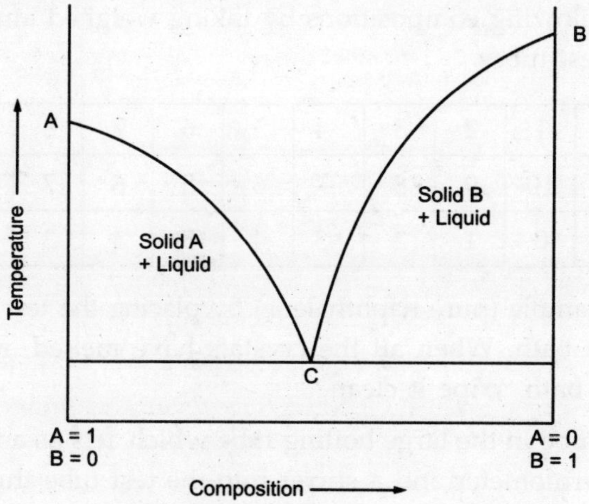

Fig. 13.4 *Simple eutectic system*

EXPERIMENT NO. 13.1

AIM **To construct a phase diagram for a binary system, naphthalene and benzoic acid and find the melting and eutectic temperatures.**

APPARATUS Boiling tube, test tubes (11), corks, thermometer (360°), stirrer, stop watch.

CHEMICALS Naphthalene, Benzoic acid.

THEORY The mixture of solids, which has the same overall composition as the liquid phase has, is called an **eutectic solid mixture**. The liquid phase of an identical composition and in equilibrium with an eutectic solid mixture, is called an **eutectic liquid mixture**. A eutectic liquid mixture has the lowest freezing point of all the possible mixtures of the components involved.

Liquid mixtures of naphthalene and benzoic acid of varying composition are allowed to cool. The temperature at which the separation of solid phase just starts (freezing point) and the temperature halt for various mixtures are noted. The temperature halt for all the mixtures will be the same because the composition of solid phase (eutectic) is same in all the cases. A graph is plotted between composition of the mixture and freezing point and eutectic point is noted.

PROCEDURE

1. Prepare the following compositions by taking weighed amounts in properly labelled test tubes.

Sample No.	1	2	3	4	5	6	7	8	9	10	11
Naphthalene	10	9	8	7	6	5	6	7	8	9	10
Benzoic acid	0	1	2	3	4	5	4	3	2	1	0

2. Melt the first sample (pure naphthalene) by placing the test tube in the hot liquid paraffin bath. When all the crystals have melted, remove the test tube from the bath. Wipe it clean.
3. Place the test tube in the large boiling tube which acts as an air jacket (Fig. 13.5). Fix a thermometer and a stirrer into the test tube through the cork.

4. Allow the temperature to fall slowly. The boiling tube can be clamped in a beaker containing water to ensure uniform cooling. Stir thoroughly and record the temperature after every half minute.
5. Note the temperature at which crystals begin to separate from the molten mass. Continue to record the temperature until it becomes constant and the whole mass solidifies.
6. Similarly, repeat the determination with other mixture.
7. Record the freezing point and temperature halt for every sample.

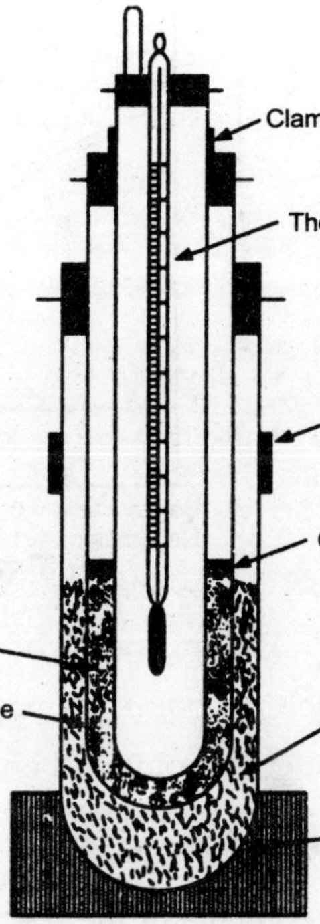

Fig 13.5

OBSERVATIONS

Sample no.	Naphthalene : Benzoic Acid	Freezing Point	Temperature Halt
1	10 : 0		
2	9 : 1		
3	8 : 2		
4	7 : 3		
5	6 : 4		
6	5 : 5		
7	4 : 6		
8	3 : 7		
9	2 : 8		
10	1 : 9		
11	0 : 10		

CALCULATIONS Plot a graph between percent composition of naphthalene and freezing temperature by taking percent composition on x-axis and temperature on y-axis. This will be the phase diagram for the above system. From the graph, find the freezing/melting temperature of the benzoic acid and naphthalene and eutectic temperature of the mixture.

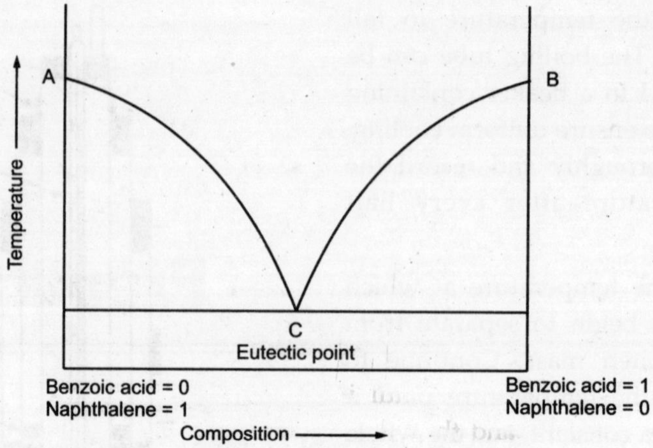

RESULT Melting point of benzoic acid =

Melting point of naphthalene =

Eutectic point of benzoic acid: naphthalene system =

PRECAUTIONS

1. The contents of the test tube are heated till all the crystals of the solid melt.

EXERCISE

1. State phase rule.
2. Define phase, component and degree of freedom.
3. How many components are present when NH_4Cl is heated in a closed vessel?
4. What is an invariant system?
5. What is meant by the term eutectic?
6. What is metastable state?
7. Give an application of eutectics?

Ans. In preparing very low melting alloys, which are used in safety devices, fire sprinklers etc.

8. What is condensed phase rule?
9. What is a phase diagram?
10. How the eutectic point of a binary system be determined?

Note. 1. Freezing point and temperature halt can also be obtained by plotting the cooling curves for various mixtures.

2. The binary mixture of any of the systems may be studied such as Benzoic acid and cinnamic acid, Naphthalene and B-naphthol, Acetamide and salicylic acid etc.

14

Preparations

EXPERIMENT NO. 14.1

AIM Preparation of aspirin.

APPARATUS Conical flasks (100 cm^3), water bath, beakers, glass rod.

CHEMICALS Salicylic acid, acetic anhydride, sulphuric acid, acetyl chloride, pyridine, alcohol.

THEORY Aspirin is an antipyretic, an analgesic and anti-inflammatory drug. It is extensively used

- in the treatment of arthritis as it is anti-inflammatory agent.
- as a medicine which may prevent heart attact by checking blood clotting in arteries.
- as a medicine for the treatment of headaches.

However it has certain side effects which include gastric problems (irritation of the mucous membrane), brain disorders in people below 18 years of age.

The chemical name of Aspirin is acetylsalicylic acid i.e. aspirin is acetyl derivative of salicylic acid. It is prepared by acetylation reaction of salicylic acid.

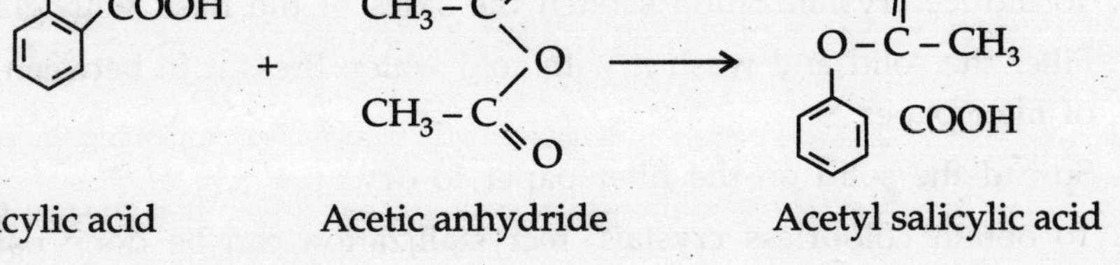

Salicylic acid　　Acetic anhydride　　Acetyl salicylic acid

Note. Acetylation is a reaction where a reactive hydrogen of hydroxy group (alcohols or phenols) or amino functional group is replaced by - $COCH_3$ group.

Acetylation reaction can be accomplished by a number of methods:

- Acetylation with acetic anhydride in presence of catalyst.
- Acetylation with acetyl chloride in presence of a base (pyridine)

MECHANISM OF ACETYLATION WITH ACETIC ANHYDRIDE IN PRESENCE OF CATALYST

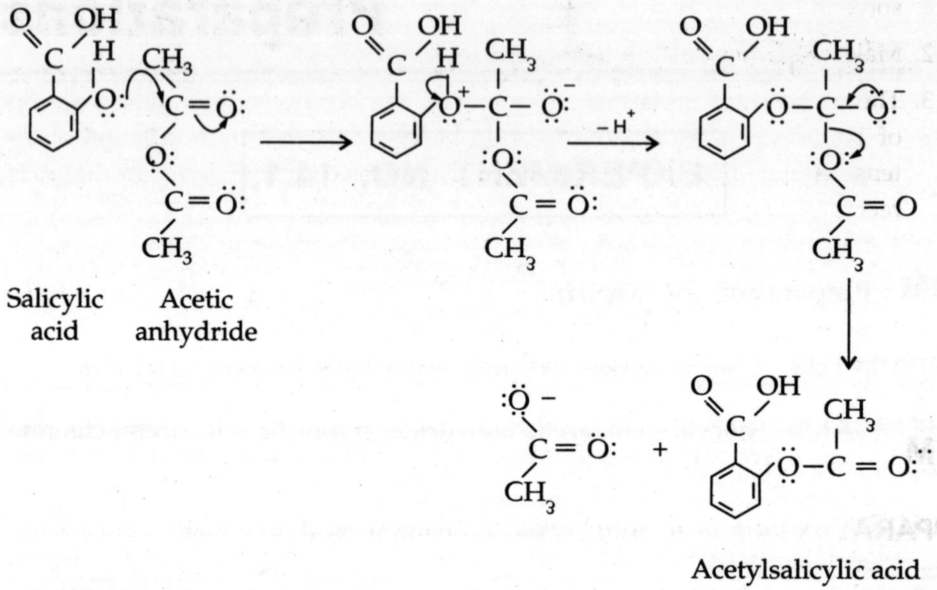

PROCEDURE

1. Take 5 g of salicylic acid in a 100 ml conical flask. To this add about 10 ml of acetic anhydride and 1-2 ml of conc. H_2SO_4.
2. Shake the contents thoroughly, the temperature will rise to 70°-80°C because of exothermic reaction. Maintain the temperature at 60°-70°C for about 15 minutes, keeping on a water bath.
3. Allow the solution to come to room temperature and then pour it in 100 ml cold water taken in a 500 ml beaker with stirring (water is added to destroy the excess acetic anhydride which gets converted to acetic acid)
4. To induce crystallization scratch the sides of the flask with glass rod.
5. Filter the solid and wash it with cold water. Press it in between the folds of filter paper.
6. Spread the solid on the filter paper to dry.
7. To obtain colourless crystals, recrystallization can be done using equal volumes of ethanol and water solvent system.
8. Determine the melting point of the recrystallized sample and report.

RESULT Yield = ...

The melting point of aspirin was found to be......

(M. pt of Aspirin = 130–135°C).

PRECAUTIONS

1. Take acetic anhydride in excess as it acts as acetylating agent as well as solvent.
2. Make sure that all the salicylic acid is dissolved.
3. The presence of unreacted salicylic acid can be checked by adding a drop of 1% ferric chloride solution and observing the colour. Formation of intense colour indicates the presence of unreacted salicylic acid. In that case recrystallization is must.

EXPERIMENT NO. 14.2

AIM **Preparation of Methyl Orange - An Azodye.**

APPARATUS Ice bath, beakers, 250 ml conical flask, test tube.

CHEMICALS N, N-Dimethyl aniline, sulphanilic acid, sodium nitrite, hydrochloric acid, acetic acid, sodium hydroxide, sodium chloride, sodium carbonate.

THEORY Dyes may be defined as the chemical substances which when applied to a fibre or a surface impart a relatively permanent colour to it.

A number of classes of dyes are known but azo dyes are the largest category of dyes. All the dyes which have one or more azo groups (– N = N –) in their structure are called azo dyes.

Preparation of an azo dye involves two stages :

(i) Diazotization

(ii) Coupling

(i) **Diazotization**. In the diazotisation step a primary aromatic amine reacts with nitrous acid (HONO) at a temperatures of 0 - 5°C to form a diazonium salt. The reaction is known as diazotisation. The nitric acid is generated in situ by sodium nitrite and a mineral acid like HCl.

In the preparation of methyl orange, sulphanilic acid is diazotized and coupled to N, N - dimethylaniline. The reaction of diazotization may be represented as follows:

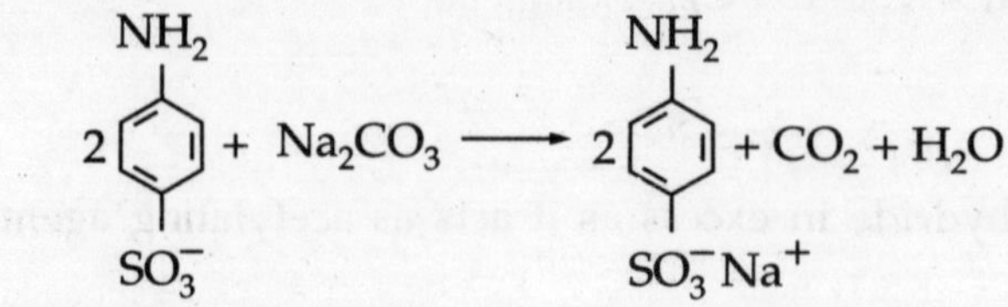

Sulphanilic acid

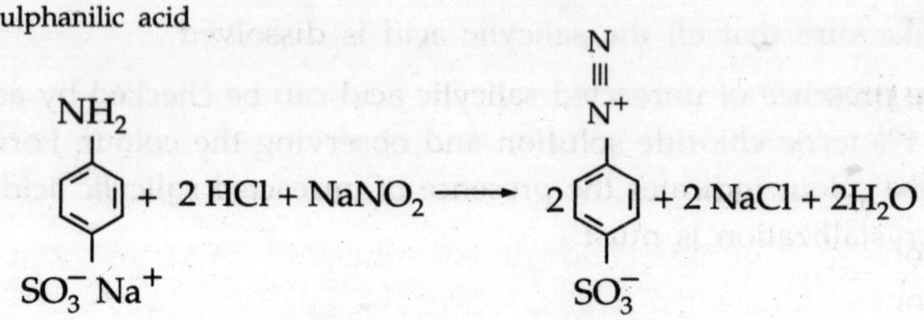

Diazotized sulphanilic acid

(ii) **Coupling.** Diazonium salts are soluble and reactive compounds. On condensation with aromatic amines in acidic medium, they produce highly coloured compounds (azo dyes). The reaction being known as coupling reaction. As diazonium salts are very unstable so the reaction is carried out at low temperature (below 5°C). These are not isolated from the solution and the coupling agent is added directly in diazonium salt solution.

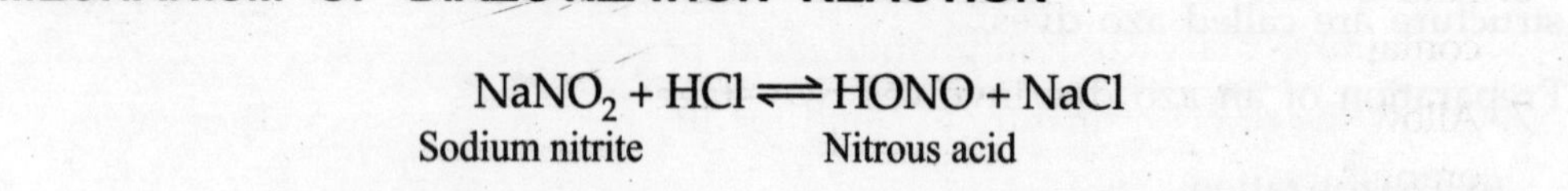

Diazotized sulphanilic acid — N,N- Dimethylaniline — Methyl orange

MECHANISM OF DIAZOTIZATION REACTION

$$\underset{\text{Sodium nitrite}}{NaNO_2} + HCl \rightleftharpoons \underset{\text{Nitrous acid}}{HONO} + NaCl$$

$$H_3O^+ + HONO \rightleftharpoons H_2O + H_2O^+\ddot{N}O \rightleftharpoons 2\,H_2O + :N^+ = 0$$

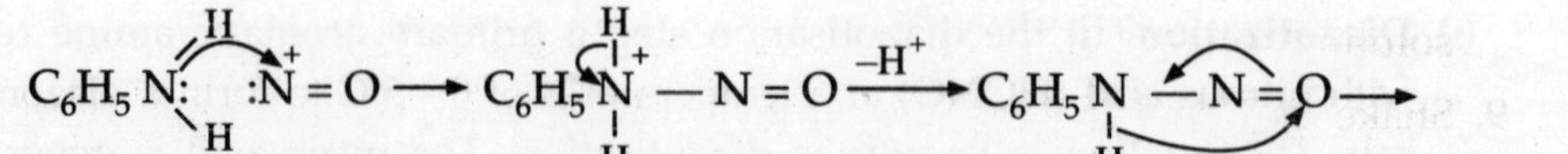

Primary aromatic amine

$$C_6H_5-\ddot{N}=\ddot{N}-OH \overset{H^+}{\rightleftharpoons} C_6H_5-\ddot{N}=\ddot{N}-\ddot{O}^+H_2$$

$$\overset{H_2O^+}{\rightleftharpoons} [C_6H_5-N^+\equiv N \leftrightarrow C_6H_5-\ddot{N}\equiv \ddot{N}^+.]$$

Diazonium ion

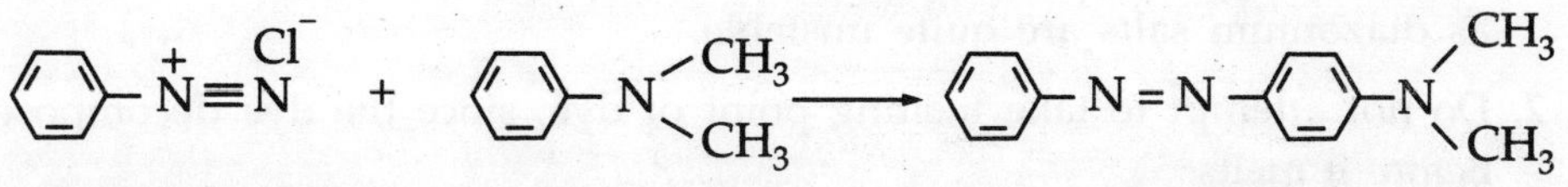

Benzene diazonium chloride | NN Dimethylaniline | A diazo compound

PROCEDURE

Diazotization

1. Dissolve 7 g of finely powdered sulphanilic acid and 2 g of sodium carbonate in 50 ml of water. Warm the mixture gently to obtain a clear solution.
2. Cool the flask under tap water and to it add a solution of 2.2 g of sodium nitrite in 10 ml of water.
3. Cool the flask in ice bath until the temperature falls to 5ºC. (Clean crushed ice can also be added in the flask). Now add very slowly, drop by drop, a solution of 8 ml of conc. hydrochloric acid in 15 ml of water, to the flask. Stir continuously. Do not allow the temperature to rise above 10º.
4. Allow the solution to stand in ice-water for 15 minutes to ensure complete diazotization. A solid material (4-diazobenzene sulphonic acid), starts separating slowly.

Coupling

5. Dissolve 4 ml of dimethylaniline in 3 ml of glacial acetic acid diluted with 10 ml of water and cool the solution in ice water.
6. Add the solution of dimethylaniline acetate, obtained in step 5, to the flask containing diazotized sulphanilic acid prepared in step 4.
7. Allow the reaction mixture to stand for 10 minutes when crystals of methyl orange start separating.
8. Add approximately 20 ml of 20% solution of sodium hydroxide to convert the dye into its yellow sodium salt. Add 5g of pure NaCl, and warm the solution to 60ºC.
9. Shake and cool the mixture in an ice bath, filter the dye under a very gentle pressure.
10. Wash with cold water, dry and recrystallize from boiling water. Weigh and report the yield.

RESULT Yield = ...g

PRECAUTIONS.

1. It is necessary to keep the temperature below 5°C in diazotization reaction as diazonium salts are quite unstable.
2. Do not attempt to take melting point of dye, since the dye decomposes before it melts.

USES OF METHYL ORANGE

- Used as a dye.
- Used as indicator for acid base titrations.

EXERCISE

1. Define 'Dye'. How are they different from coloured substances?

Ans. *Dyes may be defined as the chemical substances which when applied to a fibre or a surface impart a relatively permanent colour to it. These are distinct from coloured substances which by themselves may be coloured but may not be able to impart a colour to other things. For example, azobenzene is a coloured compound but is not a dye whereas p-aminoazobenzene is coloured and also acts as a dye.*

2. What is diazotization?
3. Why is it necessary to cool the reaction mixture to 5°C in diazotization reaction?
4. What is the mechanism of diazotization.
5. Why it is not advisable to take the melting point of the prepared dye.

EXPERIMENT NO. 14.3

AIM **Preparation of tetraammine copper (II) sulphate mono-hydrate $[Cu(NH_3)_4]\ SO_4 \cdot H_2O$.**

APPARATUS Beakers, filteration apparatus, buchner funnel, glass rod, measuring cylinder, tripod stand, watch glass, water bath, filter paper.

CHEMICALS Copper (II) sulphate, ammonia, ethanol, distilled water.

THEORY When aqueous ammonia is added to a solution of copper sulphate, cupric hydroxide is precipitated out. Cupric hydroxide dissolves in excess of aqueous ammonia solution giving a deep blue solution containing the complex, tetraammine copper(II) sulphate. The complex is precipitated from the solution by adding ethyl alcohol since it is insoluble in alcohol.

$$2Cu^{2+}(aq) + SO_4^{2-}(aq) + 2NH_3(aq) + 2H_2O(l) \rightarrow Cu(OH)_2\cdot CuSO_4(s) + 2NH_4^+(aq)$$

$$Cu(OH)_2\cdot CuSO_4(s) + 8NH_3(aq) \rightarrow 2[Cu(NH_3)_4]^{2+}(aq) + 2OH^-(aq) + SO_4^{2-}(aq)$$

$$[Cu(NH_3)_4]^{2+}(aq) + SO_4^{2-}(aq) \xrightarrow{\text{ethanol}} [Cu(NH_3)_4]SO_4\cdot H_2O(s)$$

PROCEDURE

1. Weigh out 2.0 g copper(II) sulphate and powder it. Add powdered crystals to a 250 ml beaker.
2. Prepare ammonia solution by adding 10 ml ammonia(aq) to 5 ml distilled water
3. Slowly add ammonia solution to the powder with stirring until all the copper(II) sulphate dissolves resulting in a deep blue solution. If turbidity persists, add more of NH_3 solution.
4. Add ethanol dropwise and with stirring till a deep blue precipitate is formed.
5. Heat the beaker on a water bath at 60°C, stir to dissolve the blue precipitate. Cover the beaker with a watch glass and set aside to crystallize.
6. Dark blue crystals separate after 1 hour. Filter off the crystals using a buchner funnel and wash with 2-3 ml ethanol.
7. Transfer the crystals to a watch glass and dry in a desiccator.
8. Weigh the dry crystals.

RESULT Yield = ...g

PRECAUTIONS

1. Always add ammonia in excess so that a clear blue solution is obtained.

EXERCISE

1. What do you mean by complex compound?
2. What is the difference between double salt and complex compound.
3. What is the coordination number of copper in tetrammine copper(II) sulphate?
4. What is the oxidation number of copper in the above complex.
5. What is the structure of tetrammine copper(II) sulphate?

❑❑❑

15

Analysis of Coal

Coal is regarded as a fossil fuel produced from the vegetable debris under conditions of high temperature and pressure over million of years. Chemically, coal is highly complex organic matter having varying amount of water, together with nitrogen and sulphur.

The transformation of the vegetable debris to coal takes place in two stages:

(i) **Biochemical or peat stage:** During this stage, the plant materials were at tacked by various micro organisms. The end product of this stage is peat

$$2\ (C_6H_{10}C_5)_n \rightarrow \text{Peat}$$

(ii) **Chemical stage or metamorphism** In this stage, the peat deposit buried under sedimentary deposits lose moisture and volatile components under the effect of high temperature and pressure. The peat gets enriched in carbon whereas its oxygen content decreases. The spongy peat transforms into hard brittle coal gradually. The time required for the formation of young brown coal is of the order of 10^7 years while that of most nature coals is 3×10^8 years.

CLASSIFICATION OF COAL

Coals are mainly classified on the basis of their rank, which is defined as the degree of coalification from the parent material, wood. When wood is converted into coal, there is gradual increase in the concentration of carbon and decrease in the percentage of oxygen and nitrogen. The coals are given a ranking depending upon the carbon content of the coal from wood to anthracite.

The ranking can be represented as in table 15.1.

Table. 15.1 Classification of coal according to rank

Type of coal	Percentage (dry, mineral matter free basis)					% moisture	Calorific value
	C	H	O	N	VM		cal g^{-1}
Wood	40-50	5-6	20-40	0-0.5	-	70-90	4000-4500
↓							
Peat	45-60	3.5-6.5	20-45	0.75-3	45-75	70-90	4125-5280
↓							
Brown coal	60-75	4.5-5.5	17-35	0.75-2	45-60	30-50	6600-7100
↓							
Bituminous Coal	75-90	4.0-5.5	20-30	0.75-2	11-50	10-20	6600-8800
↓							
Anthracite	90-95	3-4	2-3	0.5-2	3.8-10	1.5-3.5	8470-8800

Thus it is clear from the table that whereas carbon content and calorific value of coal increases with degree of coalification, hydrogen, oxygen, nitrogen, volatile matter, moisture decreases on increase in degree of coalification.

ANALYSIS OF COAL

The composition of coal varies according to the degree of coalification. In order to compare the coals for the use of consumer, the analysis of coal is carried out. The detailed methods of analysis and testing vary from country to country. Coal is analysed in two ways:

1. Proximate Analysis
2. Ultimate Analysis

The results of analysis are generally reported in the following ways:

- As received basis
- Air dried basis
- Moisture free basis (oven dried)
- Moisture and ash free basis

PROXIMATE ANALYSIS

Proximate analysis is an empirical but important analysis. The data varies with the procedure adopted and hence it is called proximate analysis. It gives information about the practical utility of coal. Proximate analysis of coal determines the moisture, ash, volatile matter and fixed carbon of coal.

ULTIMATE ANALYSIS

Ultimate analysis of coal is carried out to ascertain the composition of coal. For the utilization of coal for industrial purpose, it is necessary to determine the percentages of various components so as to classify the coal, to fix its price, along with putting the coal to different uses depending upon its composition. Ultimate analysis includes the estimation of carbon, hydrogen, sulphur, nitrogen and oxygen.

EXPERIMENT NO. 15.1

AIM **To determine moisture, volatile, fixed carbon and ash contents in a given coal sample by proximate analysis.**

APPARATUS Electrical or Analytical balance, electric oven, dessiccator, silica crucible with lid, tong.

Materials. Given coal sample (powdered).

(i) **Moisture Content:**

Owing to its nature and origin, coal is always associated with moisture. When wet coal is exposed to atmosphere, the external moisture evaporates but the apparently dry coal still contains some moisture. This moisture is referred to as air dried moisture.

Significance.

- Excess of moisture is undesirable in coal.
- Moisture lowers the heating value of coal and takes away appreciable amount of the liberated heat in the form of latent heat of vapourisation. For every percent of moisture present .1% of heat is lost.
- Moisture also increases the transport costs.
- Excessive surface moisture may cause difficulty in handling the coal.
- Presence of excessive moisture quenches fire in the furnace.
- Hence lesser the moisture content, better is the quality of coal as a fuel. However presence of about 5-10% moisture is desirable as it produces a uniform fuel bed and less of "fly ash".

(ii) **Volatile Matter:**

The volatile matter in coal consists of a complex mixture of gaseous and liquid products resulting from the thermal decomposition of the coal. Volatile matter does not include moisture of coal. It consists mainly of combustible gases such as H_2, CO, CH_4 and other hydrocarbons.

Significance.

- A high percent of volatile matter indicates that a large proportion of fuel is burnt as a gas or vapour or may escape unburnt.
- The high volatile matter content give long flames, high smoke and relatively low heating values.
- For efficient use of fuel, the outgoing combustible gases (volatile matter) have to be burnt by supplying secondary air. This requires a larger combustion space. A furnace with a small combustion volume or of short flame is not suitable for burning high volatile coals, since a large proportion of volatile matter will escape unburnt.
- The high volatile coals do not cake well whereas medium-volatile coals containing 20-30% of volatile matter are capable of forming hard and strong coke on carbonization.
- Coals containing less than 14% of volatile matter do not cake at all and are thus not suitable for manufacturing coke.
- However, high volatile matter content is desirable in coal gas manufacture because volatile matter in a coal denotes the proportion of the coal which will be converted into gas and tar products by heat.

(iii) **Ash.**

Coal contains inorganic mineral substances which are converted into ash by chemical reactions during the combustion of coal. Ash usually consists of silica, alumina, iron oxide and small quantities of lime, magnesia etc.

Ash can be classified as intrinsic ash and extrinsic ash. The mineral matter originally present in vegetable matter from which the coal was formed is called **intrinsic ash**. It consists of oxides of Na, K, Mg, Ca and Si. On the other hand the mineral matter, like clay, gypsum, dirt which gets mixed up during mining and handling of coal constitute the **extrinsic ash** which remains as a residue after the combustion. The extrinsic ash may consist of anhydrous $CaSO_4$, $CaCO_3$, Fe_2O_3 etc.

The minerals responsible for the formation of intrinsic ash cannot be removed by any mechanical means. Fortunately, this mineral matter content is low (about 2 to 3 percent) whereas the extrinsic ash content is relatively higher.

Significance.

- The high percentage of ash is undesirable.
- The ash does not contribute to the calorific value of coal. On the contrary it reduces the calorific value and creates many difficulties in efficient utilization of coal.
- In furnance grate, the ash may restrict the passage of air and lower the rate of combustion.
- High ash leads to large heat losses and cause problems of clinkering (forming of ash lumps).
- It has been estimated that one percent rise in coal ash is equivalent to 0.3 to 4 percent decrease in boiler efficiency.
- The composition of ash and fusion range also influences the efficiency of coal. Its composition is of considerable importance in metallurgical operations as it affects the slag and metal composition. When coal is used in boiler, the fusion temperature of ash is very significant. Generally fusion temperature lies in the range of 1000°-1700° C. Ash having fusion temperature below 1200°C is called *fusible ash* and above 1430°C is called *refractory ash*. If ash fuses at working temperature it leads to clinker formation. Apart from loss of efficiency of coal, clinker formation also leads to loss of fuel because some coal particles also get embedded in the clinkers. However, some ash is desirable since it protects the grate from direct contact with the incandescent coal, which might cause oxidation of the grate bars.

(iv) **Fixed Carbon.**

Fixed carbon content increases from low ranking coals such as lignite to high ranking coals such as anthracite. Higher the percentage of fixed carbon greater is its calorific value and better is the quality of coal. This represents the quantity of carbon that can be burnt by a primary current of air.

The percentage of fixed carbon is given by:

Percentage of fixed carbon = 100 – [% of Moisture + Volatile matter + ash]

From the above it appears that, what is left after the amount of moisture volatile matter and ash formed is subtracted, corresponds to the fixed carbon. However, the term fixed carbon is misnomer because it contains other non volatile elements also along with carbon.

Significance.

- Higher the percentage of fixed carbon, greater is the calorific value.
- The percentage of fixed carbon helps in designing the furnance and shape of the fire-box because it is the fixed carbon that burns in the solid state.

PROCEDURE

(I) Determination of Moisture (air dried):

1. Dry the coal (powdered) in air or in the moisture oven at 10 to 15°C above the room temperature.
2. Weigh silica crucible in an analytical or electrical balance accurately.
3. Transfer 1 g of air dried coal sample to the crucible and weigh it.
4. Keep the crucible in an electric oven maintained at 105 to 110°C for an hour.
5. Remove the crucible with a pair of tongs and keep it in a desiccator for cooling.
6. After it is cooled, weigh it. Note down the observations.

(II) Determination of volatile Matter

1. Take the dried sample of moisture free coal (obtained from step I) in a weighed and covered platinum crucible. The cover should have vent for escape of volatile matter.
2. Place the crucible in a muffle furnace maintained at 950 ± 20°C and heat it for exactly 7 minutes.
3. Remove the crucible with the help of tongs very carefully and bring down its temperature rapidly by placing it on cold iron plate.
4. Place the crucible in a desiccator so that it attains the room temperature.
5. Weigh it and note down the reading.

(III) Determination of Ash Content.

1. The residual coal (after the step II) is kept in a weighed silica or platinum crucible.
2. The crucible is heated in a muffle furnace maintained at 725 ± 25°C for nearly half an hour, without cover. Occasionally stir the residue with platinum wire.
3. The crucible is then cooled in a desiccator and weighed.

(IV) Determination of Fixed Carbon.

The fixed carbon is calculated as:

% Fixed carbon = [100 – (% Moisture + % Ash + Volatile matter)]

OBSERVATIONS AND CALCULATIONS

1. **Moisture**

Weight of empty crucible = a g.

Weight of crucible + coal sample = b g.

Weight of crucible + coal sample after heating = c g.

$$\% \text{ Moisture} = \frac{b-c}{b-a} \times 100$$

2. **Volatile Matter**

Weight of empty crucible = a g

Weight of crucible + sample (moisture free) = c g

Weight of crucible after the removal of volatile matter = d g

$$\% \text{ Volatile matter (on oven dried basis)} = \frac{c - d}{c - a} \times 100$$

$$\% \text{ Volatile Matter (on air dried basis)} = \frac{c - d}{c - a} \times 100$$

3. **Ash Content**

Weight of empty crucible = a g

Weight of empty crucible + coal sample = b g.

Weight of empty crucible + ash = e g.

$$\% \text{ Ash} = \frac{e - a}{b - a} \times 100$$

4. **Fixed Carbon**

100 – [% Moisture + % Volatile matter + % Ash].

RESULT The proximate analysis of the coal is as follows:

Moisture = ...%, Volatile matter = ...%, Ash = ...% and fixed carbon = ...%.

PRECAUTIONS

1. After air drying the coal the moisture analysis should be done speedly so as to prevent moisture change.

2. In the determination of volatile matter, the temperature should be raised at a slow rate.
3. If black particles appear during the determination of ash content due to the presence of unburnt carbon, continue ignition for another 15 minutes.

EXERCISE

1. Why is proximate analysis so called?
2. What is the significance of determination of moisture content, volatile matter, ash content.
3. What is fixed carbon. How is it measured.

❑❑❑

16

Lubricants

FRICTIONAL RESISTANCE

When the surface of a metal is allowed to move tangentially with respect to another metal, with which it is in contact, it experiences a resistance to its motion. This resistance is known as frictional resistance.

Actually, all the material surfaces show many irregularities in the form of peaks and valleys (of the order of 1000 A°). When the two metal surfaces come in contact with each other, the area of contact is only a small fraction of the apparent contact area, because the contact is through the peaks as shown in Fig. 16.1.

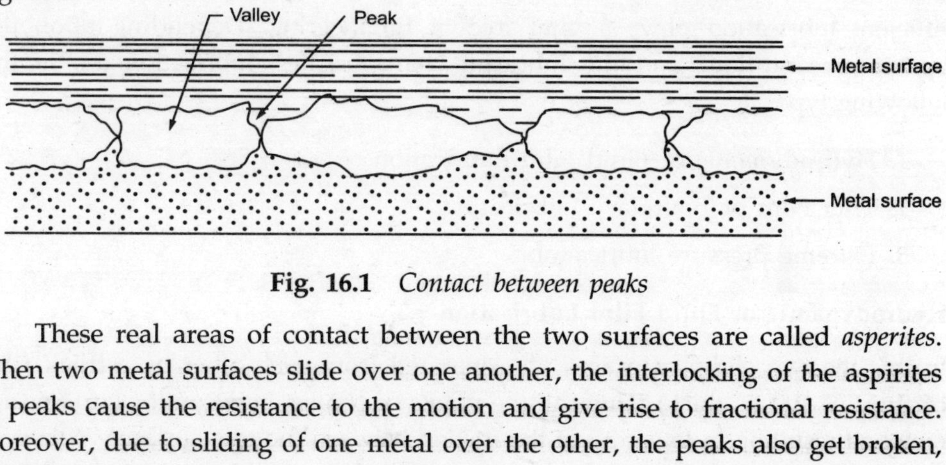

Fig. 16.1 *Contact between peaks*

These real areas of contact between the two surfaces are called *asperites.* When two metal surfaces slide over one another, the interlocking of the aspirites or peaks cause the resistance to the motion and give rise to fractional resistance. Moreover, due to sliding of one metal over the other, the peaks also get broken, resulting in wear and tear of the surfaces.

Further, the local pressure at the asperites may be sufficiently high to cause fusion of the peaks which lead to the formation of welded junction between asperites. Additional force is then required to break these junctions. This further adds to frictional resistance.

Frictional resistance results in loss of energy, a lot of wear and tear of surfaces of moving parts thereby causing loss in efficiency of machine. Sometimes, the moving parts get heated up and may also cause welding of two surfaces due to heat.

LUBRICANTS

In order to reduce frictional resistance and wear and tear of machine parts, a layer of fluid is introduced in between the moving parts. This layer of fluid is known as lubricant. The lubricant prevent direct contact of the rubbing surfaces by forming a film in between. Lubricant also act as a coolant by removing the frictional heat generated due to rubing of the surfaces.

FUNCTIONS OF LUBRICANT

- It reduces the frictional resistance.
- It protects against wear and tear.
- It takes away heat generated due to rubbing of the surface.
- It provides protection against corrosion.
- It prevents formation of welded junctions between asperites.
- It reduces the maintenance and running cost of a machine.

TYPES OF LUBRICATION

Efficient lubrication plays a vital role in Engineering. Depending upon the operating conditions and the lubricant characteristics, lubrication are of the following types:

1. Hydrodynamic or Fluid Film Lubrication.
2. Thin Film or Boundary Lubrication.
3. Extreme pressure lubrication.

Hydrodynamic or Fluid Film Lubrication

In this the two sliding surfaces are separated from each other by a **thick film of fluid** (~ 100 A° thick). Thus, all the points of contact between the surfaces are removed resulting in lowering of the friction between the two surfaces. Whatever friction is there after lubrication is due to the internal resistance between the particles of lubricant. Thus friction in such cases will depend on the viscosity, thickness of the lubricant, relative area and velocity of the sliding parts. This type of lubrication is employed in light machines like watches, clocks, guns, sewing machines, scientific instruments etc.

The most satisfactory and widely used lubricants for this type of lubrication are hydrocarbon oils which are blended with long chain polymers to maintain the required viscosity.

Thin film or boundary lubrication

In some cases under the conditions of high load, low speed or low viscosity of the lubricant, the lubricant tends to be squeezed out of position or a continuous film of lubricant cannot persist between the two surfaces. In such cases, lubrication is maintained by a very thin film of the lubricant adsorbed on the surface by physical or chemical forces or both. These adsorbed layers avoid direct metal to metal contact. Though, the moving surfaces are not completely separated but then points of contact are considerably decreased resulting in lowering of frictional resistance.

The lubricant used in such type of lubrication should have long chain and active atoms or groups capable of forming linkages with the metals on other surfaces. The boundary lubrication does not depend on the viscosity of the lubricant but depends on the load, the oiliness of lubricant (discussed later) and the nature of the underlying surface.

Vegetable and animal oils can be used for boundry lubrication but they tend to break down at high temperatures. Thus the mineral oils, which can withstand high temperatures are used but a small amount of fatty oils or acids are added to improve the oiliness.

Extreme-pressure lubrications

If the sliding surfaces are subjected to very high pressure and speed, excessive frictional heat is generated. Under such conditions, the liquid lubricants fail to stick and may decompose and even vaporize. In order to provide effective lubrication under these conditions, special additives called "extreme pressure additives" are used along with the lubricants. The active chemicals used are compounds of chlorine such as chlorinated esters, sulphur such as sulphurized fats and oils, and phosphorus such as tricresol phosphate. At prevailing high temperatures, these additives form solid surface films of chlorides, sulphides or phosphides as the case may be. These metallic compounds possess high melting points and serve as good lubricants under the conditions of extreme pressure and temperature.

CLASSIFICATION OF LUBRICANTS

A wide variety of materials have been used for lubrication purposes. Lubricants may be broadly classified as follows:

1. **Solid lubricants**. The most important solid lubricants are graphite and molybdenum disulphide. Soap stone, talc, mica etc. have also been used as solid lubricants.
2. **Semi-solid lubricants**. Greases, vaselines etc. constitute semi solid lubricants Greases are obtained by thickening a lubricating oil through the addition of a metallic soap called gelling agent. The gelling agent is usually a calcium, sodium or lithium soap.

3. **Liquid-lubricants**. Most of the liquid lubricants are of petroleum origin. Others include vegetable oils (olive oil, palm oil, castor oil etc.) animal oils (whale oil etc.), synthetic oils (silicones etc.)
4. **Emulsions**. Two types of emulsions are generally used (i) oil in water type or (ii) water in oil type.

PROPERTIES OF LUBRICANTS

The selection of the lubricant for a particular job depends on a number of factors including design of machine, operating conditions of the machine (temperature, pressure, speed of machine, etc.), and its physical and chemical characteristics. The physical and chemical properties such as viscosity, specific gravity, flash point, volatility, acid value, iodine value etc. measured in the laboratory do not indicate that how a lubricant will behave under operating conditions. Thus, it becomes imperative that these properties measured in the laboratory should be correlated to the conditions when the machine would be used. Some of the important properties of lubricants which would help an engineer to make a correct choice of the lubricant are as follows:

VISCOSITY

Viscosity is one of the most important properties of a lubricating oil. Viscosity of the lubricating oil determines its performance under operating conditions.

- If the viscosity of the liquid is too low, the fluid lubricant film cannot be maintained between the moving surfaces, as a result of which excessive wear may take place.
- If the viscosity of the lubricating oil is too high, excessive friction will take place.

Thus, in thick film lubrication the lubricant selected should be such that the selected lubricant possess a sufficiently high viscosity so that it adheres to the surface and do not squeeze out under operating conditions, yet the viscosity of the lubricant should not be too high otherwise, it will cause excessive friction.

CHANGE OF VISCOSITY WITH TEMPERATURE

Viscosity of liquids decreases with increasing temperature so that lubricating oil becomes thinner as the operating temperature is raised. It is therefore necessary that viscosity of the lubricating oil do not change much on changing temperature. Since the viscosity decreases with temperature it is impossible to select an oil having same viscosity over wide range of temperature. However, an oil whose variation with temperature is minimum can be selected.

The rate at which viscosity of a lubricant changes with temperature is measured by an arbitrary scale known as the **viscosity index** (V.I.). Viscosity index is the numerical expression of the average slope of the viscosity-temperature curve of a lubricating oil between 100°F to 210°F.

DETERMINATION OF VISCOSITY-INDEX

For determining viscosity-index, two types of references are used. One of these is **Pennysylvanian oils**. This Pennysylvanian oils consist mainly of paraffins and show a small decrease in the viscosity with increase in temperature. These oils are arbitrary given a value of 100 for V.I. The other reference taken is **naphthanic-base Gulf oils**. These consist of alicyclic hydrocarbons or naphthenes and exhibit a relatively large variation in viscosity with temperature. These are arbitrarily assigned a V.I. of O.

V.I. = 100 Pennysylvanian oils
V.I. = Zero Naphthanic-base gulf oils

For the determination of viscosity index, first the viscosity of the oil under test is determined at 100°F and 210°F let it be U and V respectively.

- Now, from the list of pennysylvanian oils, the oil having the same viscosity at 210°F as the oil under test is selected and its corresponding viscosity at 100°F is noted down. Let it be V_H.
- Similarly from the list of naphthanic base Gulf oils, the oil which has the same viscosity at 210°F as the oil under-test is selected, and its corresponding value at 100°F is noted down. Let it be V_L

$$\text{Then viscosity Index} = \frac{V_L - U}{V_L - V_H}$$

where

V_L = Viscosity at 100°F of gulf oil standard which has the same viscosity at 210°F as that of oil under test.

U = Viscosity of oil under test at 100°F

V_H = Viscosity at 100° of pennysylvanian oil standard which has the same viscosity at 210°F as that of oil under test.

Significance. Higher the V.I., lower the variation of viscosity with change in temperature.

EXPERIMENT NO. 16.1

AIM **Determination of viscosity of heavy oil by means of Redwood Viscometer (No. 1 and 2).**

APPARATUS Redwood viscometer No. 1 and No. 2, stop watch.

THEORY As is well known that viscosity, also known as coefficient of viscosity η is the tangential force per unit area required to move one horizontal plane of the fluid at unit velocity with respect to another maintained at a unit distance apart by the fluid. The units of viscosity are poise or centipoise*.

Viscosity of liquids in general is determined using Poiseuille's equation (please refer to the chapter-Viscosity). The viscosity of a lubricating oil is determined by measuring the time taken for a given quantity of oil to flow through an orifice of standard dimensions. The viscometers commonly used are: Engler's Viscometer, Saybolt Viscometer and Redwood Viscometer. Here, we will be using Red wood Viscometer and hence its description is given below:

REDWOOD VISCOMETERS

There are two types of Redwood Viscometers i.e. Redwood viscometer No. 1 and Redwood Viscometer No. 2. The two viscometers work on same principle, and have same shape and method of testing. The only difference lies in the dimensions of the discharge capillary (jet).

Redwood viscometer No. 1 is commonly used for determining the viscosities of thin lubricating oils (kerosene, mustard oil etc.) as it has smaller jet diameter of 1.62 mm with a length 10 mm. Redwood viscometer No. 2 is used for determining the viscosities of highly viscous oils (mobile oil, glycerol etc.) and has jet diameter of 3.8 mm and length 50 mm.

Viscometer	Diameter	Length of the Jet
RW 1	1.62 mm	10 mm
RW 2	3.8 mm	50 mm

Redwood viscometer consists of the following parts:

(i) **Oil cup:** Oil cup is a silver-plated brass cylinder having height of 90 mm and a diameter of 46.5 mm. The bottom of the cup is provided with a

* **Note** For the definition and units of viscosity please refer to the chapter 'Viscosity'.

central hole in which an agate jet with bore diameter 1.62 mm and length 10 mm is fitted. This jet can be opened or closed by a valve rod. The valve rod comprises of a small silver plated brass ball fixed to a stout wire. To indicate the level to which the cylinder is to be filled with oil, a pointer is fixed on the inner side of the cylinder. The pointer is a stout, tapered, upward-pointing wire. A thermometer is inserted in the cup through the lid of the cup.

(ii) **Heating bath:** A cylindrical copper bath containing water surrounds the oil cup. It is provided with a tap for emptying the water and has a long side tube projecting outwards, for heating the water by means of a gas burner. A thermometer is also inserted in the copper bath to know the temperature of water. Heating bath is also provided with a stirrer for stirring the water to keep the temperature of water uniform throughout. The stirrer is provided with a circular shield to prevent any water splashing into the cylinder.

(iii) **Spirit Level:** A spirit level used for levelling the apparatus vertically is also provided in the lid of the cup. The entire apparatus is supported on three legs, provided at the bottom with levelling screws.

(iv) **Kohlrausch flask.** A specially shaped flask known as Kohlrauch flask is provided to receive the oil from the outlet. The flask is of 50 ml capacity.

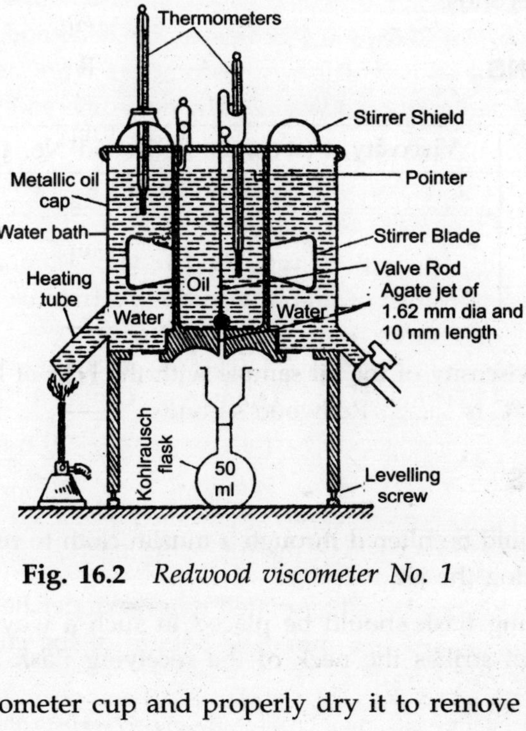

Fig. 16.2 *Redwood viscometer No. 1*

PROCEDURE

1. Clean the viscometer cup and properly dry it to remove any trace of the solvent.

2. Level the viscometer with the help of levelling screws.
3. Fill the outer bath with water for determining the viscosity at 80°C and below. For higher temperatures, the bath is filled with an oil having suitable viscosity at the test temperature.
4. Place the ball valve on the jet to close it and pour the test oil into the cup upto the tip of the indicator.
5. Place a clean dry Kohlrausch flask immediately below and directly in line with discharging jet.
6. Insert a thermometer and a stirrer in the cup and cover it with a lid.
7. Heat the water filled in the bath slowly with constant stirring. When the oil in the cup attains a desired temperature stop the heating.
8. Lift the ball valve and start the stop watch. Oil from the jet flows into the flask.
9. Stop the stopwatch when lower meniscus of the oil reaches the 50 ml mark on the neck of the receiving flask. (Replace the ball valve in position immediately so as to prevent overflow).
10. Record the time taken for 50 ml of the oil to collect in the flask.
11. Repeat the experiment to get the reproducible results and report the mean value in seconds.

OBSERVATIONS

S.No.	Viscosity, t (seconds, Redwood No. 1)
1.	
2.	
3.	Mean = ...s

RESULT The viscosity of the oil sample with the help of Redwood viscometer No. 1 at ....°C is ....Redwood seconds.

PRECAUTIONS

1. The oil should be filtered through a muslin cloth to remove solid particles that may clog the jet.
2. The receiving flask should be placed in such a way that the oil stream from the jet strikes the neck of the receiving flask and does not cause foaming.
3. After each reading the oil should be completely drained out of the receiving flask.

EXERCISE

1. How does the viscosity of liquid vary with temperature?
2. What is Viscosity index?
3. How the Viscosity Index of a lubricating oil improved.

Ans. *V.I. of lubricating oils can be improved by adding certain polymers, which are only partially soluble in the oils. At low temperatures the solubility of the polymers is low and hence the effect of polymer on the viscosity of the lubricant is also very low. At high temperatures, the solubility of the polymer in lubricating oil is quite high and hence the increased concentration of the polymer raises the viscosity of the oil. This compensates for the decrease in the viscosity of the oil itself.*

4. What is the difference between Redwood Viscometer No. 1 and Redwood Viscometer No. 2?
5. Name a few lubricants with high viscosity index?

Ans. *Silicones, polyglycol ethers, Diesters or triesters.*

CLOUD POINT AND POUR POINT

Cloud Point Cloud point is the temperature at which the oil becomes cloudy or hazy in appearance when it is cooled in a standard apparatus at a standard rate.

Lubricating oils which are derived from petroleum, mostly contain dissolved paraffin wax and other asphaltic or resinous impurities. The amounts of these impurities may vary depending upon the refining process used. At low temperatures i.e. when the oil is cooled down, these impurities tend to separate out of the oil in the form of minute crystals and hence result in cloudiness of the oil.

Pour Point. Pour point is the temperature at which the oil just ceases to flow when cooled at a standard rate in a standard apparatus.

If, after the cloud point, the oil is cooled further, more and more of impurities will crystallize and a stage is reached when the oil solidifies and stops flowing. Thus pour point can also be defined as the lowest temperature at which an oil will flow or pour under prescribed conditions, when it is cooled undisturbed at a fixed rate.

Significance. Cloud point and pour point indicate the suitability of a lubricant in cold conditions.

- The lubricating oils used in capillary feed systems or wicking arrangement should have low cloud points, otherwise at low temperatures the crystals of impurities will separate out clogging the capillary or wick interstices.
- In the fuel intake system of diesel engine, the lubricants having high cloud points may clog the filter screens.

- A high pour point leads to the solidification of the lubricant that may cause jamming of the machine.
- Pour point also gives an idea of the lowest temperature at which an oil can be transferred by pouring.
- Pour point determines the suitability of the lubricant for low temperature installations which may be required to start and operate at sub-zero temperatures. Examples refrigerator plants and air craft engines.

EXPERIMENT NO. 16.2

AIM **To determine the cloud and pour point of lubricating oil.**

APPARATUS Cloud and Pour Point apparatus.

THEORY When an oil is cooled slowly, the temperature at which it becomes cloudy or hazy is called cloud point, while the temperature at which the oil ceases to flow or pour is called pour point.

The cloud and pour point of a lubricating oil is determined with the help of 'cloud and Pour Point Apparatus'. It consists of the following parts:

(i) **Flat bottomed tube.** The flat bottomed tube is about 3 cm in diameter and 12 cm high and is fitted with a thermometer.

(ii) **Air jacket.** The tube is surrounded by the air jacket.

(iii) **Cooling bath.** The jacket is surrounded in turn by a cooling bath containing freezing mixture. The freezing mixture can be crushed ice and salt or ice and $CaCl_2$ or solid CO_2 and petrol, depending upon the cooling temperature required. A thermometer is inserted in the cooling bath to determine the temperature.

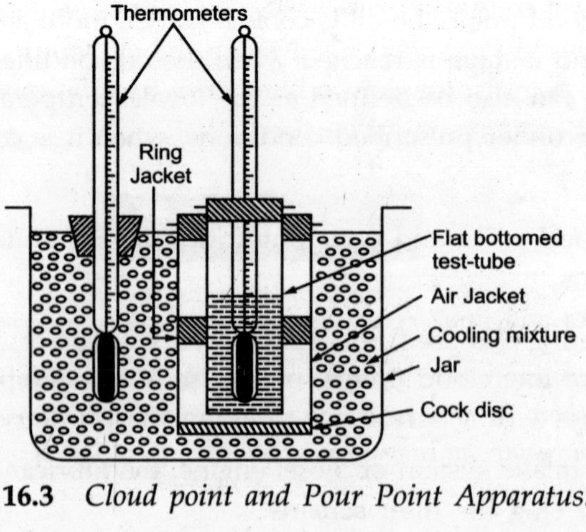

Fig. 16.3 *Cloud point and Pour Point Apparatus.*

PROCEDURE

1. Dry the lubricating test oil by shaking it with a small amount of anhydrous sodium sulphate and filter it through lintless filter paper.
2. Fill the oil upto the mark inside the flat bottomed glass tube and fit the tube with a cork into which a thermometer is inserted. The bulb of the thermometer should dip inside the oil.
3. Keep the tube in the cooling air jacket surrounded by the cooling bath containing the cooling mixture. The temperature of the oil will start falling down.
4. After every 1°C fall in temperature of oil, take out the test jar from the jacket and inspect for cloudiness. If no cloudiness appears, immediately replace it in the jacket.
5. Record the temperature at which such inspection first reveals a distinct cloudiness in the oil near the bottom of the jar.
6. Continue the cooling and take out the tube after every 3°C fall of temperature and tilt it just enough to see any movement or flow of oil. If the movement or flow is there, replace it immediately in the jar.
7. Continue the test until no movement of oil is observed when the test jar is held in a horizontal position for exactly 5 seconds. Note down the temperature and report it as Pour Point.

RESULT The cloud point of the given lubricant oil is ...... and the pour point of the given lubricating oil is ....

PRECAUTIONS

1. The complete operation of the removal and replacement of the test jar should be completed within 3 seconds.
2. When the separation of wax starts, care should be taken not to disturb the mass of the oil.
3. If the pour point is very low, a number of freezing mixtures with decreasing lower temperature should be used.

EXERCISE

1. Define cloud and Pour Point of a liquid lubricant.
2. How can the pour point of an oil be lowered?

Ans. *Wax pour point be lowered by dewaxing or adding suitable pour point depressant. Viscosity pour point can be lowered by lowering the viscosity of the oil which is achieved by removing the viscous constituents of the oil.*

3. What is the significance of pour point and cloud point determination.

4. What are pour point depressant?

Ans. *Pour point depressants are the materials which when added to oils containing wax get adsorbed on the surface of wax crystals and thus reduce the size of wax crystals.*

FLASH POINT AND FIRE POINT

Flash Point. The flash point of a lubricating oil is defined as the minimum temperature at which the oil gives off enough vapours that ignite for a moment when a tiny flame of standard dimensions is brought near it. The flash immediately disappears for want of more vapours i.e. the temperature is not enough for the vapours to be formed at high rate.

Fire Point. The fire point of an lubricating oil is defined as the lowest temperature at which the vapours of oil burn continuously for atleast five seconds, when a tiny flame is brought near it.

At slightly higher temperature than the flash point i.e. at fire point the heat from the flash becomes sufficient to evaporate more liquid and maintain combustion. The fire point is usually 5 to 40°C higher than the flash point.

Significance. The flash and fire points do not have any bearing with the lubricating property of the oil, but these are important from the following point of view:

- The knowledge of Flash and Fire point is helpful in providing safegaurds against fire hazards during their storage, transportation, handling and use.
- A lubricant selected for a particular job should have a flash point which is reasonably above its working temperature. This ensures safety.
- To ensure safety of illuminating oils like kerosene etc., the flash point of these should be reasonably above the average maximum temperature of a country.
- Liquids having flash point less than 140°F are called **flammable** and those having flash points above 140°F are called **combustible liquids**.
- Low flash point indicates the possibility of formation of gas-locks in fuel pipes or Spark Ignition Engines.
- Flash and Fire points have also been used to detect solvent contamination.

EXPERIMENT NO. 16.3

AIM **To determine Flash Point and Fire Point of an oil by Pensky Marten's Apparatus.**

APPARATUS Pensky Marten's flash and fire point apparatus.

THEORY **Flash Point** is the lowest temperature at which the lubricating oil gives off enough vapours that ignite for a moment when a tiny flame is brought near it. **Fire point** is the lowest temperature at which the vapours of the oil burn continuously for at least five seconds when a tiny flame is brought near it.

The determination of flash and fire points can be made by using the following apparatus.

(i) Cleverland Flash Point apparatus

(ii) Abel's closed cup apparatus

(iii) Pensky-Marten's apparatus.

In this experiment we will be using Pensky-Marten's apparatus, the description of which is given below.

Pensky Marten's Flash and Fire Point Apparatus. It consists of the following parts:-

(i) **Oil Cup:** The oil cup made up of brass is about 5 cm in diameter and 5.5 cm deep. It has a marking inside to indicate the level upto which the cup is filled. The lid of the cup is provided with four openings. Through one opening, a thermometer is inserted while a stirrer passes through the other, carrying two brass blades. The third opening is for introducing a test flame and the fourth one is for introducing the air.

(ii) **Shutter:** The lid is provided with a shutter which can be moved to introduce the flame when required.

(iii) **Flame exposure device:** It is a device for introducing a standard flame. It is connected to the shutter in such a way that when the shutter is turned, openings of the test flame and air are opened and the flame exposure device dips into the opening over the surface of the oil.

(iv) **Air bath:** The oil cup is supported by its flange over air bath which is heated by a gas burner.

(v) **Pilot Burner:** Pilot burner lights the test flame which gets extinguished when it is introduced into the opening for the test. The moment test flame returns to the original position pilot burner lights it.

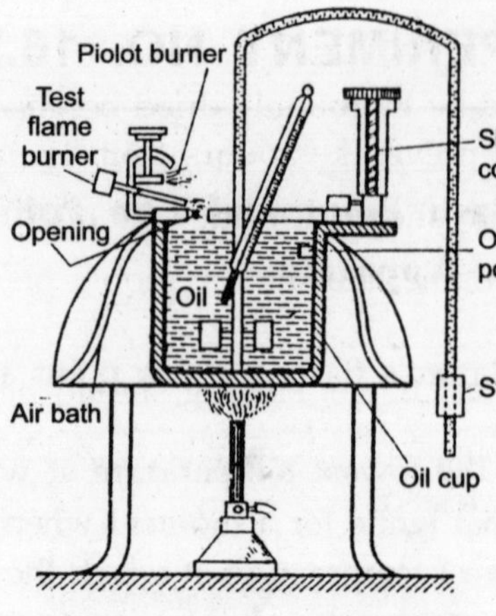

Fig. 16.4 *Pensky Marten's apparatus*

PROCEDURE

1. Clean and dry all parts of the apparatus.
2. Fill the oil cup with the test oil upto the mark.
3. Fix the lid on the top through which are inserted a thermometer and a stirrer. Ensure that the flame exposure device is fixed on the top.
4. Light the test flame and adjust it to about 4 mm in diameter.
5. Heat the apparatus by a gas burner so that the temperature of the test oil increases by about 5 to 6° per minute while the stirrer is rotated at approximately 60 revolutions per minute.
6. At every 1°C rise of temperature, introduce the test flame for a moment into the oil vapour. This is done by operating the shutter. On moving the knob of the shutter, the test flame is lowered in the oil vapours through the opening. The moment, knob is released the test flame springs back to its original position and the opening is closed.
7. When the test flame causes a distinct flash in the interior of the cup, note the temperature which represents the Flash point.
8. Further heat the test oil at the rate of 1°C/minute and continue applying the test flame as before.
9. The temperature at which the vapours of the oil give a clear and distinct blue flash for five seconds is recorded as the Fire Point of the oil.

RESULT The flash Point of given oil sample = ...°C

The Fire Point of given oil sample = ...°C

PRECAUTIONS

1. The apparatus should be thoroughly dried because the moisture affects the flash point as steam prevents vapours from igniting.
2. The thermometer bulb should dip into the oil.
3. For introducing the flame, the shutter should be drawn open slowly and closed quickly.
4. While applying the test flame, stirring should be discontinued.
5. Samples of very viscous materials may be warmed until they are reasonably fluid before they are tested.

EXERCISE

1. Define flash point and fire point of a lubricating oil.

2. What should be the flash point of a good lubricant?

Ans. *Flash point must be atleast above the temperature at which the lubricant is to be used, otherwise there is every risk of fire during use.*

3. In Pensky Marten's apparatus, what is the purpose of the air jacket surrounding the oil cup?

Ans. *Air being a bad conductor of heat ensures very slow and uniform temperature change whether the oil is being heated or cooled.*

4. What are the factors that affect the flash and fire point.

Ans.
- *Presence of moisture*
- *Vapour pressure*
- *Apparatus used.*
- *Frequency of application of test flame*
- *Rate of heating the test oil etc.*

5. How moisture is removed from the lubricating oil.

Ans. *By absorption through dehydrating agent such as Na_2SO_4, anhydrous $CaCl_2$ etc.*

6. Fatty oils do not have sharp or distinct flash points?

Ans. *These may undergo thermal decomposition with the evolution of vapours of decomposition products which may give irregular flashes (freaky flashes) before the true flash point.*

7. What is the difference between Abel's Apparatus and Pensky Marten's Apparatus for flash point determination?

Ans. *In Abel's apparatus, heating is done by water bath and the temperature rises very slowly where as in Pensky Marten's Apparatus, the heating is done by gas burner directly.*

Abel's apparatus is employed for oils having flash point upto 120°F whereas Pensky Marten's apparatus is employed for oils having flash point above 120°F.

ANILINE POINT

Aniline point of an oil is defined as the minimum temperature at which the oil is completely miscible with an equal volume of freshly distilled aniline. Alternatively, Aniline point is the minimum equilibrium solution temperature for equal volumes of aniline and the lubricating sample.

Since like dissolves like therefore aniline is readily soluble in those lubricants which are rich in aromatic and naphthenic compounds. Whereas the lubricants which are rich in paraffins, the dissolution takes place only if the temperature is increased.

Significance.

- Aniline point is a measure of aromatic content of the lubricating oil.
- A lubricating oil with low aniline point have a high aromatic content and will attack the rubber seals used in the system. This is because the aromatic compounds have a tendency to dissolve natural and certain types of synthetic rubbers.
- A higher Aniline point means low percentage of hydrocarbons and hence is desirable.
- Thus aniline point is used as an indication of possible deterioration of rubber sealings etc.

EXPERIMENT NO. 16.4

AIM **Determination of Aniline Point of an lubricating oil.**

APPARATUS Aniline point apparatus (Test tube and jacket, rubber , thermometer, stirrer, heating bath).

CHEMICALS Freshly distilled aniline and anhydrous sodium sulphate.

THEORY Aniline point of an oil is defined as the lowest temperature at which equal volumes of freshly distilled aniline and oil are miscible. For the determination of Aniline Point, equal volumes of aniline and sample are taken in a test tube. The mixture is heated till a homogeneous solution is obtained. The tube is allowed to cool at a uniform rate. The temperature at which the two phases (oil and aniline) separate out is recorded as the aniline point. The separation of the two phases is indicated by the appearance of cloudiness in the medium.

PROCEDURE

1. Clean and dry the Aniline Point Apparatus. (Dry it at 100 - 110°C in electric oven)
2. Take 5 ml of pure aniline and dry it over KOH pellets, filter it.
3. Take 5 ml of the test oil and dry it by shaking with anhydrous Na_2SO_4 and filter it.
4. The aniline and the sample (5 ml each) are transferred to a pyrex glass tube (2.5 × 15 cm). The tube is fitted with a cork holding a stirrer and a thermometer. The bulb of the thermometer should dip 5 mm above the bottom of the tube.
5. This tube is inserted into an outer air jacket (4 × 17.5 cm) which is made of heat resistant glass.

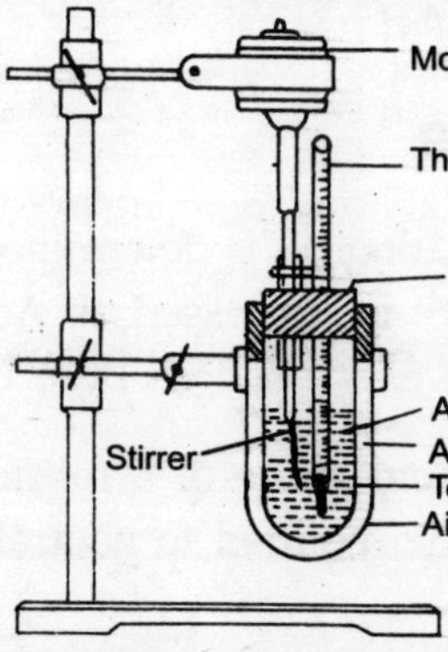

Fig. 16.5 *Aniline Point Apparatus*

6. The aniline and sample mixture is stirred to get a homogeneous solution. If these are not completely miscible at room temperature, the jacket holding the tube is immersed in a hot bath.
7. The temperature of the bath is raised at a constant rate and the stirring is continued till the two liquids are completely miscible.
8. Discontinue heating at this point (you can remove the jacket from the hot bath). Allow the temperature of the solution to fall at a rate 1°C per minute.
9. The temperature at which the cloudiness or haziness appears due to the separation of the two phases is noted down as Aniline Point of the oil.

RESULT Aniline Point of the oil is ......°C.

PRECAUTIONS

1. No moisture should be adhering to the apparatus. It should be absolutely dry.

2. In hot bath, water should never be used as aniline is hygroscopic.
3. As Aniline is highly toxic, it should not be sucked into the pipette by mouth. A pipette provided with a rubber suction bulb should be used.
4. Precaution should be taken during stirring that no air bubbles should be formed.

EXERCISE

1. Define aniline point?
2. What is the relation between aniline point and aromatic content.
3. Is these any relation between Aniline Point and ignition quality of a diesel fuel.

Ans. *High aniline point of a diesel fuel means low aromatic content. And low aromatic content means less knocking in diesel fuel.*

NEUTRALIZATION NUMBER

The acidity or alkalinity of lubricating oil is determined in terms of neutralization number. Determination of acidity is expressed as Acid Number or Acid Value and determination of basicity is expressed as Base Number or Base Value.

Acid Number or Acid Value. Acid value of a lubricating oil is the number of milligrams of potassium hydroxide required to neutralize all the free acid present in 1 g of the oil.

Base Number or Base Value. Base value of a lubricating oil is the number of milligrams of a titrating acid required to neutralize all the basic constituent present in 1 g of the oil.

Out of the two, determination of acidity is more common. Fatty oils consist mostly of glycerol or other esters of higher fatty acids and in some cases, considerable amounts of free acids are present in the lubricating oil if these are not refined carefully i.e. acids and bases used in refining process have not been completely removed.

Significance

- The small intrinsic acidity may not be harmful in itself but the degree to which it increases in used oils is usually taken as a measure of deterioration of the oil due to oxidation or contamination.
- The acid content of the fatty oils increases with time due to hydrolysis with moisture and is therefore a rough indicator of age of the oil.
- The deterioration in the flavour of edible oils with time is due to the increase in free acid concentration formed by hydrolysis or oxidation.

- Periodic determination of Acid number is thus helpful in deciding the state, where the lubricant needs replacement.

However the presence of free acid in lubricating oils is very rare and indicates contamination. A lubricating oil must possess acid value **less than 0.1**.

EXPERIMENT NO. 16.5

AIM **Determination of Acid Value (Acidity) of an oil.**

APPARATUS Burette, pipette, conical flask, beakers water bath.

CHEMICALS N/10 KOH, Neutral ethyl alcohol, phenolphthalein indicator.

THEORY The acid value of a lubricating oil is defined as number of milligrams of KOH required to neutralize the free acid present in 1 g of the oil. For the determination of Acidity, a known weight of lubricating oil is dissolved in a suitable solvent and is titrated with a standard solution of KOH using phenolphthalein as indicator.

$$RCOOH + KOH \rightarrow RCOOK + H_2O$$

$$H^+ + {}^-OH \rightarrow H_2O.$$

PROCEDURE

1. Weigh the empty beaker in which the oil is to be taken.
2. Transfer about 150 drops of the lubricating oil (about 5 g) into the beaker and weigh again. Find out the exact weight of the oil taken by subtracting the weight of empty beaker from it.
3. Add 50 ml of the solvent (neutral alcohol) to the oil and shake vigorously to dissolve the oil. Transfer the contents into a 250 ml conical flask.
4. Heat the flask over a water bath for 30 minutes. Cool the flask to room temperature.
5. Add 10-12 drops of phenolphthalein indicator.
6. Titrate with the standard N/10 KOH solution until a faint permanent pink colour appears at end point. Note down the volume of alkali used. Let it be y ml.
7. Run a blank titration on the same volume of hot solvent. Let the volume of KOH used be z ml.

OBSERVATIONS AND CALCULATIONS

Initial weight of the beaker = x_1 g

Final weight of the beaker and oil = x_2 g.

Weight of the oil = $x_2 - x_1$ g

Volume of N/10 KOH used in the test oil = y ml

Volume of N/10 KOH used in the blank titration = z ml

Therefore volume of N/10 KOH used for titrating the acid in the oil = $(y - z)$ ml.

$$\text{Acid value} = \frac{\text{ml of KOH used} \times \text{Normality} \times 56}{\text{weight of oil taken}}$$

$$= \frac{(y-z)}{\text{weight of oil}} \times \frac{1}{10} \times 56 = \frac{(y-z)\times 5.6}{\text{weight of oil}}$$

where 5.6 represents the amount of KOH in mg present per ml of N/10 KOH solution. (1000 ml of 1 N KOH ≡ 56 g of KOH)

RESULT The acid value of the given oil is ....mg.

PRECAUTIONS

1. KOH should be standardized just before the titration.
2. If the oil is coloured (red or black) phenolphthalein is not a satisfactory indicator. In that case use p-naphthol-benzoin as indicator where the end point will be from orange to green.
3. If the pink colour fades repeatedly after subsequent additions of alkali, the titration should be completed rapidly and the first appearance of pink colour should be taken as the end point.

EXERCISE

1. Define Acid number. How is it determined?
2. When phenolphthalein is used as an indicator, the pink colour at the end point sometimes fades away after a few seconds. Explain

Ans. *Sometimes the oil contain the materials which are attacked by the alkali. If this extra drop is consumed by the reactable materials, pink colour disappears.*

3. What do you mean by base value?
4. What is the significance of determination of acid value?

SAPONIFICATION VALUE/NUMBER

Saponification value or number of an oil is the number of milligrams of potassium hydroxide required to saponify 1 g of oil. (Saponification is the hydrolysis of an ester with KOH to give alcohol and sodium or potassium salt of the acid).

Mineral oils are the mixture of hydrocarbons and hence do not react with KOH and are not saponifiable. On the other hand vegetable and animal oils are the mixtures of glyceryl esters of fatty acids and require large amount of alkali to get hydrolyzed. Their saponification values are very high.

Significance

- The determination of saponification value helps us to ascertain whether the oil under reference is mineral or vegetable oil or a compounded oil containing vegetable and mineral oil.
- Each of the oil has its own specific saponification number. Any deviation from this value in a given sample indicates the probability and extent of adulteration.

EXPERIMENT NO. 16.6

AIM **Determination of saponification value of an oil.**

APPARATUS Condensor, water bath, round bottomed flask, burette, pipette, conical flask.

CHEMICALS Standard HCl solution (N/2), standard KOH solution (N/2) and phenolphthalein.

THEORY Saponification number is the number of milligrams of potassium hydroxide required to saponify 1 g of oil. The saponification value of an oil is determined by refluxing a known amount of sample with known excess of standard KOH solution. The alkali consumed is then determined by titrating the unconsumed KOH against standard HCl using phenolphthalein as indicator.

$$RCOOH + KOH \rightarrow RCOOK + H_2O$$

$$H^+ + \overline{O}H \rightarrow H_2O.$$

PROCEDURE

1. Transfer about 5 g of accurately weighed oil sample in a conical flask.
2. Add 50 ml N/2 alcoholic KOH and 50 ml of alcohol into the flask through a pipette.
3. Add the same amount of alcohol and N/2 KOH in another flask for blank determination.
4. Fit the two flasks with water condensers and reflux the contents on a water bath for a minimum period of 1 hour.
5. Cool the contents slightly and disconnect the condenser. Rinse the condenser with a small amount of distilled water into the condenser.
6. Add 10-12 drops of phenolphthalein indicator and titrate the contents of the two flasks with N/2 HCl until the pink colour just disappears.

OBSERVATIONS AND CALCULATIONS

Weight of the empty beaker = w_1 g

Weight of the beaker and oil = w_2 g

Weight of the sample taken = $w_2 - w_1$ g.

Volume of alcoholic KOH added to both the flasks = 50 ml.

Volume of N/2 HCl used in sample determination = V_1 ml

Volume of N/2 HCl used in blank determination = V_2 ml

Volume of N/2 HCl equivalent to KOH used in saponifying $(w_2 - w_1)$ g of oil $= (V_2 - V_1)$ ml.

$$\text{mg of KOH present in 1000 ml of N/2 KOH} = \frac{1}{2} \times 56 \text{ g} = 28 \text{ g} = 28 \times 1000 \text{ mg}$$

$$\text{mg of KOH present in 1 ml of N/2 KOH} = \frac{28}{1000} \times 1000 = 28 \text{ mg}$$

$$\text{mg of KOH present in } (V_2 - V_1) \text{ ml of N/2 KOH} = (V_2 - V_1) \times 28 \text{ mg}$$

$$\text{Saponification value of the sample} = \frac{(V_2 - V_1) \times 28}{\text{Amount of oil}}.$$

RESULT The saponification value of the given oil is ....mg.

PRECAUTIONS

1. Boiling should be done at a slow rate.
2. To hasten the process of saponification, the sar ɔle flask should be occasionally shaken during refluxing.
3. The alcoholic solution of KOH should be freshly prepared.

EXERCISE

1. What is saponification value?
2. Why mineral oils are not saponifiable?
3. What is the significance of determination of saponification number?

IODINE VALUE

Iodine value or iodine number is the number of grams of iodine equivalent to the amount of Iodine Monochloride (ICl) absorbed by 100 g of the oil.

As is well known, oils and fats are mixtures of mixed glycerides and can be represented as:

CH_2OOCR
$CHOOCR'$
CH_2OOCR''

These are formed by glycerol and monobasic fatty acids. The acids present in combination with glycerol have all even number of carbon atoms (C_4 to C_{22}) and may be unsaturated or saturated. Example, $C_{15}H_{31}COOH$ (palmitic acid) $C_{17}H_{35}COOH$ (stearic acid), $CH_3(CH_2)_4$ $CH=CHCH_2CH=CH(CH_2)_7$. COOH (linoleic acid) etc.

The oils having predominance of short chain saturated fatty acids or long chain fatty acids with a considerable degree of unsaturation are liquids at room temperature and are called oils. Others are solid at room temperatures and are called fats.

The degree of unsaturation is reported in terms of Iodine Number.

Significance

- As each oil has a specific iodine value so the determination of iodine value can be utilized to detect the extent of contamination in any specific oil.
- It can be used to classify oils based on iodine value because unsaturation confers drying characteristics on oils.

Iodine Value	Type of oil	Examples
> 150	Drying oil	Linseed oil, tung oil
100 to 150	Semidrying oil	Castor oil (dehydrated), soyabean oil
< 100	Non Drying oil	Coconut oil, olive oil

- When a thin film of an oil with a high degree of unsaturation (high Iodine Value) is spread on a surface, it takes up oxygen at the double bond sites and gets oxidized and polymerized into a hard, tough, infusible mass. Thus these oils having high I.V. are used in varnishes and paints.
- To use oil as lubricant its Iodine Value should be low so that it does not deteriorate.

EXPERIMENT NO. 16.7

AIM **Determination of Iodine Value of an oil.**

APPARATUS Burette, pipette, conical flask, beakers. etc.

CHEMICALS N/10 $Na_2S_2O_3$ solution, Wij's solution, (solution of ICl in glacial acetic acid) Potassium iodide solution (10%), chloroform or carbon tetrachloride, starch solution (freshly prepared).

THEORY Iodine Value is the number of grams of iodine equivalent to the amount of iodine monochloride (ICl) absorbed by 100 g of the oil.

Determination of iodine value is done through iodometric titration*. A known weight of the oil is dissolved in $CHCl_3$ or CCl_4 and treated with a known excess of Wij's solution. One molecule of ICl adds to the double bond.

After the completion of the reaction, KI is added and unreacted ICl liberates I_2 from it

$$ICl + KI \rightarrow I_2 + KCl.$$

The liberated iodine is titrated against $Na_2S_2O_3$ using starch as indicator.

PROCEDURE

1. Remove the moisture present in the oil sample by straining it through a filter paper.
2. Weigh about 5 g of the sample oil and transfer it to a conical flask. Add 25 ml of CCl_4 and 25 ml of Wij's solution to it. In another flask take the same volumes of CCl_4 and Wij's solution for blank determination.
3. Close the flask and mix the contents by swirling. Keep in dark for about one hour at a temperature below 30°C. The glass stopper may be moistened with KI solution.

4. Now add 20 ml of KI solution to both the flasks and 100 ml of distilled water.
5. Titrate the liberated iodine with N/10 $Na_2S_2O_3$ solution till the colour of the solution becomes pale yellow.
6. Now add the indicator, i.e. starch solution, the solution will turn blue.
7. Continue the titration till the blue colour disappears.
8. Note down the volume of $Na_2S_2O_3$ used, in both the cases, i.e. test solution and blank.

OBSERVATIONS AND CALCULATIONS

Weight of the weighing tube = w_1 g

Weight of the weighing tube and oil = w_2g

Weight of the sample = $(w_2 - w_1)$g

Volume of $Na_2S_2O_3$ used to titrate excess iodine in sample determination = x ml.

Volume of $Na_2S_2O_3$ used to titrate total iodine in blank titration = y ml.

Volume of N/10 $Na_2S_2O_3$ equivalent to iodine (or ICl) consumed by $(w_2 - w_1)$g of oil = $(y - x)$ ml.

Iodine present in (y – x) ml of N/10 iodine solution = $(y - x) \times 127 \times \frac{1}{10} \times \frac{1}{10^3}$ g

(Equivalent weight of iodine = 127)

$$\therefore \text{ Iodine Value } = \frac{(y-x)\times 127}{(w_2 - w_1)} \times \frac{1}{10} \times \frac{1}{10^3} \times 100 = \frac{(y-x)}{(w_2 - w_1)} \times 1.27$$

RESULT The iodine value of the given oil is...........

PRECAUTIONS

1. Wij's solution should never be sucked through mouth. It should be added with the help of burette.
2. Wij's solution should be added in about 100% excess than the expected requirement.
3. Wij's solution should be added in the two flasks simultaneously.

Note. (For details on iodometric titration, please refer to chapter-1, iodometric titrations.)

EXERCISE

1. What do you mean by iodine value?
2. What is Wij's solution?
3. Why is glass stopper moistened with KI solution?

Ans. *To prevent the leakage of I_2 vapour.*

4. What is the significance of determination of Iodine Value?

Note to the Lab Assistant. For the preparation of various solutions ($Na_2S_2O_3$ starch solution etc.) refer to the chapter on iodometry. (students are also advised to go through the chapter on iodometry in volumetric titrations for complete understanding of such type of titrations).

Wij's solution. Wij's solution is available commercially or it can be prepared by dissolving 1.6 g of commercially available iodine monochloride (ICl) in 100 ml of A.R. glacial acetic acid. Wij's solution is stored in amber coloured stoppered bottles in a dark place below 30°C

❑❑❑

17

Chromatography

Number of chemical reactions yield the *mixture* of products rather than a *single* product. These products must be separated and analysed. Chromatography is one of the techniques which is used to separate and identify the chemical constituents present in mixture.

DEFINITION

Chromatography is a technique of separation and identification of chemical constituents present in a mixture. The separation is brought about by distributing the sample into two phases. One of these phases consists of *stationary phase* of large surface area and the other being a *mobile phase* which percolates through the first phase. The stationary phase may be solid or liquid and the mobile phase may be liquid or a gas.

Different substances, under similar conditions, when passing through a material of a different phase are absorbed at different rates. Thus separations are based on differences in absorption or migration rates among the sample components.

CLASSIFICATION

The chromatography can be classified in various ways.

1. On the basis of physical status of mobile phase, it is of two types:

   (a) Liquid Chromatography (Mobile phase is liquid)

   (b) Gas chromatography (Mobile phase is gas)

2. On the basis of physical states of stationary phase and the principle involved it is of following types:

(a) Adsorption chromatography. In this type of chromatography, the stationary phase consists of a solid adsorbent and the mobile phase is liquid or gas containing the mixture. The selective adsorption of the components of the mixture occurs on the surface of the solid.

(b) Partition chromatography. In this type of chromatography, the stationary phase is a liquid or a liquid supported on an inert solid. The moving phase can be liquid or a gas. The separation of different components is based upon their selective partition between the moving phase and a thin liquid film (stationary phase) held on an inert solid.

(c) Ion Exchange chromatography. The stationary phase is an ion exchanger and the distribution of the components of the mixture is based on ion exchange principle. This type of chromatography is applicable to ionic species.

(d) Gel Permeation chromatography. In gel chromatography, the stationary phase is gel and the separation is based on its sieving action.

(e) Electrophoresis. Electrophoresis is carried out using strips of moistened filter paper between the ends of which a potential difference is applied. The separation of the mixture takes place because of the different speeds with which organic ions move under the influence of an electric field towards cathode or anode.

3. On the basis of the type of stationary phase, it may be classified as:

(a) Column chromatography. In this type of chromatography, the stationary phase is formed into a column in a cylindrical tube.

(b) Paper chromatography. In this form of partition chromatography, the stationary phase is the absorbed water present in the filter paper and the moving phase is a solvent saturated with water. Filter paper used is made of highly purified cellulose $(C_6H_{10}O_5)_n$.

(c) Thin layer chromatography. In this, the stationary phase is thin layers of certain fine solid particles, such as silica gel and alumina, supported on glass plates or plastic plates.

THIN LAYER CHROMATOGRAPHY

Thin layer chromatography (TLC) is an efficient, fast and simple method of separating the mixtures. In this technique a thin layer of silica gel or alumina is spread on a glass or plastic plate and hence the name. The basic principle of separation is the *selective absorption* of the components of the mixture on the stationary phase, when they are carried through it by the flow of mobile phase. The steps involved are as follows:

(i) Preparation of thin layer plates.

(ii) Application of the material on the plate.

(iii) Development of the plate.

(iv) Visualization

(v) Calculation of R_f values.

(i) Preparation of thin layer plates. The plates are usually prepared using glass plates or plastic plates of the size 20 × 20 cm or 5 × 20 cm. A thin layer of silica or alumina is spread on the glass or the plastic plate. The prepared thin layer on glass is often called a *chromaplate*. Chromaplates are prepared by applying a uniform layer of silica or alumina in the form of a thin aqueous paste. The solid adsorbent (silica or alumina) is usually mixed with a small amount of 'binder' such as calcium sulphate.

The binder helps in proper adherence of the adsorbent onto the plate. Special applicators are available commercially to spread the adsorbent. The thickness of the adsorbent layer is generally 0.25 mm. The paste or slurry of the adsorbent is prepared in water.

(ii) Application of the material on the plate. The material is dissolved in an appropriate solvent and deposited on the thin layer with the help of a capillary. The capillary is first dipped in the sample solution, the solution goes up the capillary. The sample is applied on the plate by touching the capillary to the adsorbent surface at about 2 cm from the edge. Position of spot is marked with a pencil so that the Rf values may be calculated after separation.

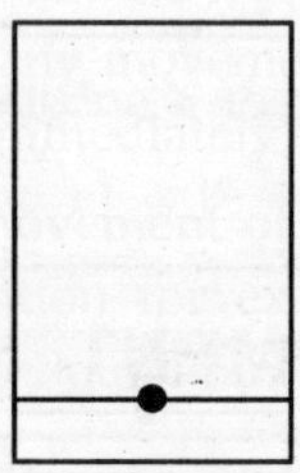

Fig. 17.1 *Thin layer plate with the spot of sample*

(iii) Development of the plates. When the spot is dry, the chromaplate is placed in a jar (developing chamber) containing small amount of solvent. The jar must be saturated with solvent vapour before the plate is placed in it. This can be achieved by lining the jar with filter paper dipping into the solvent. Wide-mouthed, capped bottles are used as developing chambers in the laboratories. The plate is kept in such a way that lower end of the plate dips 1-2 cm below the surface of the solvent. The spot should be above this level. The top of the jar is firmly closed. The solvent rises through the adsorbent layer, *and the components of the mixture ascend at different rates depending on their affinities for the adsorbent*. When the solvent has nearly reached the top of the adsorbent layer (3/4 of the TLC plate), the components should be well separated. The relative distance travelled by the components can be increased by using a solvent of higher or lower polarity. Typical solvents in order of increasing elutive* power:

*Elution: Elution is a process in which solutes are washed through a stationary phase by the movement of a mobile phase.
Eluent It is a solvent used to carry the components of a mixture through a stationary phase.

Petroleum < benzene < chloroform < ether < ethanol

(iv) Visualization. The plate is removed from the tank, the position of the solvent front is marked with a pencil and the solvent is allowed to evaporate from the plate. If the components of the mixture are coloured, they can be detected by simple visual examination. However, if they are colourless, they are located by (a) keeping the plate in a closed dry tank containing crystals of iodine. The spots show up as brown spots (b) spraying the plate with conc. H_2SO_4 and then heating the plate to about 200°C. The separated spots of organic compounds will render visible due to highly sensitive charring effect (c) Irradiation with UV light to locate the substance which give flourescence on exposure to UV.

(v) Calculation of R_f Values. R_f value (Retention Factor) is defined as the *ratio of the distance travelled by the component at its point of maximum concentration to the distance travelled by the solvent.*

$$R_f \text{ Value} = \frac{\text{Distance moved by component}}{\text{Distance moved by the solvent front}}$$

The R_f value is characteristic of a substance for a given solvent.

EXPERIMENT NO. 17.1

AIM **Determination of R_f value of amino acids by thin layer chromatography (TLC) and identify the amino acids present.**

APPARATUS TLC jar, capillaries, measuring cylinder, Glass plates

CHEMICALS Silica gel, Butanol-1, propanol-1, conc. ammonia solution, Amino acids (L-Lysine, L-leucine, L-Alanine), Ninhydrin reagent, unknown sample of amino acids.

THEORY One of the important applications of TLC is the separation of amino acids. Basic principle of TLC is given in discussion part.

The sample is spotted at one end of glass plate coated with thin layer of adsorbent. The plate is placed in a covered jar containing the solvent for developing. The solvent rises up by capillary action and the solute is distributed between the stationary (absorbent) phase and the mobile phase. A solute which is more strongly adsorbed will migrate slowly up the TLC plate. The sample is subsequently separated. Spraying the dried chromotogram with ninhydrin forms the coloured zones of different amino acids. The various amino acids are identified by the calculation and comparison of R_f values.

PROCEDURE

Preparation of TLC plate

1. Take a clean glass plate. Mix about 10 g of Silica Gel G with 20 ml of water. Shake well to remove any lumps.
2. Pour this slurry on the glass plate and spread it out with the help of applicator.
3. Let the plate dry for ten minutes and put the plate at 110°C in a oven for atleast 30 minutes to activate the adsorbent.

Application of the amino acid

4. Prepare the solutions of the given amino acids by dissolving 5 mg of each separately in 0.5 ml of distilled water. Mix one drop of each solution to provide a mixture of the three amino acids and dilute the remainder of the three solutions to 1 ml separately.
5. Apply each of the above solution and the unknown solution, with the help of capillary to a silica gel plate. The spot of the solution should not be larger than 2 mm in diameter. Allow the spot to dry.

Developing

6. Prepare the developing solution by mixing 70 ml of propanol-1 with 30 ml of conc. ammonia.
7. Line the developing jar with filter paper moistened with the solvent prepared above in step 6.
8. Place the prepared plate in the jar and introduce sufficient developing solvent by means of a pipette so that the lower edge of the absorbent layer is immersed in the solvent. Cover the jar and allow the solvent (developer) to ascend along the plate.
9. When the developer ascends to the required height (3/4 of the plate), mark the solvent front and dry the plate at 100°C for about 10 minutes.

Visualization

10. Spray the chromatogram (TLC plate) with ninhydrin reagent (2% solution of ninhydrin in butanol) and heat at 110°C for 5-10 minutes.
11. Mark the centre of each spot. Measure the distance travelled by each component and the solvent. Determine the Rf value.

OBSERVATIONS AND CALCULATIONS

S.No.	Amino Acid	Distance of Amino acid from the line of application (dm)	Distance of solvent (ds)	$R_f = \frac{dm}{ds}$
1.	L-lysine			
2.	Leucine			
3.	Alanine			
4.	Unknown mix.			
	(i)			
	(ii)			
	(iii)			

Compare the R_f value of amino acids present in the unknown mixture with the R_f values of pure amino acids and thus identify the amino acids present in the unknown mixture.

RESULT The amino acids present in the unknown mixture are

(i) .... with R_f value = ...

(ii) .... with R_f value = ...

PRECAUTIONS

1. The plate prepared should be of uniform thickness.
2. The spot of the compound should not exceed 2 mm in diameter.
3. Different spots should be atleast 5 mm apart.
4. The developing solvent in the tank must be sufficient so that plate dips to the depth of about 1 cm.

Instructions for the Lab. Assistant

Provide the following

1. **Unknown solution** of amino acids - Dissolve any two amino acids (15 mg in 1 ml of distilled water) separately and mix them. Warm if particular amino acid is not soluble in cold.

Note. Ninhydrin is the 2 hydrate of indane-1,2,3 trione (or triketohydrindene hydrate) having the formula

2. Provide the solution of pure amino acids.
3. **Ninhydrin Reagent.** Take 100 ml of butanol-1 and 100 ml of distilled water in separatory funnel. Shake gently and allow it to form layers. Transfer the upper organic layer to a spraying bottle and to this add 0.2 g of ninhydrin and shake well.
4. Silica Gel G is silica gel containing gypsum as binding agent.

EXPERIMENT NO. 17.2

AIM **To separate a mixture of dyes using thin layer chromatography.**

APPARATUS Same as previous experiment.

CHEMICALS 'Silica Gel G', any three dyes (azo-benzene, p-methoxy azobenzene, Sudan yellow, Sudan red, p-amino azobenzene, p-dimethylamino azobenzene, p-hydroxy azobenzene), benzene.

THEORY Similar to previous experiment number : 17.1.

PROCEDURE Proceed as in the previous experiment number : 17.1.

(i) The solutions of the dyes is prepared by dissolving each dye in benzene (20 mg in 10 ml)

(ii) The developing solvent used in this case would be benzene

(iii) No spraying agent would be used, as coloured spots would be obtained.

Calculate the R_f value of the components in pure solutions and in the mixture provided.

PAPER CHROMATOGRAPHY

Paper chromatography (PC) is a simple technique to separate complex mixtures of metal ions, amino acids, sugar, dyes and drugs. It is a form of partition chromatography in which the stationary phase is the absorbed water, always present in filter paper. The support is the paper itself and the mobile phase is a solvent previously saturated with water.

Paper chromatography is a form of *partition chromatography but adsorption phenomenon is also involved. Filter paper used is made of highly purified cellulose which, being a polyhydroxy compound of high molecular weight, absorbs and retains water molecules. The stationary phase is not the filter paper, but a very thin film of the solvent, usually water adhering to the surface of the filter paper. Thus the paper chromatography will not work in dry climates unless the paper is exposed to air of high humidity, before being placed in the developing solvent.

A drop of a solution of the mixture (or drops of individual components of the mixture) is applied to the paper by means of the capillary tube and the paper dried. The paper is then suspended in a suitable, closed container so that it can be irrigated with an organic liquid or a mixture of liquids. The bottom of the paper should immerse in the solvent. The capillary action of the paper will cause the solvent to rise steadily up the paper. During this process, the solvent which now contains the mixture in solution, is continuously extracted by the retained water molecules in the paper.

The extent to which a particular solute has moved, relative to the extent to which the solvent has moved, known as R_f value, will be a function of the partition coefficient of the solute between the solvent and the film of retained water.

When the ascending solvent has travelled the required distance, the paper is removed, the position of solvent-front marked and the paper is dried. The position of various solutes, if they are coloured, appear as coloured spots. If the spots are not coloured, the location of the spots is determined by (i) spraying the paper with a chemical reagent which produces insoluble coloured derivatives with the solutes (eg. ninhydrin for amino acids) (ii) or by UV light in which the solute exhibit flourescence. The R_f value is calculated.

$$R_f = \frac{\text{Distance moved by the substance}}{\text{Distance moved by the solvent}}$$

In paper chromatography, two types of techniques are used.

(i) **Ascending technique.** In this the solvent ascends through the filter paper as described above.

(ii) **Descending technique.** In this the solvent is contained in a small trough at the top of developing tank and travels downwards. The mixture is applied at a point external to the trough.

---

* Partition is the process of distribution of one or more solutes between two immiscible phases. Distribution coefficient K_D is given by :

$$K_D = \frac{\text{conc. of solute in solvent A}}{\text{conc. of solute in solvent B}}$$

EXPERIMENT NO. 17.3

AIM **Separation of metallic ions by paper chromatography.**

APPARATUS What man No. 1 filter paper sheets, spotting capillaries, developing tank with cover.

CHEMICALS Potassium chromate, Lead nitrate, Silver nitrate, mercurous nitrate, Nitric acid, unknown solution, (or nitrates of metallic ions which are to be separated).

THEORY The general theory of paper chromatography is given in the discussion part.

The separation of inorganic compounds is based upon their different solubility in the organic solvents. In some cases, the solubility is due to the formation of soluble complex between the compound and the solvent whereas in other cases complex forming agents are deliberately added.

PROCEDURE

Preparation of solutions

1. Prepare 1 ml aqueous solution of lead nitrate, silver nitrate and mercurous nitrate by dissolving about 10 mg crystals of each separately in a small test tube. Add a few drops of HNO_3 to prevent hydrolysis.

Application of the sample

2. Cut whatman No. 1 filter paper strip of 25 cm square.
3. On the strip draw a line about 2-3 cm from one end and put four crossings on this line at atleast 5 cm apart. This end will be the bottom and development will start from this end.
4. Apply the solutions of Pb^{2+}, Ag^{2+} and Hg_2^{2+} and the unknown mixture on the crosses, using separate capillary for each.

Development

5. Curl the filter paper and place it in a jar containing sufficient amount of distilled water that lower end of the paper dips in below the spots. Alternatively the filter paper can be suspended in the chromatographic jar

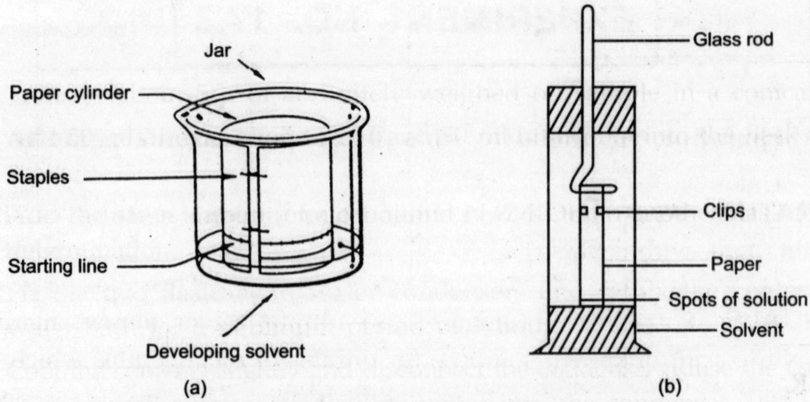

Fig. 17.1 *Apparatus for paper chromatography.*

7. Allow the developer (solvent) to rise along the paper until it reaches near the upper end of the paper.

Drying and locating the spots

8. Remove the paper and mark the solvent front with the help of a pencil.
9. Dry the paper.
10. Dip the paper into the solution of potassium chromate. Different coloured spots due to metallic ions will be observed. Encircle the coloured spots.
11. Calculate the R_f values and compare the R_f values of individual cation with that of R_f values in unknown mixture and on its basis identify the cations.

OBSERVATIONS AND CALCULATIONS

The coloured spots of the metallic ions have the following colours.

Pb^{2+} = Yellow

Ag^{+} = Orange red

Hg^{2+} = Orange.

Calculate the R_f value as

$$R_f = \frac{\text{Distance of the solute from the base line}}{\text{Distance of the solvent front}} = \frac{d_{\text{met}\cdot\text{ions}}}{ds}$$

Sample	d_{met}·ions	d_s	$\frac{d_{met \cdot ions}}{ds}$
Ag^+			
Hg_2^{2+}			
Unknown mixture			
1.			
2.			

The R_f values of the components of the mixture resembles with .... and ......

RESULT The metallic cations present in the given mixture are ... and .....

PRECAUTIONS

1. The spots of the mixture and the components should be applied on a smaller area. Larger spots result in poor separation.
2. The solvent should be below the spots of the components and the mixture.
3. Before placing the paper in the chromatographic jar the spots should be dried.

Instructions to the Lab Assistant

Provide the following:

1. Solution of unknown mixture: Dissolve any two nitrates of the above compounds in water.
2. K_2CrO_4 solution: Dissolve 24.25 g of K_2CrO_4 in distilled water (.25 M solution).

SEPARATIONS OF DIFFERENT MIXTURES OF INORGANIC IONS BY PAPER CHROMATOGRAPHY

As is discussed in the previous experiment, a number of inorganic ions can be separated using the same procedure. However the locating reagent and the solvent (developer) is different for different mixtures. Some of these are given below

---

Notes. When the paper is dipped in K_2CrO_4 solution, the coloured spots which are formed are due to the formation of chromates of lead, mercurous and silver.

2. Lead moves with the mobile phase at the fastest rate and hence appears on the uppermost area followed by mercurous. The slowest moving is silver.

Ions to be separated	Solvent Developer	Locating Reagent	Remarks
1. Co, Ni and Zn (aqueous solutions of the nitrates of these metals are mixed)	Mix acetone (87 ml) conc. HCl (8 ml) and water (5 ml)	Dissolve 50 mg of trisodium pentacyanoamine ferroate in 20 ml of water and also dissolve 10 mg of rubeanic acid in 10 cm^3 ethanol. Mix the two.	Expose the dried chromatogram to NH_3 vapour to neutralize any acid, before placing in locating reagent
2. Fe, Mn, Co and Ni (aqueous solutions of nitrate)	Acetone (90 volume) 1 M HCl (10 volume)	Ammonical solution of dimethyl glyoxime	With locating reagent cobalt give dark brown, iron gives brown, manganese give red brown, and nickel gives red coloured spots
3. Na, K and Li	Absolute methanol	20% alcoholic solution of violuric acid	Coloured spots with locating reagent of Na, K and Li would be yellow, brown and violet respectively
4. Hg^{2+}, Cu^{2+}, Cd^{2+} and Bi^{3+}	n-Butyl alcohol and HCl	Conc. solution of dithizone in chloroform	Colours of metal ions produced with dithizone are Cu^{2+} - brown, Cd^{2+} - Purple, Hg^{2+} - Pink, Bi^{3+} - Brown
5. Cl^-, Br^- and I^- (Na salts of these are dissolved in water)	n-butanol (40 ml), pyridine (20 ml) and 1.5 M ammoniacal solution (40 ml)	Ammonical silver nitrate solution	

COLUMN CHROMATOGRAPHY

Column chromatography is a technique based on the use of a column of a solid absorbent (stationary phase), hence the name column chromatography. The separation of the components of the mixture may involve any physical principle like adsorption, partition or ion exchange.

The success of a separation by column chromatography depends on the choice of the stationary and mobile phases.

(i) Selection of the stationary phase (Adsorbent) There are number of adsorbents which can be used in column chromatography. The choice of the adsorbent is dependent upon the following factors:

- The adsorbent must not react with the substance under test.
- It should be insoluble in the solvent to be used.

- It should not catalyze the decomposition of the substances.
- The particle size of the adsorbent should be uniform.

(ii) Selection of the mobile phase (Solvent) The separation in a column involves adsorption, partition and ion exchange phenomenon, the choice of solvent depends upon these properties and other factors like polarity and solubility.

Different solvents are used for placing the solute in the column, developing the chromatogram and eluting the adsorbed material. For example, a relatively non polar solvent is used to place the mixture to be separated on the column, whereas polar solvents are used for eluting strongly adsorbed components and non polar solvents are used for weakly adsorbed components.

The solvents in increasing order of polarity are Petroleum ether < carbon tetrachloride < cyclo hexane < carbon disulphide < ether < acetone < benzene < esters < chloroform < ethanol < methanol < water < pyridine < organic acids etc.

The basic steps involved in the column chromatography are as follows:

1. Preparation of the column
2. Application of the sample
3. Development of the column.
4. Recovery of the pure constituents.

1. **Preparation of the column:** A glass column or a burette of 20 cm in length and 7-8 mm diameter is used as a column. The column is packed with the required adsorbent (alumina Al_2O_3; activated charcoal, calcium carbonate, starch etc.) as the stationary phase. For packing the adsorbent, the glass tube is first cleaned and dried. It is clamped to a stand. A small wad of cotton wool is used as the column support. Half of it is filled with the eluting solvent and a thin slurry of powdered packing material is poured into the tube. The slurry can be prepared by mixing the absorbent (silica, alumina etc.) with the requisite amount of the solvent. The stopper is opened and solvent is allowed to run slowly from the column so that the adsorbent particles settle. The column is tapped gently with a glass rod. The slurry is again added and the process is continued till all the adsorbent is added into the tube and the column of desired length is obtained. The column is covered with the layer of the solvent. The column can also be packed by dry powdered form of the adsorbent with continual tapping and then the solvent is added. But in this method air bubbles are liable to be trapped in the column.

Note. It is customary to use a relatively non-polar solvent in the beginning. The eluting solvent must be significantly less polar than the components of the mixture. If it is more polar, then the components will remain in the mobile phase and little separation will result. Moreover, the mixture should be soluble in the solvent, otherwise it will remain permanently adsorbed on the adsorbent and no separation will take place.

2. **Application of the sample.** The mixture or the sample is dissolved in a minimum amount of the solvent (usually the same solvent used for preparing the column is taken). Liquid mixtures may be applied directly to the column. The solution of the sample is added in one lot at the top of the column.

3. **Development of the column.** As the solution of the sample penetrates into the column, these solutes are adsorbed. When the solution has just completely entered the column, fresh solvent is added at the top. The solvent flows down and redissolves the solutes and carries them to lower sections of the column where they are readsorbed in amounts according to their adsorption coefficients. The least tightly held material tends to move ahead more rapidly. The various components will pass through with different rates and separate out into coloured bands or zones. This is called the **development** of the column and the whole column with bands of separated mixture is known as **chromatogram**.

4. **Recovery of the pure constituents.** If the components are coloured their positions on the column is readily apparent as a series of coloured bands. Sometimes the positions of the bands can be detected by their fluorescence in ultraviolet light or by the use of an indicator or marking substance. In other cases, the addition of the solvent is continued. By successively changing the solvents in order of their increasing polarity, the components are allowed to pass through the column with the solvent and collected separately. This is called **elution**. The solvent used for washing a particular band is known as **eluent**.

EXPERIMENT NO. 17.4

AIM **Separation of plant pigments (chlorophylls and carotenoids) from green leaves by column chromatography.**

APPARATUS Chromatography column, glass wool, cotton wool, beaker, conical flask, china dish, separatory funnel, graduated cylinder etc.

CHEMICALS Methanol, petroleum ether, chloroform, acetone, activated alumina, spinach (green leaves).

THEORY For theory of column chromatography, please refer to the discussion part.

Green leaves mainly contains chlorophylls and carotenoids hydrocarbons. The carotenoids are non-polar and are weakly adsorbed by alumina and can be eluted with non-polar solvents like hydrocarbons. On the other hand chlorophylls having number of polar sites per molecule are strongly adsorbed by alumina and their elution requires polar solvent like chloroform.

PROCEDURE

Preparation of the Extract:

1. Take 5-10 g of spinach leaves (or leaves of a green plant) and cut it up into fine pieces. Grind the leaves in a mortar and add 50 ml of methanol. Grind again.
2. Filter and discard the filterate. Grind the residue with methanol (20 ml) and petroleum ether (20 ml) again. Filter the mixture through the cotton wool.
3. Grind the residue with methanol (20 ml) and petroleum ether (70 ml) in the mortar and filter. The collective filterate of steps (2) and (3) is taken in the separating funnel.
4. Shake the separating funnel gently and allow the layers to settle.
5. The bottom layer is water-methanol layer and the upper layer is of petroleum ether containing organic pigments. Discard the bottom layer and wash the upper layer with water to remove any dissolved methanol.
6. Transfer the upper layer containing the extract to a dry conical flask and to this add anhydrous sodium sulphate. Shake the flask and leave it over for about 10 minutes to remove any water present.
7. Transfer the extract to a clean, dry test tube and concentrate it to small volume over a hot bath.

Preparation of the column:

8. Thoroughly clean the chromatographic tube and place a small wad of cotton wool as the column support.
9. Prepare a thin slurry of activated alumina and pack the column in the usual manner (refer to discussion)

Loading of the extract:

10. Fix the column in a stand vertically and add 2 ml of the extract prepared above on the top of it.

11. Allow it to run evenly into the adsorbent until a green zone 3-4 mm deep is formed at the top of the column.

Development and Recovery of the constituents:

12. Add the developer (solvent) which is a mixture of petroleum ether and benzene (1 : 9 v/v).
13. The carotenoid, being least adsorbed by the adsorbent are collected in a conical flask, till the solvent dropping off the column is almost colourless.
14. The chlorophylls are then eluted by adding chloroform to the column.
15. Concentrate both the fractions separately on a water bath, dry and weigh.

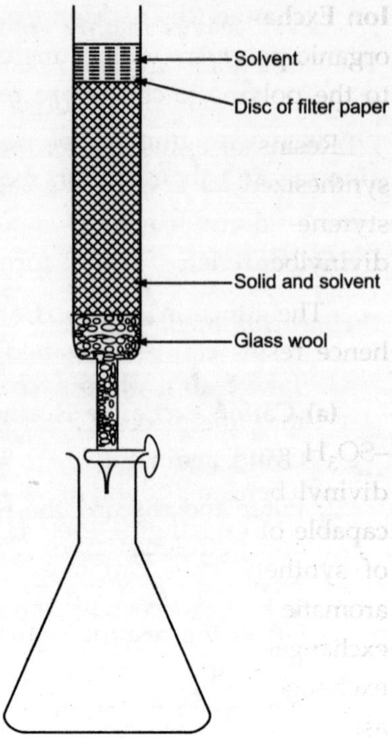

Fig 17.2 *Chromatographic column*

RESULT The plant pigments (chlorophylls and carotenoids) are separated.

PRECAUTIONS

1. The extract from green leaves should be completely free from water since the presence of a polar substance can alter the course of development.
2. The various fractions should be collected carefully.

ION EXCHANGE CHROMATOGRAPHY

Ion exchange may be defined as the reversible interchange of like ions between a solid phase (ion-exchange material) and a liquid phase in which there is no permanent change in structure. Thus, Ion Exchange Chromatography may be regarded as a special example of liquid-solid chromatography wherein strong ionic attractions replace relatively weak polar adsorption forces. In this technique, substances to be separated are previously adsorbed on the ion exchangers and then released by eluting or washing the column with a solution containing a solute which is more tightly held by the adsorbent than the component to be separated.

Ion Exchange Resins Ion exchange resions consist of cross- linked, long chain organic polymers with a microporous structure. The functional groups attached to the polymeric chains are responsible for ion exchange.

Resins of suitable chemical compositions and physical properties may be synthesized for specific ion-exchange applications. The resins usually consist of styrene- divinylbenzene copolymer. Other polymers like methacrylic acid-divinylbenzene or phenol formaldehyde polymers are also used.

The functional groups attached to the resin may be acidic or basic and hence resins can be classified as:

(a) Cation exchange Resins: Resins containing acidic groups like –COOH or $-SO_3H$ group are termed as cation exchange resins. These are usually styrene divinyl benzene copolymers which on sulphonation or carboxylation become capable of exchanging their H^+ ions with the cations of the solution. A number of synthetic-resin cation exchangers which contain sulphonated phenolic or aromatic hydrocarbons have been prepared. These resins are called H-form cation exchangers. It is also possible to convert the H-form cation exchanger to Na-form exchanger by using NaCl. Schematically a cation exchanger can be represented as:

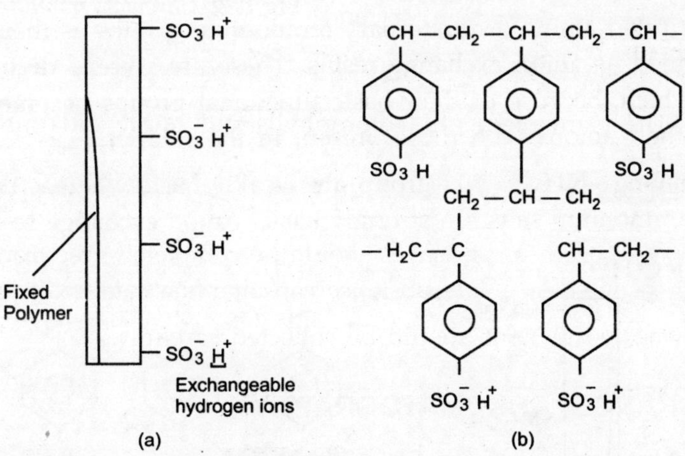

Fig. 17.3 *Representation of Cation Exchanger.*

Resins having SO_3H group are termed as strongly acidic resins whereas having COOH group are termed as weakly acidic resins. The cation exchange resins can be represented as RH^+.

Table 17.1 gives some of the examples of commercial cation Exchange resins.

Table 17.1 Commercial Cation Exchange Resins

Types of Resin	Functional Group	Nature of Resin	pH Range	Commercial Name	Applications
Strongly acid ons	$-CH_2SO_3^-H$	Sulphonated Polystyrene	1-14	Amberlite IR-100 IR-105 IR-109	Inorganic separations, fractionation of cati-
	$-SO_3^-H^+$	Sulphonated Polystyrene	1-14	IR-112 IR-120	Peptides, amino acids, lanthanides, vitamin B
Weekly acidic	–COOH	Carboxylic	5–4	Amberlite IR-6 IR-50	Biochemical separations, amino acids, antibiotics
	$-PO(OH)_2$	Phosphonated Polystyrene	5–4	Duolite C-63	Organic bases, transition elements

(b) **Anion exchange Resins.** Resins containing basic functional groups like amine, substituted amine or quaternary ammonium groups as their hydroxide salts are termed as anion exchange resins. These are styrene divinyl benzene complexes which, because of their basic functional groups become capable of exchanging their anions with those present in the solution.

Resins having $-NH_2$, = NH group are weakly basic whereas those having quaternary ammonium salts are strongly basic. Anion exchange resins are represented as $R\overline{O}H$.

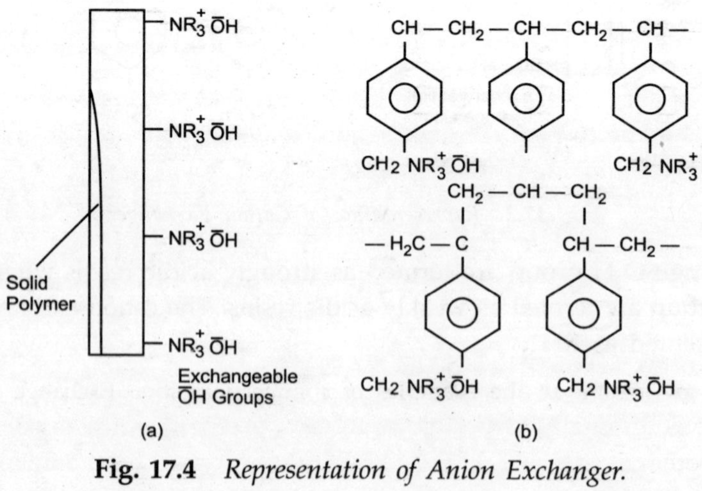

Fig. 17.4 *Representation of Anion Exchanger.*

Table 17.2 gives the examples of some of the commercially available anion exchangers.

Table 17.2 Commercial Anion Exchange Resins

Types of Resin	Functional Group	Nature of Resin	pH Range	Commercial Name	Applications
Strongly basic	$-CH_2NR_2OH$	Quaternary Polystyrene	0-12	IRA-410	Halogens, fatty acids,,
	$-CH_2N$	Ammonium Polystyrene	0-12		Alkaloids, vitamin B complex
	$-(CH_2COOH)_2$	Chelating amino diacetic acid	0-12	DowexA-I	Fractionation of anions
	$-CH_2NH_2$		0-12	Dowex IR-4B	Fatty acids
Weekly basic	$-N(C_2H_5)_2$	Polyamine polystyrene	0-9	De-Acedite -E Amberlite IR-4B	Amino acids, vitamins
	$-OH$	Phenol formaldehyde	0-9	Duolite A-1 to 7	Fractionation of anionic compounds of metals

Action of Ion Exchangers: The action of ion exchangers can be illustrated by taking the example of softening of hard water in which Ca^{2+} and Mg^{2+} ions present in water are replaced by Na^+ ions and SO_4^{2-} and Cl^- are exchanged by $\overline{O}H$ ions.

Process: The ion exchange unit consists of two tanks. Cation exchange resins and anion exchange resins are placed in first and second tank respectively. First, hard water is made to pass through cation exchange resin which exchanges its H^+ ions for Ca^{2+} and Mg^{2+} ions in water.

$$2RH^+ + CaSO_4 \rightarrow R_2Ca^{2+} + H_2SO_4$$

$$2RH^+ + MgSO_4 \rightarrow R_2Mg^{2+} + H_2SO_4$$

$$2RH^+ + CaCl_2 \rightarrow R_2Ca^{2+} + 2HCl$$

$$2RH^+ + MgCl_2 \rightarrow R_2Mg^{2+} + 2HCl.$$

The anions originally associated with Ca^{2+} or Mg^{2+} will be associated with H^+ ions. In otherwords HCl, H_2SO_4, H_2CO_3 are formed in amounts equivalent to the anions present in raw water. Thus water after passing through cation exchanger is free of Ca^{2+} and Mg^{2+} ions. This water is then passed through another column having anion exchange resin. Here the anions like SO_4^{2-}, Cl^- etc present in water are exchanged for $\overline{O}H$ ions of the resin.

$$R\overline{O}H + Cl- \rightarrow RCl^- + \overline{O}H$$

$$2R\overline{O}H + SO_4^{2-} \rightarrow R_2SO_4^{2-} + 2\overline{O}H$$

$$2R\overline{O}H + CO_3^{2-} \rightarrow R_2CO_3^{2-} + 2\overline{O}H$$

The water coming out of the anion exchanger is completly free from cations and anions responsible for hardness. It is known as deionized water or demineralized water. It is as pure as distilled water.

H^+ released from cation exchanger and $\overline{O}H$ released from anion exchanger combine to form unionized water.

$$H^+ + \overline{O}H \rightarrow H_2O.$$

Regeneration: After some time the cation and the anion exchangers get exhausted and stop working.

The cation exchanger is regenerated by first back washing and then passing a 2% H_2SO_4 solution through it. The regeneration reactions can be represented as:

$$R_2Ca^{2+} + H_2SO \rightarrow 2RH^+ + CaSO_4$$

$$R_2Mg^{2+} + H_2SO_4 \rightarrow 2RH^+ + MgSO_4.$$

The column is then rinsed with soft water to remove $CaSO_4$ and $MgSO_4$. However if HCl is used for regeneration, $MgCl_2$ and $CaCl_2$ are formed.

$$R_2Ca^{2+} + HCl \rightarrow 2RH^+ + CaCl_2$$

$$R_2Mg^{2+} + 2HCl \rightarrow 2RH^+ + MgCl_2.$$

HCl is ideally suited for regeneration since $CaCl_2$ and $MgCl_2$ are more soluble.

The anion exchanger is regenerated by back washing and then passing a solution of NaOH. The Na_2SO_4 or NaCl so formed, are rinsed with soft water and washing are led to sink.

$$R_2SO_4^{2-} + 2\ NaOH \rightarrow 2R\overline{O}H + Na_2SO_4$$

$$R_2Cl^- + NaOH \rightarrow R\overline{O}H + NaCl.$$

The generated exchanger is ready for further use.

Ion Exchange capacity. Capacity of ion exchanger is defined as the number of ions which can be exchanged per gram of the dry resin. Ion exchange capacity can be expressed in milliequivalents per gram of the exchanger.

Most of the cation exchangers are available in their Na^+ form. Before determining the capacity of resin they are converted into their H^+ forms because then the determination will only involve a simple acid base titration using phenolphthalein indicator. The conversion of Na^+ to H^+ forms is achieved by simply keeping the dry resin in 1 M HCl solution for about one hour.

$$R\ Na_2^+ + 2H^+ \rightarrow RH_2^+ + 2Na^+$$

Similarly, the anion resins are converted into the $\overline{O}H$ form by keeping the resin in 1 M NaOH solution for about one hour.

The capacities of weakly acidic and basic ion exchangers are functions of pH. The ion exchange capacity for some resins is given in milliequivalent per gram of resin.

Resin	Total Exchange Capacity m eq/g of dry resin
Anionic	
(i) De - Acedite FF (Cl^- form)	9–10
(ii) Amberlite - IRA - 410 (Cl^- form)	3.0
(iii) Amberlite - IR 4 B ($\overline{O}H$ form)	5.2–9.1
Cationic	
(iv) Zeo carb (Na^+ form)	4.5–5.0
(v) Zea carb (H^+ form)	9–10
(vi) Dowex - 50 (Na^+ form)	4-9–5.0

Technique of Ion-Exchange Chromatography

The technique and the apparatus used in ion exchange chromatography is same as for column chromatography. The only difference is that in this the adsorbent (ion exchanger) can be regenerated whereas in column chromatography it is discarded.

EXPERIMENT NO. 17.5

AIM **To determine the Ion Exchange capacity of a given cation exchanger.**

APPARATUS A cation exchange resin column (a burette with a glass wool plug at the lower end having dimensions 20 cm length and 7-8 mm diameter) conical flask, burette, pipette, measuring flask, pH paper.

CHEMICALS 0.1 N NaCl, 0.1 N NaOH solution, 0.1 N HNO_3 solution, phenolphthalein.

THEORY The cation exchange resins consist of cross-linked, long chain organic polymers with microporous structure having –COOH or SO_3H as functional groups (refer to the discussion part) are commercially available as their sodium salts and can be represented as $RSO_3^- Na^+$. These are first converted to their H^+ form by passing .1N HCl solution or .1 N HNO_3 solution

$$RSO_3^- \; Na^+ \xrightarrow{HNO_3} RSO_3^- \; H^+$$

The H^+ ion of the resin is then exchanged for Na^+ ions by passing a solution of NaCl

$$RSO_3^- \; H + NaCl \rightarrow RSO_3^- Na^+ + HCl$$

The hydrogen ions are released in equivalent amounts of sodium ion. The HCl so released is titrated against standard NaOH solution. The exchange capacity of a cation exchanger then is measured by determining the number of milligram equivalents of sodium ion which are absorbed by 1 g of the resin in the hydrogen form.

PROCEDURE

1. Take a weighed quantity of the given cationic resin sufficient to produce 20-30 cm long column and stir it with 100 - 200 ml of distilled water to form a slurry.
2. Fill the chromatography column (or a burette having a glass wool plug) with the slurry and allow it to settle down, as described in column chromatography.

3. Pass 250 ml of 0.1 N HNO_3 through the column to convert the resin into H^+ form.

4. Wash the column with distilled water several times till the pH paper shows no change of colour, when tested with the washings (the reddish colour of pH paper indicates that some HNO_3 is still present and washing is not complete).

5. Add .1 N solution of NaCl (250 ml) and allow it to pass through the column at a rate 2 ml/minute. Collect the effluent in a 250 ml conical flask.

6. Check the effluent (solution) frequently with pH paper when it shows no change in colour, it suggests that replacement is complete. The collection of effluent is stopped at this stage.

7. The solution (acid) collected in the conical flask is titrated against .1 N NaOH using phenolphthalein as indicator. Record the volume of .1 N NaOH used for titration.

OBSERVATIONS AND CALCULATIONS

Normality of NaOH= .1 N

Volume of NaOH consumed = V ml

Weight of resin packed in the column = wg

Capacity of cation exchange resin

$$= \frac{\text{Volume of NaOH (V)} \times \text{Normality of NaOH}}{\text{W}}$$

$$= \frac{\text{V} \times 0.1}{\text{W}} \text{ m eq/g}$$

RESULT The capacity of cation exchanger is .... m eq/g

PRECAUTIONS

1. For preparing all the solutions and rinsing the glasswares, only distilled water should be used.

2. The rate of effluent coming out the chromatographic column should be adjusted to 3-4 ml/min

EXPERIMENT NO. 17.6

AIM **Determine the ion-exchange capacity of a given anion exchange resin.**

APPARATUS Same as in previous experiment.

CHEMICALS 0.1 N NaOH solution, 0.1 N NaCl, pH paper, 0.1 N HCl, phenolphthalein.

THEORY The anion exchange resins consist of cross linked, long chain organic polymers with microporous structure having basic groups like amine or substituted amines as their hydroxides. Commercially available anion exchangers are available as Cl^- form which are converted into their $\overline{O}H$ form by passing .1 N NaOH solution through the resin.

$$RCl^- + NaOH \rightarrow R\overline{O}H + NaCl$$

The $\overline{O}H$ ions of the anion resin are the exchanged for Cl^- ions by passing a solution of NaCl.

$$R\overline{O}H + NaCl \rightarrow RCl^- + NaOH$$

The $\overline{O}H$ released in equivalent of chloride ions. The exchange capacity of the resin is then determined by determining the number of milliequivalents of chloride ions which are absorbed by 1 g of the resin in hydroxide form.

PROCEDURE

1. Take weighed quantity of anionic resin sufficient to produce 20-30 cm long column and stir it with 100-200 ml of distilled water to form a slurry.
2. Fill the chromatography column (or a buretle plugged with glass wool) with the slurry and allow it to settle down by letting the tap open.
3. Pass 250 ml of 0.1 N NaOH solution through the column to convert the resin into $\overline{O}H$ form.
4. Wash the resin with distilled water several times and the washings are checked with pH paper for completion of the process.
5. Now add .1 N NaCl in the column slowly so that the rate of the effluent coming out of the resin column should not exceed 3-4 ml/min. Collect the effluent in a clean conical flask.

6. Check the subsequent washings with filter paper for complete replacement of $\overline{O}H$ inside the resin. When pH paper shows no colour, it indicates that the replacement is complete. The collection of effluent is stopped at this stage.
7. Titrate the collected solution against N/10 HCl adding 2-3 drops of phenolphthalein indicator. Disappearance of pink colour marks the end point. Note the volume of the HCl used as V ml.

OBSERVATIONS AND CALCULATIONS

Normality of HCl = .1 N

Volume of HCl consumed = V ml

Weight of the anionic resin taken in the column = Wg

Capacity of the anion exchange resin

$$= \frac{\text{Volume of HCl (V)} \times \text{Normality of HCl solution}}{\text{W}}$$

$$= \frac{V \times 0.1}{W} \text{ m eq/g}$$

RESULT The capacity of anion exchange resin is .... m eq/g

PRECAUTIONS Same as in previous experiment.

EXPERIMENT NO. 17.7

AIM **Separation of cobalt and nickel using an anion exchanger and their estimation.**

APPARATUS Beakers, chromatographic column (burette) glass rod, burette, pipette, conical flasks.

CHEMICALS Anion exchange resin (De-Acidite FF), 9 M HCl, 0.1 M EDTA, murexide indicator, xylenol orange indicator, ammonia solution, KOH, hexamine, NH_4Cl solution.

THEORY The separation of the above mixture is based upon the fact that cobalt forms a stable monovalent complex anion in 9 M HCl. This cobalt complex anion can be extracted from the solution by a strongly basic anion-exchange resin, such as De-Acidite FF. The nickel present in the mixture is not retained by the resin and is washed out of the column by 9 M HCl.

PROCEDURE

1. Prepare a slurry of the resin De-Acidite FF (Chloride form) in 100 ml of distilled water in a beaker. Allow the particles to settle and decant the supernatant liquid.
2. Transfer the slurry to a chromatographic column (or a burette plugged with cotton). Pack the column as discussed in the discussion part.
3. Now pass 50 ml of 9 M HCl through the column and drain to almost bed level.
4. Introduce the solution containing the mixture of Co and Ni with a little 9 M HCl from the top of the column.
5. Pass 25 ml of 9 M HCl through the column in one lot. Collect the eluate containing nickel in a 250 ml conical flask at the rate of 2 ml per minute. Pass 3 more lots of 25 ml each of 9 M HCl through the column and collect the eluate in the same conical flask.
6. Concentrate the eluate to a small volume on wire gauze in order to remove excess acid.

Estimation of Nickel

7. Test the solution obtained in step 6 for acidity using litmus paper. If acidity still persists, add KOH to neutralize it.
8. Dilute the solution to 50 ml and add 3 drops of freshly prepared murexide indicator.
9. Add 10 ml of NH_4Cl and introduce concentrated ammonia dropwise until the pH is about 7 as indicated by yellow colour of the solution.
10. Titrate it with 0.1 M EDTA solution until the end point is reached i.e. from yellow to voilet.

Estimation of Cobalt

11. After the nickel has been eluted from the column, pass 200 ml of water at the rate of 3-4 ml per minute through the chromatographicc olumn.
12. The anionic anionic chloro complex will be decomposed. Collect the effluent in a small beaker.

13. Concentrate the effluent to a small volume and neutralize with potassium hydroxide. Adjust the pH of the solution to about 6 by ading powdered hexamine. Add distilled water to make up the volume to 50 ml.
14. Add the indicator (xylenol orange). Warm to about 60ºC and titrate with standard 0.01 M EDTA until the colour changes from red to orange-yellow.

OBSERVATIONS

1. Volume of .1 M EDTA used for titration = V_1 ml
2. Volume of .01 M EDTA used for titration = V_2 ml

CALCULTAIONS

1. 1 ml of 0.1 M EDTA ≡ 5.871 mg of Ni

   V_1 of 0.1 M EDTA ≡ 5.871 × V_1 mg of Ni = x mg of Ni
2. 1 ml of 0.01 M EDTA ≡ .5891 mg of Co

   V_2 of 0.01 M EDTA ≡ .589 × V_2 mg of Co = y mg of Co

RESULT x mg of Ni and y mg of Co is recovered from the given mixture.

EXERCISE

1. Define the term chromatography.
2. What are the different types of chromatographic techniques?
3. Define partition chromatography.

Ans. *It is a technique in which mixtures of substances are separated by means of partition between a moving solvent and a stationary liquid which is held on a suitable support. The difference in partition coefficient of individual solutes in mobile and stationary phase is basis of separation.*

Based on thin layer chromatography

4. What is the principle of thin layer chromatography?
5. How the solid layers adhere to the glass plates?

Ans. *With the help of binding agents like $CaSO_4$ etc.*

6. Define chromatogram, developer, adsorbent.
7. What is R_F value? How it is determined?
8. What are different materials used as adsorbent in thin layer chromatography?

9. What is the structure of ninhydrin.
10. What is silica Gel G?
11. In what ways TLC is superior to other chromatographic techniques?

Ans. *It is simple, having short development time and is highly sensitive.*

Based on paper chromatography

11. What is the paper chromatography?
12. What is R_F value?
13. Differentiate between ascending and descending techniques?
14. Choice of solvent is an important criteria for separation in paper chromatography. Explain.
15. State the factors on which the rate of flow of substance depend?

Ans. *The rate of flow of substance depend on the size, nature of the substance, solvent density and diffusion coefficient.*

Based on column chromatography

16. What is the principle of column chromatography?
17. How the column is prepared in the laboratory?
18. What do you mean by chromatogram?
19. How the pure constituents are recovered?
20. What is elution?
21. Why it is advisable to start elution with the non polar solvent in the beginning?

Based on Ion Exchange Chromatography

22. Define ion exchange chromatography?
23. What are ion exchangers? How these are classified?
24. What is the principle of ion exchange chromatography?
25 What is ion exchange capacity?
26. Name some of the commercially available ion exchangers.
27. What is regeneration? What chemicals are used for regeneration for cationic and anionic exchangers respectively.

❑❑❑

18

Adsorption

The term refers strictly to the existence of a higher concentration of the molecules of a substance at the surface of a liquid or solid phase than present in the bulk.

The surface of a liquid is in a state of strain due to strong inward pull, due to which the liquid surface tends to contract. This property arises due to the presence of unbalanced forces on the surface. Similar to it, surface of a solid also has unsatisfied valence or residual forces. When a solid surface is exposed to a gas or a liquid, molecules of the gas or solution phase accumulate or adsorbed on the surface. The substance that is adsorbed on the surface is called *Adsorbate* and the solid on whose surface the molecules are adsorbed is called *Adsorbent.*

Adsorption is a surface phenomenon and at a given temperature and pressure depends upon the surface area of the adsorbent. Solids in finely divided state, because of large surface area are good adsorbents. *Adsortion is an exothermic process.*

There are two categories of adsorption depending upon whether the association between the gas and the solid surface is physical or chemical in character. The two types are-

(a) **Physical adsorption.** Physical adsorption is also referred to as Vander Waals adsorption in which gas, molecules or atoms are held to the solid surface by weak Vander Waals forces. It is characterized by low enthalpy of adsorption of the order of 20 kJ mol^{-1}, which is of the same order as the heat of condensation. This small enthalpy change is insufficient to lead to breaking of bonds. Physical adsorption is usually observed at low temperatures or on relatively inert surfaces.

(b) **Chemisorption.** In chemisorption, a term for chemical adsorption, the molecules are adsorbed to the surface by forming chemical bonds. These bonds may be covalent or ionic in nature. The enthalpy of chemisorption is much greater than that of physiosorption of the order of 200 kJ mol^{-1}. The chemisorbed molecules may get completely dissociated into atoms and the existence of these dissociated molecules on the surface is one reason why solid surface catalyze the reactions. Chemisorption of hydrogen ion on nickel can be represented as-

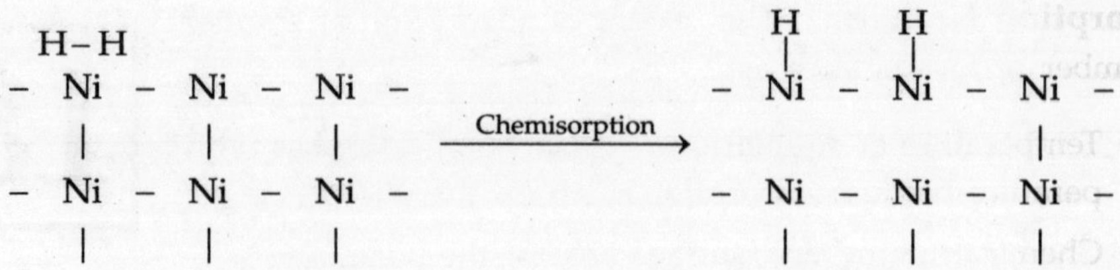

The difference between the physical adsorption and chemical adsorption is given in Table 18.1

Table 18.1

Physical Adsorption	Chemisorption
1. It involves the physical forces of attraction, like Vander Waals forces etc.	It involves the chemical forces of attraction like covalent bonds etc.
2. It is characterized by low enthalpy of adsorption of the order of 20 kJ mol^{-1}.	It is characterized by high enthalpy of adsorptionof the order of 80-200 kJ mol^{-1}.
3. The adsorption is reversible.	It is irreversible.
4. Physiosorption are non specific. Every gas is adsorbed to more or less extent on all solid surfaces.	Chemisorptions are specific in nature A gas will be chemisorbed on such solids only with which it can combine chemically
5. Multilayers may be formed in physical adsorption.	Only monolayer is formed in chemisorption.
6. Physical adsorption occurs rapidly at low temperature and decreases with increasing temperature.	Chemisorption usually occurs at high temperatures.
7. The activation energy of desorption is very low as weak Vander Waals forces are involved.	The activation energy of desorption is high as adsorbate molecules are held by comparatively strong valence forces.
8. Some example of physical adsorption are- (a) Adsorption of various gases on charcoal (b) Adsorption of nitrogen on mica.	Some examples of chemical adsorption are- (a) Adsorption of O_2 on tungsten. (b) Adsorption of H_2 on nickel.

Adsorption isotherms. The extent of adsorption at equilibrium depends on a number of factors such as:

- Temperature at equilibrium - Adsorption decreases with increase in temperature because adsorption is an exothermic process.
- Chemical nature and surface area of the adsorbent
- Concentration of the adsorbate.

At a given temperature, the variation in amount of gas adsorbed with change in pressure is given by an empirical relation suggested by Freundlich and is known as Freundlich adsorption isotherm. The relationship is,

$$\frac{x}{m} = k\,C^{1/n} \qquad ...(i)$$

where x is the amount of the adsorbate

m is the amount of the adsorbent

C is the equilibrium concentration of the adsorbate at a particular temperature, and k and n are constants.

Taking logarithms of the terms in Freundlich isotherm we get,

$$\log \frac{x}{m} = \log k + \frac{1}{n} \log C \qquad ...(ii)$$

If a graph is plotted between log $\frac{x}{m}$ as ordinate versus log C as abscissa, a straight line would result. The slope of this line will be equal to 1/n and log k will be given by the intercept which the straight line makes on the ordinate.

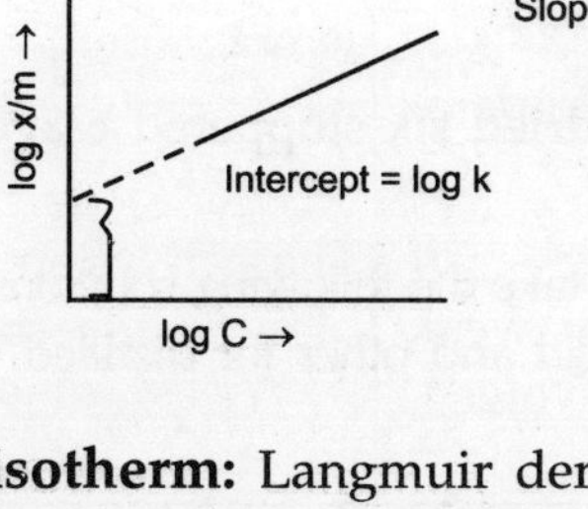

Langmuir adsorption isotherm: Langmuir derived another relationship between the amount adsorbed and the equilibrium concentration of the solute. According to it,

$$\frac{x}{m} = \frac{\alpha C}{1 + \beta C}$$

where α and β are constants.

Taking inverse we get

$$\frac{1}{x/m} = \frac{1+\beta C}{\alpha C}$$

or

$$\frac{C}{x/m} = \frac{1+\beta C}{\alpha}$$

or

$$\frac{C}{x/m} = \frac{1}{\alpha} + \frac{\beta C}{\alpha}$$

This equation can be verified by plotting $C/(x/m)$ against C, which would result in straight line. The slope will give β/α and the intercept will give $1/\alpha$.

EXPERIMENT NO. 18.1

AIM **Study the adsorption of oxalic acid from solution on activated charcoal and examine the validity of Freundlich isotherm.**

APPARATUS Burette, pipettes (25 ml, 20 ml, 10 ml) six stoppered bottles or conical flasks with corks, funnel, watch glass, thermometer, glazed paper.

CHEMICALS Activated charcoal, .25 M oxalic acid and .1 M NaOH or 0.02 M $KMnO_4$

THEORY As given in the discussion part.

PROCEDURE

1. Take well cleaned and dried six stoppered reagent bottles and label them from 1 to 6.
2. With the help of a burette take the following solutions in these bottles. (set up two burettes, one for oxalic acid and other for distilled water to take the respective volumes.)

Bottle No.	1	2	3	4	5	6
Oxalic acid (.25 M) (ml)	50	40	30	20	10	0
Distilled water (ml)	50	60	70	80	90	100

Mix the solutions well.

3 Place all the six bottles in a water bath and set them aside to acquire the temperature of the bath. Measure the temperature of the bath.

4. Take six glazed pieces of paper and weigh 2 g of activated charcoal on each of these.

5. To each of the bottles add two grams of weighed charcoal in quick succession.

6. Shake well and place the bottles in a thermostat at room temperature for about 30 minutes. Shake the contents from time to time.

7. Filter the contents of the bottles into labelled and dried bottles through a small dry filter paper. Reject first 5-10 ml of the filterate (As initial filterate will have a slightly reduced concentration of the acid due to adsorption of the acid by the dry filter paper used.)

8. Titrate the filtered solution of each bottle against M/10 NaOH using phenolphthalein as indicator (titration can also be done using 0.02 M $KMnO_4$) Take atleast two readings for each filterate.

9. Titrate the stock solution (given) of oxalic acid with .1 M NaOH.

OBSERVATIONS

Room temperature = t°C.

Weight of activated characoal in each bottle = 2 g.

Volume of oxalic acid solution for titrating in each case = 10 ml.

Bottle No.	Initial Reading	Final Reading	Volume of 0.1 M NaOH	Mean used (ml)
1.	(i)			
	(ii)			
2.	(i)			
	(ii)			
3.	(i)			
	(ii)			
4.	(i)			
	(ii)			
5.	(i)			
	(ii)			
6.	(i)			
	(ii)			

CALCULATIONS

(i) Initial concentration (C_i) of solutions 1 to 6

Bottle No.		Initial Conc. (Ci)
1.	$\frac{50}{100} \times .25$ M	
2.	$\frac{40}{100} \times .25$ M	
3.	$\frac{30}{100} \times .25$ M	
4.	$\frac{20}{100} \times .25$ M	
5.	$\frac{10}{100} \times .25$ M	
6.	—	Zero

(ii) Equilibrium concentration of the acid solution in each bottle (molarities of solution 1 to 6 after adsorption).

The molarity of oxalic acid after absorption is calculated for each solution using the following equation

$$M_{\text{oxalic acid}} = \frac{2 \times .1\, M_{NaOH} \times \text{Volume of NaOH}}{\text{Volume of oxalic acid taken for titration}}$$

($\because$ 2 moles of NaOH reacts with one mole of oxalic acid).

The values are entered in the table as C.

(iii) Amount of oxalic acid adsorbed (x) is calculated using the equation-

x = Decrease in the amount of oxalic acid in solution

= Decrease in conc. of oxalic acid × Volume of solution in liters

$= (C_i - C) \times \frac{100}{1000} = C_i - C/10$

Bottle No.	Initial conc. before adsorption (C_i)	Equilibrium conc. of acid after adsorption (C)	Amount of acid adsorbed, $x = (C_i - C)/10$
1			
2			
3			
4			
5			
6			

We can calculate x/m for each bottle and find the value of log x/m, where m is the amount of the adsorbent. The log of C is also noted in each case.

Amount of adsorbent m = 2 g

Bottle No.	Log C	log x/m	C/(x/m)
1			
2			
3			
4			
5			
6			

A graph is plotted with log x/m as ordinate and log C as abscissa. It will be a straight line which proves the Freundlich adsorption isotherm. To prove the validity of Langmuir adsorption isotherm, the values of $C/(x/m)$ are plotted as ordinate against C as abscissa. The straight line will prove the validity of Langmuir isotherm.

PRECAUTIONS

1. For filteration small papers should be used.
2. First 5 ml of each bottle should be discarded.

EXPERIMENT NO. 18.2

AIM **To study the adsorption of acetic acid on active charcoal and to verify the Freundlich and Langmuir isotherms.**

APPARATUS Burette, pipettes (25 ml, 20 ml, 10 ml) six stoppered bottles or conical flasks with corks, funnel, watch glass, thermometer, glazed paper.

CHEMICALS Activated charcoal, .5 M acetic acid, .1 M NaOH.

THEORY As given in the discussion part.

PROCEDURE

1. Take well cleaned and dried six stoppered reagent bottles and label them form 1 to 6.
2. With the help of a burette take the following solutions in these bottles. (set up two burettes, one for acetic acid and other for distilled water to take the respective volumes.)

Bottle No.	1	2	3	4	5	6
Acetic Acid .5 M (ml)	50	40	30	20	10	0
Distilled water (ml)	0	10	20	30	40	50

Mix the solutions well.

3. Place all the six bottles in a water bath and set them aside to acquire the temperature of the bath. Measure the temperature of the bath.
4. Take six glazed pieces of paper and weigh 2 g of activated charcoal on each of these.
5. To each of the bottles add two grams of weighed charcoal in quick succession.
6. Shake well and place the bottles in a thermostat at room temperature for about 30 minutes. Shake the contents from time to time.
7. Filter the contents of the bottles into labelled and dried bottles through a small dry filter paper. Reject first 5 – 10 ml of the filterate (As initial filterate will have a slightly reduced concentration of the acid due to adsorption of the acid by the dry filter paper used.)
8. Titrate the filtered solution of each bottle against M/10 NaOH using phenolphthalein as indicator. Take atleast two readings for each filterate.
9. Titrate the stock solution (given) of acetic acid with .1 M NaOH.

OBSERVATIONS

Room temperature = t°C

Weight of activated characoal in each bottle = 2 g.

Volume of acetic acid solution for titrating in each case = 10 ml.

Bottle No.	Initial Reading	Final Reading	Volume of 0.1 M NaOH used (ml)	Mean
1.	(*i*)			
	(*ii*)			
2.	(*i*)			
	(*ii*)			
3.	(*i*)			
	(*ii*)			
4.	(*i*)			
	(*ii*)			
5.	(*i*)			
	(*ii*)			
6.	(*i*)			
	(*ii*)			

CALCULATION

(*i*) Initial concentration (C_i) of solutions 1 to 6

Bottle No.	Initial Conc. (C_i)
1.	$\frac{50}{50} \times .5\,M$
2.	$\frac{40}{50} \times .5\,M$
3.	$\frac{30}{50} \times .5M$
4.	$\frac{20}{50} \times .5M$
5.	$\frac{10}{50} \times .5M$
6.	Zero

(*ii*) Equilibrium concentration of the acid solution in each bottle (molarities of solution 1 to 6 after adsorption).

The molarity of acetic acid after absorption is calculated for each solution using the following equation

$$M_{\text{acetic acid}} = \frac{.1\,M_{NaOH} \times \text{Volume of NaOH}}{\text{Volume of acetic acid taken for titration}}$$

The values are entered in the table as *C*.

(*iii*) Amount of acetic acid adsorbed (x) is calculated using the equation-

$$x = \text{Decrease in the amount of acetic acid in solution}$$

$$= \text{Decrease in conc. of acetic acid} \times \text{Volume of solution in liters}$$

$$= (C_i - C) \times \frac{50}{1000} = \frac{C_i - C}{20} \text{ moles}$$

Bottle No.	Initial conc. before adsorption (C_i)	Equilibrium conc. of acid after adsorption (C)	Amount of acid adsorbed, x (C_i – C)/20
1			
2			
3			
4			
5			
6			

We can calculate x/m for each bottle and find the value of log x/m.

The amount of adsorbent $m = 2$ g

The log of C is also noted in each case.

Bottle No.	Log C	log x/m	C/(x/m)
1			
2			
3			
4			
5			
6			

A graph is plotted with log x/m as ordinate and log C as abscissa. It will be a straight line which proves the Freundlich adsorption isotherm. To prove the validity of Langmuir adsorption isotherm, the values of $C/(x/m)$ are plotted as ordinate against C as abscissa. The straight line will prove the validity of Langmuir isotherm.

PRECAUTIONS

1. For filteration small papers should be used.
2. First 5 ml of each bottle should be discarded.
3. Do not use wet filter paper as it may dilute the solution.

EXPERIMENT NO. 18.3

AIM **To study the adsorption of iodine from alcoholic solution by charcoal.**

APPARATUS

CHEMICALS Activated charcoal, 0.5 M solution of iodine in ethanol, standard 0.2 M sodium thiosulphate solution.

THEORY As given in discussion part.

PRECEDURE

1. Take well cleaned and dried six stoppered reagent bottles and label them from 1 to 6.

2. Prepare 250 ml of .5 M solution of iodine in ethanol by dissolving about 32 g of iodine in ethanol and making the solution to 250 ml.
3. With the help of a burette take the following solutions in these bottles. (set up two burettes, one for iodine solution and other for distilled water to take the respective volumes.)

Bottle No.	1	2	3	4	5	6
Iodine solution .5 M	50	40	30	20	10	0
Pure ethanol (ml)	0	10	20	30	40	50

Mix the solutions well.

4. Place all the six bottles in a water bath and set them aside to acquire the temperature of the bath. Measure the temperature of the bath.
5. Take six glazed pieces of paper and weigh 2 g of activated charcoal on each of these.
6. To each of the bottles add two grams of weighed charcoal in quick succession.
7. Shake well and place the bottles in a thermostat at room temperature for about 3 hours. Shake the contents from time to time.
8. Filter the contents of the bottles into labelled and dried bottles through a small dry filter paper. Reject first 5 – 10 ml of the filterate.
9. Titrate the filtered solution of each bottle against .2 M $Na_2S_2O_3$ solution using starch indicator. Take atleast two readings for each filterate.
10. Titrate the stock solution (given) of iodine with .2 M $Na_2S_2O_3$.

OBSERVATIONS

Room temperature = t°C

Weight of activated characoal in each bottle, m = 2 g.

Volume of iodine solution taken for titrating in each case = 10 ml.

Bottle No.	Initial Reading	Final Reading	Volume of 0.2 M $Na_2S_2O_3$ used (ml)	Mean
1.	(*i*)			
	(*ii*)			
2.	(*i*)			
	(*ii*)			
3.	(*i*)			
	(*ii*)			
4.	(*i*)			
	(*ii*)			
5.	(*i*)			
	(*ii*)			
6.	(*i*)			
	(*ii*)			

CALCULTIONS

(*i*) Initial concentration (C_i) of solutions 1 to 6

Bottle No.		Initial Conc. (C_i)
1.	$\frac{50}{50} \times .5\,M$	
2.	$\frac{40}{50} \times .5\,M$	
3.	$\frac{30}{50} \times .5\,M$	
4.	$\frac{20}{50} \times .5\,M$	
5.	$\frac{10}{50} \times .5\,M$	
6.		Zero

(*ii*) Equilibrium concentration of the iodine solution in each bottle (molarities of solution 1 to 6 after adsorption).

The molarity of iodine after absorption is calculated for each solution using the following equation

$$M_{\text{iodine}} = \frac{2 \times .2\,\text{M Na}_2\text{S}_2\text{O}_3 \times \text{Volume of Na}_2\text{S}_2\text{O}_3}{\text{Volume of iodine taken for titration}}$$

($\because$ 2 moles of $Na_2S_2O_3$ reacts with one mole of iodine).

The values are entered in the table as C.

(*iii*) Amount of iodine adsorbed (x) is calculated using the equation -

$$x = \text{Decrease in the amount of iodine in solution}$$

$$= \text{Decrease in conc. of iodine} \times \text{Volume of solution in liters}$$

$$= (C_i - C) \times \frac{50}{1000} = C_i - C \times \frac{1}{20}$$

Bottle No.	Initial Conc. of iodine before adsorption (C_i)	Equilibrium conc. of iodine after adsorption (C)	Amount of iodine adsorbed. $(C_i - C)/20$
1			
2			
3			
4			
5			
6			

We can calculate x/m for each bottle and find the value of log x/m. The log of C is also noted in each case.

Bottle	Log C	log x/m	$C/(x/m)$
1			
2			
3			
4			
5			
6			

A graph is plotted with log x/m as ordinate and log C as abscissa. It will be a straight line which proves the Freundlich adsorption isotherm. To prove the validity of Langmuir adsorption isotherm, the value of $C/(x/m)$ are plotted as ordinate against C as abscissa. The straight line will prove the validity of Langmuir isotherm.

PRECAUTIONS

1. For filteration small papers should be used.
2. First 5 ml of each bottle should be discarded.

EXERCISE

1. Define adsorption.
2. What is the difference between physical adsorption and chemisorption ?
3. What is effect of increase of temperature on the adsorption of a gas on solid surface ?

Ans. *In case of physical adsorption, rise in temperature decreases the adsorption, but in chemisorption, it first increases and then decreases.*

4. What is Freundlich relationship for adsorption?

❑❑❑

19

Environmental Chemistry

EXPERIMENT NO. 19.1

AIM **To determine the alkalinity of a given water sample.**

Refer to Chapter 2, Wate ranalysis, Page - 104.

EXPERIMENT NO. 19.2

AIM **Determination of dissolved oxygen in the given sample of water.**

Refer to Chapter 2, Wate ranalysis, Page - 116.

EXPERIMENT NO. 19.3

AIM **To determine Biological Oxygen Demand (BOD) of a given sample of water.**

Refer to Chapter 2, Wate ranalysis, Page - 122.

EXPERIMENT NO. 19.4

AIM **To find the chemical oxygen demand (COD) of a waste water sample using $K_2 Cr_2 O_7$.**

Refer to Chapter 2, Wate ranalysis, Page - 126.

EXPERIMENT NO. 19.5

AIM **Determination of pH, conductivity and turbidity in some drinking water samples and preparation of report.**

EXPERIMENT 19.5(A)

AIM **Determination of pH in some drinking water samples.**

APPARATUS pH meter or potentiometer, glass electrode, thermometer, beaker, stirrer

THEORY One of the frequently used test in water chemistry is to determine the pH of water. The water supplied for drinking purposes has to undergo many processes such as neutralization, precipitation, coagulation etc. All these processes are pH dependent. pH of natural water lies in the range of 4.4 to 8.5.

For determining the pH of a solution different methods are employed. The most widely used method is by using pH meter. The value of pH, the logarithm of the reciprocal of the hydrogen ion concentration in solution, is determined by measuring the difference in potential between two electrodes immersed in a sample solution.

The method is based on the determination of the potential difference between an electrode pair consisting of a glass electrode sensitive to the difference in the hydrogen ion activity in the sample solution and the internal filling solution, and a reference electrode, which is supposed to have a constant potential independent of the immersing solution. These days ,***combined electrodes*** are used in which the reference electrode is combined with the glass electrode. The measured potential difference is compared with the potential obtained when both electrodes are immersed in a solution or buffer with known pH or hydrogen ion concentration. The pH is defined by the formula:

$$pH_{(sample)} = pH_{(reference)} + (E_{(sample)}) - E_{(reference)}\ F/RT1n10$$

where E are the electrode potentials, R is the universal gas constant, T the absolute temperature and F is the Faraday constant.

PROCEDURE

1. ***Calibration*** Calibrate the electrode as per the following procedure:

(a) Connect the pH meter to the power supply. Switch on the instrument.

(b) Take a standard buffer solution of pH 7 in 100 mL beaker. Note the temperature of the buffer solution.

(c) Set the temperature control of the pH meter to the temperature of the buffer solution.

(d) Remove the combination electrode from the storage solution, wash it with distilled water and blot dry with soft tissue paper.

(e) Connect the combination electrode to the input socket and dip the electrode in 7pH buffer solution.
(f) Set the Function Selector Switch to pH position and adjust with 'Calibrate" control till the digital display shows the pH 7.
(g) Now move the function switch to 'Stand by' position.
(h) Remove the electrode from the buffer solution and wash it with distilled water. Dry with tissue paper.
(i) Dip the electrode in the buffer solution having pH 4.
(j) Set the "temperature" to the temperature of the solution.
(k) Set the Function Selector Switch to pH position and adjust with 'Slope correction" control till the digital display shows the pH 4.

2. *pH measurements:*

(a) Connect the combination electrode to the input socket after washing it with distilled water.
(b) Dip the combined electrode in the solution under test.
(c) Set the Temperature knob to the temperature of the solution.
(d) Set the 'Function Selector Switch' to pH position.
(e) Note the pH of the solution.
(f) Repeat the same procedure for other water samples

OBSERVATION Drinking water samples:

Sample	*pH*
1	
2	
3	
4	

RESULT The pH of the given water samples are: ....

SIGNIFICANCE

1. Knowing 'pH' value is very important parameter for analysis of water/ waste water treatment, its suitability for domestic use and for irrigation. Certain chemicals and biological processes work only at a particular pH of 6.5 to 8.5 has no direct adverse effect on health. However lower value will produce sour taste and higher value above 8.5 a bitter taste. Higher value of pH encourages the scale formation of trihalomethanes which are causing cancer in human beings. The water for domestic consumption should have pH between 6.5 to 8.5.
2. Corrosion of water mains is the main problem associated with acidic waters. Acidic/alkaline waters cannot be used for construction purposes also. If pH is less, algae will die, fish cannot reproduce and it causes acidity, corrosion, irritation of mucous membrane, tuberculosis and other health problems in humans.

PRECAUTIONS

1. Immediately after testing, the electrodes should be washed off with a gentle stream of distilled H_2O.
2. For storage, after cleaning, the electrodes are suspended in distilled water and the system is protected against evaporation. The drying out of the electrodes should be avoided.

EXPERIMENT 19.5(B)

AIM **Determination of conductivity of some drinking water samples.**

THEORY Conductivity is a useful test in raw and finished water for a quick determination of minerals. In drinking water total dissolved solids are made up of primary inorganic salts, with small concentration of organic matter. Contributing ions are mainly CO_3^{2-}, HCO_3^{2-}, Cl^-, SO_4^{2-}, NO_2^-, K^+, Na^+, Ca^{2+}, Mg^{2+}. These ions are added to the water mainly through its contact with rocks and soil with minor contributions from pollution including urban runoff from industries.

Raw and potable water normally register specific conductance from 50 to 500μΩ/cm whereas mineralized water register values over 500 and even over 1000μΩ/cm.

PROCEDURE

1. Wash the conductivity cell thoroughly with distilled water.
2. Rinse the conductivity cell repeatedly with KCl solution.
3. Take the sufficient volume of KCl solution in a beaker and note down its temperature.
4. Connect the instrument to the mains and switch on the instrument using the power switch. Connect the electrode leads in the input sockets at the rear of the instrument.
5. Set the 'Function' switch to 'check' position and adjust the display to 1.000 with CAL control at the back panel.
6. Dip the conductivity cell in KCl solutin and adjust the temperature knob of the conductivity bridge at the temperature of the KCl solution.
7. Move the 'Function' switch to 'conductance' position and range position to appropriate range.
8. Adjust the cell constant knob so that the display reads the known value of theKCl solution at that temperature.
9. Bring the 'Function switch' to 'cell constant' position and read the value of the cell constant from the display window.
10. Take 100 ml of the water sample and note its conductivity.
11. Similarly note the conductivity of all the samples.

OBSERVATIONS

S.No	*Sample*	*Conductivity*
1	I	
2	II	
3	III	
4	IV	

RESULT The conductivities of drinking water samples are:

EXPERIMENT 19.5(C)

AIM **Determination of turbidity in the given water sample.**

APPARTUS Laboratory Nephelometer, sample cells.

THEORY Turbidity in water is caused by suspended and colloidal mater such as clay, silt finely divided organic and inorganic matter, and plankton and other microscopic organisms. Turbidity is an expression of the optical property that causes light to be scattered and absorbed rather than transmitted with no change in direction or flux level through the sample. Correlation of turbidity with the weight or particle number concentration of suspended matter is difficult because the size, shape, and refractive index of the particles affect the light-scattering properties of the suspension. When present in significant concentrations, particles consisting of light-absorbing materials such as activated carbon cause a negative interference. In low concentrations these particles tend to have a positive influence because they contribute to turbidity. The presence of dissolved, color-causing substances that absorb light may cause a negative interference. Some commercial instruments may have the capability of either correcting for a slight color interference or optically blanking out the color effect.

PRINCIPLE The nephelometric method is based on a comparison of the intensity of light scattered by the sample under defined conditions with the intensity of light scattered by a standard reference suspension under the same conditions. The higher the intensity of scattered light, the higher the turbidity. Formazin polymer is used as the primary standard reference suspension. The turbidity of a specified concentration of formazin suspension is defined as 4000 NTU.

INTERFERENCE Turbidity can be determined for any water sample that is free of debris and rapidly setting coarse sediment. Dirty glassware and the presence of air bubbles give false results. "True color," i.e., water color due to dissolved substances that absorb light, causes measured turbidities to be low. The effect usually is not significant in treated water.

Unit of measurement: The nephelometric measurements are reported as nephelometric turbidity units (NTU).

PROCEDURE (a) *Preparation of stock primary standard formazin suspension:*

1. *Solution I:* Dissolve 1.000 g hydrazine sulfate, $(NH_2)_2.H_2SO_4$, in distilled water and dilute to 100 mL in a volumetric flask.

   CAUTION Hydrazine sulfate is a carcinogen; avoid inhalation, ingestion, and skin contact. Formazin suspensions can contain residual hydrazine sulfate.
2. Solution II: Dissolve 10.00 g hexamethylenetetramine, $(CH_2)_6N_4$, in distilled water and dilute to 100 mL in a volumetric flask.
3. In a flask, mix 5.0 mL Solution I and 5.0 mL Solution II. Let stand for 24 h at 25 ± 3°C. This results in a 4000-NTU suspension. Transfer stock suspension to an amber glass or other UV-light-blocking bottle for storage. Make dilutions from this stock suspension. The stock suspension is stable for up to 1 year when properly stored.

(*b*) *Nephelometer calibration:* Follow the manufacturer's operating instructions. Run at least one standard in each instrument range to be used. Make certain the nephelometer gives stable readings in all sensitivity ranges used.

(*c*) *Measurement of turbidity:* Gently agitate sample. Wait until air bubbles disappear and pour sample into cell. Read turbidity directly from instrument display.

RESULT The turbidity of given water sample is _____ NTU.

EXPERIMENT NO. 19.6

AIM **Determination of total residual chlorine in water**

Refer to Chapter 2, Wate ranalysis, Page - 129.

EXPERIMENT NO. 19.7A

AIM Determination of pH of some soil samples.

APPARATUS pH meter or potentiometer, glass electrode, thermometer, beaker, stirrer.

THEORY Soil acts as an electrolyte and hence its pH can be measured. The pH is measured to determine the degree of acidity or alkalinity in the soil sample. The pH of soil is an important Physico-chemical characteristic because it influences:

- The suitability of a soil for crop production.
- The availability of nutrients in the soil.
- The microbial activity in the soil.
- The lime and the gypsum requirement.
- The soil's physical properties like structure, permeability etc.

For determining the pH of a solution different methods are employed. The most widely used method is by using pH meter. The value of pH, the logarithm of the reciprocal of the hydrogen ion concentration in solution, is determined by measuring the difference in potential between two electrodes immersed in a sample solution.

The method is based on the determination of the potential difference between an electrode pair consisting of a glass electrode sensitive to the difference in the hydrogen ion activity in the sample solution and the internal filling solution, and a reference electrode, which is supposed to have a constant potential independent of the immersing solution. These days , *combined electrodes* are used in which the reference electrode is combined with the glass electrode. The measured potential difference is compared with the potential obtained when both electrodes are immersed in a solution or buffer with known pH or hydrogen ion concentration. The pH is defined by the formula:

$$pH_{(sample)} = pH_{(reference)} + (E_{(sample)}) - E_{(reference)}\ F/RT1n10$$

where E are the electrode potentials, R is the universal gas constant, T the absolute temperature and F is the Faraday constant

pH interpretation:

S.No.	*pH*	*Interpretation*
1.	Below 6.5	Acidic – requires liming
2.	6.5 – 8.7	Normal – No treatment, optimum for most of the crops
3.	Above 8.7	Alkaline – Requires reclamation with gypsum and by growing the salt toleration crops.

PROCEDURE

1. *Soil Sample:*

(a) Take the soil sample with the help of clean spatula and crush it if the lumps are there.

(b) Take around 50 gm of the sample in a beaker and add 100 mL of distilled water. Stir at least four times over a 30 minutes period or continuous shaking for 10 min on mechanical shaker. This time is allowed for the soil and water to reach equilibrium

(c) Shake the soil suspension and filter it.

2. *Calibration:* Calibrate the electrode as per the following procedure:

(a) Connect the pH meter to the power supply. Switch on the instrument.

(b) Take a standard buffer solution of pH 7 in 100 mL beaker. Note the temperature of the buffer solution.

(c) Set the temperature control of the pH meter to the temperature of the buffer solution.

(d) Remove the combination electrode from the storage solution, wash it with distilled water and blot dry with soft tissue paper.

(e) Connect the combination electrode to the input socket and dip the electrode in 7pH buffer solution.

(f) Set the Function Selector Switch to pH position and adjust with 'Calibrate" control till the digital display shows the pH 7.

(g) Now move the function switch to 'Stand by' position.

(h) Remove the electrode from the buffer solution and wash it with distilled water. Dry with tissue paper.

(i) Dip the electrode in the buffer solution having pH 4.

(j) Set the "temperature" to the temperature of the solution.

(k) Set the Function Selector Switch to pH position and adjust with 'Slope correction" control till the digital display shows the pH 4.

3. pH measurements:

(a) Connect the combination electrode to the input socket after washing it with distilled water.

(b) Dip the combined electrode in the soil solution under test.

(c) Set the Temperature knob to the temperature of the solution.

(d) Set the 'Function Selector Switch' to pH position.

(e) Note the pH of the solution.

(f) Repeat the same procedure for other soil samples

OBSERVATIONS Soil Samples:

	Sample	*pH*
1		
2		
3		
4		

RESULT The pH of the given soil samples are: ....

PRECAUTIONS

1. Immediately after testing, the electrodes should be washed off with a gentle stream of distilled H_2O.

2. For storage, after cleaning, the electrodes are suspended in distilled water and the system is protected against evaporation. The drying out of the electrodes should be avoided.

EXPERIMENT NO. 19.7B

AIM **To determine the conductivity of the given soil sample.**

APPARATUS Beaker, magnetic stirrer/glass rod, conductivity meter

THEORY Electrical conductivity (EC) is the reciprocal of resistance of a conductor, 1 cm long and 1 cm^2 in cross sectional area and is expressed in mhos/cm. In the case of soils, the electrical conductivity of a water extract is proportional to its salt concentration. Therefore the electrical conductivity measurements are used in soil-testing to estimate soil salinity. Since most soil solutions have a conductivity much less than unity, the measurements are expressed in terms of milli mhos / cm ($EC \times 10^3$) (the unit for expressing the conductivity of irrigation waters is micromhos / cm = $EC \times 10^6$)

The conductivity of a saturation extract is generally recommended for appraising soil salinity in relation to plant growth. In the range of Electrical conductivity, suitable for plant growth, the following equations describe the relationship between the osmotic pressure (O.P.) and the total soluble salts of soil solutions and conductivity measurements.

$$OP = 0.36 \times EC \times 10^3 \quad \ldots (1)$$

Percentage of soluble salts in soil

$$= 0.064 \times EC \times 10^3 \times SP / 100 \quad \ldots (2)$$

However the difficulty of obtaining saturation extracts to find out saturation percentage (SP) of the soil sample makes the latter prohibitive for routine use.

The salinity estimation is based on the conductivity of dilute extracts of suspensions which are more convenient for rapid determinations. *For the soil sample, soil:water ratios of 1:2 is used for better result.*

Soils generally differ in their salt content, which affects their ability to grow crops. The soils containing an excess of soluble salts are called saline. Excess salts interfere with water and nutrient uptake, and in saline soils all crops may not grow successfully. Various crops differ in their tolerance to the presence of salts in the soils. The electrical conductivity of the soil is measured with a conductivity meter known as 'salt bridge'.

PROCEDURE

1. Weigh 50 gm of soil and add to it 100 ml of distilled water.
2. Stir the suspension continuously stirring for 20 min.

3. Filter the solution using Whatmann Filter paper.
4. Connect the instrument and calibrate the cell constant of the cell and with 0.01 M KCl. Put the temperature probe also in the KCl solution.
5. After calibration, dip the conductivity cell in the filtered solution.
6. Turn the knob towards the manual read and note the conductivity of the solution.

Interpretation: The following approximate values can be used as a guide when using a 1:2 soil water extract for conductivity measurement.

Table 19.1 Guidelines for the use of soil water extract conductivity

S.No.	*Conductivity mhos/cm^{-1}*	*Relative level of salts*	*Plant response*
1.	0.0 – 0.2	Low	No injury to any plants. May indicate lack of nutrients.
2.	0.2 – 0.8	Medium	Optimum level for most plants. Usually indicates well fertilized soils. Very sensitive plants may be injured.
3.	0.8 – 1.5	High	Injury to salt sensitive plants Germination seeds or seedlings likely to be injured.
4.	1.0 – 5.0	Excessive	Definitely injurious to most plants of any age.

RESULT The conductivity of the given soil sample is __________

PRECAUTIONS

1. Immediately after testing, the electrodes should be washed off with a gentle stream of distilled H_2O.
2. For storage, after cleaning, the electrodes are suspended in distilled water and the system is protected against evaporation. The drying out of the electrodes should be avoided.

EXPERIMENT NO. 19.8

AIM **To estimate the soil moisture content in the given soil sample.**

APPARATUS Watch glass, Drying oven, Spatula, Dessicator

THEORY Water plays a very significant role in soil growth relationship. It forms major part of plant itself, essential for physiological activities, acts as a solvent and nutrient carrier and maintains turgidity of the plants. Infact water is a regulator of physical, chemical and biological activities in the soil.

Water in a soil may be measured in a number of ways: Gravimetric, Volumetric method, Neutron scattering, Soil moisture tension and Electrical conductivity. However gravimetric method is the simplest and the most widely used method. In this the weighed soil sample is placed in an oven at 105°C and it is dried to constant weight. The weight difference is considered to be water present in soil sample.

Percent moisture = (Loss in weight /Oven dry weight of soil) × 100

PROCEDURE

1. Weigh the empty watch glass.
2. Put 10gms soil sample on the watch glass and weigh immediately to avoid loss in moisture content.
3. Place the watch glass in a preheated oven at 105°C for one hour.
4. Keep the watch glass in a Dessicator for cooling for atleast half an hour.

Now weigh the watch glass along with the oven dry soil.

OBSERVATIONS

Weight of empty watch glass = (x)

Wt. of watch glass + moist soil = (y)

Wt. of watch glass + oven dry soil = (z)

CALCULATIONS

Moisture content in soil = $(y - z)$

Weight of oven dry soil = $(z - x)$

Percentage moisture in soil = $(y - z)/(z - x) \times 100$.

RESULT The amount of moisture present in soil sample is ……%.

EXPERIMENT NO. 19.9

AIM Determination of Total dissolved solids in water/effluent sample.

THEORY Dissolved solids" refer to any minerals, salts, metals, cations or anions dissolved in water. Total dissolved solids (TDS) content of drinking water varies between 20-1000 mg/l, and it comprises mainly of inorganic salts (principally calcium, magnesium, potassium, sodium, bicarbonates, chlorides and sulfates) and some small amounts of organic matter and dissolved gases.

TDS in drinking-water originate from natural sources, sewage, urban run-off, industrial wastewater, and chemicals used in the water treatment process, and the nature of the piping or hardware used to convey the water, i.e., the plumbing.

Table 19.2 Total Dissolved Solids (TDS) in some water samples.

Water Type	*TDS (mg/l)*
Sea water	35,000
Recommended for potable water	< 1000
Irrigation water	< 500 no restriction on use, 500-2000 moderate restriction on use, > 2000 severe restriction on use.

PROCEDURE

1. Take the empty weight of evaporating dish/crucible.
2. Measure accurately 100ml of well mixed water/effluent sample and filter the sample through the whatman filter paper.
3. Transfer filtrate to a preweighed evaporating dish/crucible and evaporate to dryness on a hot plate.
4. Dry for at least 1h in oven at 180°C.
5. Cool in a desiccator and take the final weight. Calculate the total dissolved solids (TDS) using the given equation.

CALCULATION

mg of total dissolved solids/L = (A-B) x 1000/ Sample volume, ml

where, A = weight of dried residue + dish, mg, and

B = weight of dish, mg.

RESULT The amount of total dissolved solids in a given water sample is.....mg/l.

SIGNIFICANCE An elevated TDS indicates the following:

1. The concentration of the dissolved ions may cause the water to be corrosive, salty or brackish taste, result in scale formation, and interfere and decrease efficiency of hot water heaters; and
2. Many contain elevated levels of ions that are above the Primary or Secondary Drinking Water Standards, such as: an elevated level of nitrate, arsenic, aluminum, copper, lead, etc.

EXPERIMENT NO. 19.10

AIM (a) **To determine λ_{max} (wave length of maximum absorp-tion) of solution of $KMnO_4$ using a spectrophotometer**

(b) **Verify Beer's law and apply it to find the concentration of the given unknown solution.**

Refer to Colorimetric and Spectrophotometric Analysis, Chapter - 6, page - 191.

EXPERIMENT NO. 19.11

AIM **To determine the concentration of ferrous ions in a given sample spectrophotometrically by O-phenan-throline method.**

Refer to Colorimetric and Spectrophotometric Analysis, Chapter - 6, page - 196.

EXPERIMENT NO. 19.12

AIM **Determine the concentration of iron in the given sample of water by spectrophotometric method (colorimetric method) using KCNS as colour developing agent.**

Refer to Colorimetric and Spectrophotometric Analysis, Chapter - 6, page - 197.

EXPERIMENT NO. 19.13

AIM **To determine the concentration of particulate matter in the ambient air using High Volume Sampler.**

APPARATUS High Volume Sampler, Glass fibre filter (size : 203 × 254 mm)

THEORY Particles orginate from numerous sources; both natural and industrial. Also, they are emitted as primary pollutants or formed in the atmosphere as secondary pollutants from precursor gases. Natural sources include wind-blown dust, sea-spray, volcanoes, and forest fires. Combustion of fossil fuels, whether in power stations or motor vehicle engines, is a major anthropogenic source of dust particles. This produces not only visible smoke, but also finer particles that result from atmospheric reactions of gases emitted by fossil fuel combustion. SO_2 and NO_x can react in the atmosphere to produce sulfate and nitrate containing particles. Many other activities also generate particulate pollution: smelting and processing of metals, cement manufacture, agricultural activities, quarrying, construction, etc. Particulates can vary considerably in size and chemical composition depending on their origin. They can extend in diameter from fine particles < 0.1 μm to large particles > 100 μm. Generally, larger particles (>2 μm) originate from natural sources such as wind-blown dust and sea-spary.

Table 19.3 Definition of terms relating to atmospheric aerosol

Term	*Definition*
Aerosol	Suspension of small particles or liquid droplets in a gas
Smoke	Ash, soot, and grit emitted from combustion processes
Haze	Suspension of small particles (less than a few microns in diameter causing reduction of visibility
Mist	Suspension of droplets causing reduction of visibility (visibility > 1 km)
Fog	Suspension of water droplets causing great reduction in visibility (visibility < 1 km)

Both particle size and composition play an important role in determining the health effect of particulate matter. Particles > 10 μm in diameter do not penetrate the respiratory system as they are removed in the nasal region. Particles < 10 μm can be deposited in the respiratory system and are called *inhalable.* These particles can penetrate beyond the larynx into the tracheo-bronchial region. Particles < 2.5 μm can penetrate into the pulmonary region consisting of bronchioles and alaveoli, and they are called *respirable* or *fine* particles. Because of the potential health effects of the smaller particles, most air-quality monitoring stations nowadays measure the inhalable fraction (< 10 μm), which is termed PM_{10}, using specially designed sampling devices. In the past most stations measured *total suspended particulates* (TSP), also called *suspended particulate matter* (SPM). Most air-quality standards are defined in terms of PM_{10}, and more recently $PM_{2.5}$. WHO no longer specifies guidelines for particulate matter and it considers it a no-threshold pollutant.

Particles can have harmful effects on their own or they may act synergistically to enhance the toxic effects of other pollutants. High levels of SO_2 and particulates can lead to increased mortality, morbidity, and aggravation of respiratory and cardiovascular disease. Sulfuric acid aerosols have the highest irritant potential. Atmospheric particles also contain many toxic metals (*e.g.* lead, cadmium, that can result in various illnesses, and carcinogenic compounds, such as PAH. An association has been found in major US cities between particulate matter and mortality and morbidity at levels of PM_{10} below the prescribed NAAQS. In addition, high concentrations of atmospheric aerosol during smog and haze episodes can significantly reduce visibility. This can lead to increased risk of traffic accidents, closure of airports, *etc.* An increase in the total particulate load in the atmosphere could affect the earth's radiation balance with consequent effects on weather and climate. Particles can also cause soiling of materials and increase corrosion rates. Toxic substances present in atmospheric dust are washed out and deposited at the earth's surface, and thus introduced into terrestrial and aquatic ecosystems. There they can have harmful effects on living organisms. Humans can be exposed to these compounds through the food chain and the water supply. Major constituents of airborne particles and their sources are given in Table 19.4.

Each of these compounds may contribute between 5 and 5% of the total mass of the particles, depending on the size. Various transition metals (*e.g.* Pb, Mn, Fe, Co, Cu, Zn, Ni, Sn, Cd) are also found in atmospheric particles. Generally, each of the trace metals contributes < 1% of the total particulate mass. Anthropogenic sourcs of trace metals include metal manufacturing industries, fossil fuel combustion, and incineration. Typical ranges of concentrations measured in air are given in Table 19.5.

Sampling and Analysis The most common method of sampling dust particles in the air is by means of filtration using High-volume samplers. Hi-Vol samplers suck air through a large filter paper (203 × 254 mm) at a flow rate (1–2 m^3 min^{-1}). This improves accuracy and allows for the analysis of many compounds in the dust. The sampler is mounted in a shelter of heavy gauge aluminium, so that the filter is parallel to the ground and the gabled roof protects the filter from direct precipitation (Fig. 19.3). The filter is generally changed every 24 or 48 h, weighed, extracted, and analysed. Longer sampling periods can be used if required. The high volume (Hi-Vol) sampler is efficient for particles < 100 μm and it is used to determine the concentration of TSP, also called SPM.

Table 19.4 Major constituents of atmospheric particles and their sources

Component	*Major source*
Carbonaceous material	Smoke emissions from combustion
Na^+, K^+, Ca^{2+}, Mg^{2+}	Sea salt, wind-blown dust
SO_4^{2-}	SO_2 emissions from coal combustion
NO_3^-	NO_x emissions from combustion
NH_4^+	NH_3 emissions from animal waste
Cl^-	Sea salt, HCl from coal combustion
Insoluble minerals	Wind-blown dust

Table 19.5 Typical concentrations of some trace metals in air

Element	*Concentrations range* ($\mu g\ m^{-3}$)	
	Urban	*Background/Rural*
Fe	0.1–10	0.04–2
Pb	0.1–10	0.02–2
Mn	0.01–0.5	0.001–0.1
Cu	0.05–1	0.001–0.1
Cd	0.0005–0.5	0.0001–0.1
Zn	0.02–2	0.003–0.1
V	0.02–0.2	0.001–0.05

The High Volume Sampler is usually operated at ground level. In normal usage it is never operated more than 15 meters above ground level. In order to obtain a representative sample, the Sampler should not be positioned near a wall or other obstructions that would prevent free air flow. In excessively turbulent conditions or in the presence of strong surface winds or otherwise bad weather, the sampling rate is likely to decrease rapidly and perhaps in a non-linear fashion due to filter choking. If the sampler is operating in the vicinity of a source day-to-day variations in the measurement are expected due to varying meteorological conditions and changing atmospheric phenomena, like wind speed and direction, disperal, diffusion, etc.

High Volume Sampler

PROCEDURE Leave the glass fibre filter paper (203 × 254 mm) to equilibrate in dessicator for 24 h before and after sampling.. The relative humidity should be in the range of 20–45%. Weigh on an analytical balance sensitive to 0.1 mg. You may roll the filter paper loosely for weighing if necessary. Place the filter paper in the sampling assembly and sample for 24 or 48 h. Measure the flow rate of the Hi-Vol sampler according to the manufacturer's instructions at the start and at the end of the sampling period, and calculate the mean flow rate. The effective filtration area is 178 × 229 mm (you may have to measure this for your particular case). After sampling, the filter paper should be folded along the long axis with the exposed surface facing inward. Filter papers can be stored in envelopes. After determining the mass concentration filter papers can be stored for up to 1 year for chemical analysis.

From the difference in weight of the filter paper after and before sampling, the duration of sampling and the average flow rate, determine the concentration of TSP in the air:

$$\text{TSP } (\mu\text{g m}^{-1}) = \frac{[\text{Final weight (mg)} - \text{initial weight (mg)}] \times 1000]}{\text{Flow rate } (\text{m}^3 \text{ min}^{-1}) \times \text{sampling period (min)}}$$

RESULT The total suspended particulate matter in the ambient air is —— $\mu g/m^3$.

❑❑❑

APPENDICES

APPENDIX I

Handling of Data

Analytical chemistry involves the separation, identification and determination of the amounts of the components in a given sample. However, it is impossible to perform a chemical analysis that is totally free of experimental errors.

Errors may be due to number of reasons such as, improper calibration of the apparatus, different conditions of measurement, degree of accuracy to which our eyes can read etc. Therefore the measurements are uncertain to a certain degree. However, the result of the measurement taken from the experiments should be reported in such a way that they involve minimum uncertainity.

The terms commonly used in reporting the measurements in chemical analysis are;

1. Precision. Generally, repeated measurements are taken for a sample, under identical conditions. The results of measurements are usually not identical. The values may differ among themselves as well as from the true value. **Precision is defined as the degree of agreement between repeated measurements.** In otherwords, precision tells the reproducibility of the results. For example, the volume of the acid used in a acid-base titration by two students is reported as follows:

(i) 6.20 ml, 6.21ml, 6.19 ml

(ii) 6.18 ml, 6.21 ml, 6.24 ml

Evidently, the first set of measurement is more precise than the second because in first set the three values for replicate measurement are close as compared to second set.

2. Accuracy. Accuracy is the degree of agreement between the measured value and the theoretical value. It is expressed by the error.

Precision	Accuracy
1. Precision describes the agreement among several results that have been measured in the same way.	Accuracy measures the agreement between a result and its true value of a given quantity.
2. Precision is expressed in terms of deviation from the mean.	Accuracy is expressed in terms of absolute error or relative error.
3. Precision relates to the reproducibility of the results.	Accuracy relates to the reliability.

Error is of two types - (i) Absolute error or (ii) relative error.

Absolute Error. The absolute error of a measurement is the difference between the measured value and true value. It can have positive or negative sign.

$$E = \text{Observed value} - \text{True value}$$

Relative Error. The relative error of a measurement is the absolute error divided by the true value.

$$E_r = \frac{\text{Observed Value} - \text{True Value}}{\text{True Value.}}$$

Two types of errors are introduced because of number of reasons and are accordingly classified as:

Random Errors. Random errors are the errors that affect the precision of the measurement. A random error means that the measurement has an equal probability of being high or low from the true (accepted) value.

Systematic Error. This type of error occurs in the same direction each time i.e. either the observed value is low or it is high from the true or accpeted value. Hence systematic error affect the accuracy of the result.

To minimize the errors, the reulsts of the experiments are always reported in **scientific notation** having **significant figures**. The two terms are discussed in the following sections.

SCIENTIFIC NOTATION

The numbers less than one or greater than 10 are expressed in scientific notation. The following rules are followed for expressing the numbers in scientific notation.

(i) The number is written with a decimal after the first non-zero unit. This is called the coefficient term.

(ii) The number is multiplied by 10 raised to the number of spaces, the decimal has to be moved to bring it after the first non zero unit. If the decimal is moved to the left, the power (exponent) to which 10 is raised is positive. On the other hand if the decimal is moved to the right the power to which 10 is raised is negative.

895. The first nonzero unit is 8. The decimal is assumed to be present after five and hence has to be moved two places to the left. Hence the above number is expressed as scientific notation as 8.95×10^2.

Similarly, the scientific notation of some more numbers is as follows,

Number		Scientific Notation
4365	=	4.365×10^3
.0082	=	8.2×10^{-3}
.0593	=	5.93×10^{-2}
80900	=	8.09×10^4

SIGNIFICANT FIGURES

The final result for an experiment involves a number of mathematical operations such as adding, subtracting, multiplying or dividing. Thus, it becomes important that the final result is reported in such a way that the uncertainity associated with it is known correctly. A simple way of indicating the uncertainity associated with a given result is the use of significant figures.

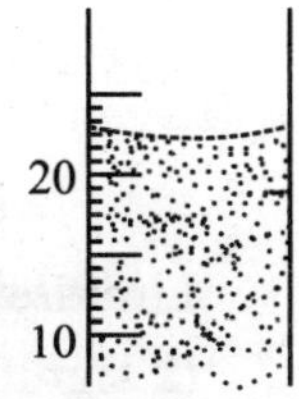

Fig. 1

The significant figures in a number are all the digits whose values are known with certainity with one additional digit whose value is uncertain. For example, when the level of the liquid is observed in the burette shown in Fig. 1, it is seen that the level of the liquid is greater than 20.3 and less than 20.4. The volume of the liquid delivered can be reported as 20.34 ml. In this number, the first three digits viz. 20.3 are certain and fourth is estimated and hence uncertain. Using the significant figure convention, the volume is reported as 20.34 ml.

Rules for determining the number of significant figures:

(i) Non zero integers are always counted as significant figures. Example, the number 43 has two significant figures whereas 3.82 has three significant figures.

(ii) It zero is placed in between two non-zero digits, it is significant. Example, the number 40082 and 5.609 have 5 and 4 significant figures.

(iii) Trailing zeroes i.e. the zero at the right end of the number are significant only if the number contains a decimal point. The number 100 has only one significant number whereas the numbers 1.00×10^2 has three significant figures. If hundred is written as 100, then it also contains three significant digits.

(iv) Zero is not significant when it is used to mark the position of the decimal point in a number less than one. Example, 0.3625 and 0.0003625 both have four significant figures.

(v) Universal constants and quantities obtained by calculations are known as exact numbers. Exact numbers are assumed to contain an infinite number of significant figures.

Significant Figures and Mathematical Operations:

1. For multiplication or division the result has the same number of significant digits as the least precise measurement used in calculation. In otherwords, the answer should be rounded off so that it contains the same number of significant digits or the original number with the smallest number of significant digits. For example,

$$8.35 \times 6.3 = 52.605 \xrightarrow{\text{corrected}} 52.6$$

↑

Limiting term

(2 significant digits)

2. For addition or subtraction, the result has the same number of decimal places as the least precise measurement used in the calculation. Example,

$$42.41 + 16.8 + 3.002 = 62.212 \xrightarrow{\text{corrected}} 62.2$$

↑

Limiting term

(one decimal place)

Hence for multiplication and division, significant figures are counted. For addition and subtraction, number of decimal places are counted.

Rules for rounding off

1. Rounding off is not done before the calculation is complete. The extra digits are carried through to the final result and then round off.
2. If the digit to be removed is less than five, the preceding digit remains the same. For example, 8.22 rounds off to 8.2.
3. If the digit to be removed is equal to or more than 5 than the preceding digit is increased by 1. For example, 8.36 rounds off to 8.4.

❑❑❑

APPENDIX II

Preparation of Indicator Solutions

Indicator	*Preparation of Solution*
Diphenylamine	Dissolve 1g of substance in 100 ml of conc. H_2SO_4.
Eriochrome Black T	Dissolve 1g of pure dye in 100 ml of rectified spirit or methanol.
Ferroin	Dissolve 3.7125 g of O-phenanthroline monohydrate with 1.6375 of $FeSO_4.7H_2O$ and make up the volume to 1 litre.
Ferric Alum	Dissolve 25 g of ferric alum in 100 ml. of hot distilled water. Add 10 ml of nitric acid and boil until yellow coloured solution is obtained.
Phenol phthalein	Dissolve 1 g of phenolphthalein in 100 ml of 50% alcohol.
Potassium chromate	Dissolve 5 gm of potassium chromate in 100 ml of distilled water.
Potassium ferricyanide	Dissolve 55 g of substance in 1 litre of distilled water.
Starch solution	Dissolve 1g of powdered starch with distilled water to form a paste and add 100 ml of boiled water to it with constant stirring, boil for about five minutes and then cool.
Methyl orange	Dissolve 50 mg of methyl orange in distilled water and make the volume to 100 ml with distilled water. Filter.
Methyl red	Dissolve 50 mg of methyl red in 500 ml of hot distilled water and cool.
Murexide	Prepare saturated solution of murexide in distilled water.

❑❑❑

APPENDIX III

Atomic Masses of the Elements

Element	Symbol	At. No.	At wt.
Aluminium	Al	13	26.98
Arsenic	As	33	74.92
Barium	Ba	56	137.34
Calcium	Ca	20	40.08
Carbon	C	6	12.011
Chlorine	Cl	17	35.453
Chromium	Cr	24	51.996
Copper	Cu	29	63.54
Fluorine	F	9	18.998
Gold	Au	79	196.967
Hydrogen	H	I	1.008
Iodine	I	53	126.904
Iron	Fe	26	55.847
Lead	Pb	82	207.19
Magnesium	Mg	12	24.312
Manganese	Mn	25	54.938
Mercury	Hg	80	200.59
Nitrogen	N	7	14.007
Phosphorus	P	15	30.974
Potassium	K	19	39.102
Silver	Ag	47	107.870
Sodium	Na	11	22.990
Sulphur	S	16	32.064
Tin	Sn	50	118.69
Zinc	Zn	30	65.37

APPENDIX IV

Equivalent Weights of Substances used in Volumetric Analysis

Titration and Substances used	*Formula*	*Eq. Weights*
Acid - Alkali Titrations:		
Sodium carbonate	Na_2CO_3	MW/2 = 53.06
Sodium bicarbonate	$NaHCO_3$	M.W/1 = 84.00
Potassium carbonate	K_2CO_3	M.W/2 = 69.00
Potassium bicarbonate	$KHCO_3$	M.W/1 = 100.00
Sodium hydroxide	NaOH	M.W/1 = 40.00
Barium hydroxide	$Ba(OH)_2.8H_2O$	M.W/1 = 157.75
Oxalic acid (hydrated)	$H_2C_2O_4.2H_2O$	M.W/2 = 63.03
Hydrochloric acid	HCl	M.W/1 = 36.46
Sulphuric acid	H_2SO_4	M.W/2 = 49.04
Redox Titrations:		
Potassium permanganate	$KMnO_4$	M.W/5 = 31.606
Potassium dichromate	$K_2Cr_2O_7$	M.W/6 = 49.03
Ferrous sulphate (hydrated)	$FeSO_4.7H_2O$	M.W/1 = 278.00
Ferrous sulphate (anhydrous)	$FeSO_4$	M.W/1 = 152.00
Ferrous Ammonium sulphate	$FeSO_4.(NH_4)_2SO_4.6H_2O$	M.W/1 = 392.12
Oxalic acid (hydrated)	$H_2C_2O_42H_2O$	M.W/2 = 63.04
Iron (ferrous)	Fe	At. W/I = 55.84
Iodine Titrations:		
Iodine	I_2	M.W/2 = 126.92
Hypo (Sodium thiosulphate Hydrated)	$Na_2S_2O_35H_2O$	M.W/1 = 248.19
Anhydrous sodium thiosulphate	$Na_2S_2O_3$	M.W/1 = 158.09
Arsenious Oxide	As_2O_3	M.W/1 = 49.45
Copper sulphate (hydrated)	$CuSO_4.5H_2O$	M.W/1 = 249.68
Argentometric Titrations		
Silver nitrate	$AgNO_3$	M.W/1 = 169.89
Silver	Ag	At.W/1 = 107.88
Sodium chloride	NaCl	M.W/1 = 58.45
Chloride ion	Cl^-	At.W/1 = 35.46
Potassium sulphocyanide	KCNS	M.W/1 = 97.17
Ammonium sulphocyanide	NH_4CNS	M.W/1 = 76.12

APPENDIX V

Weights of Substances Required to Prepare 1 Litre of a Solution of Given Normality in Volumetric Analysis

S. No.	Substance	*Mol.wt.*	*Eq. wt.*	*N/10*	*N/20*	*N/30*
1.	Sodium hydroxide, NaOH	40	40	4.0000	2.0000	1.3333
2.	Anhydrous sodium carbonate Na_2CO_3	106	53	5.000	2.6500	1.7667
3.	Oxalic acid ($H_2C_2O_4.2H_2O$)	126.08	63.04	6.3040	3.1520	2.1013
4.	Potassium dichromate, $K_2Cr_2O_7$	294.2	49.03	4.9030	2.4515	1.6343
5.	Potassium permanganate $KMnO_4$	158.03	31.606	3.1606	1.5803	1.0535
6.	Mohr's salt of Ferrous ammonium sulphate	392.10	392.10	39.2100	19.6050	13.0700
7.	Copper sulphate $CuSO_4.5H_2O$	249.71	249.71	24.9710	12.4855	8.3237
8.	Sodium thiosulphate, $Na_2S_2O_3.5H_2O$	248.20	248.20	24.8200	12.4100	8.2733
9.	Silver nitrate, $AgNO_3$	169.30	169.90	16.9900	8.4950	5.6633
10.	Ammonium thiocyanate, NH_4CNS	76.10	76.10	7.6100	3.8050	2.5367

❑❑❑

APPENDIX VI

Strengths of acids and bases

(A) STRENGTH OF AQUEOUS SOLUTIONS OF CONCENTRATED ACIDS AND BASES.

Compound	Molecular formula	Molecular weight	Approximate			Volume to make 1 litre of approx 1 M solution (m1)
			% of solute by weight	Specify gravity	Molarity	
Hydrochloric acid	HCl	36.5	35	1.18	12	84
Nitric acid	HNO_3	63.0	70	1.42	16	63
Sulphuric acid	H_2SO_4	98.0	96	1.84	18	56
Acetic acid	CH_3COOH	60.0	100	1.05	17.5	58
Phosphoric acid	H_3PO_4	98.0	> 85	1.70	14.7	68
Sodium hydroxide	$NaOH$	40.0	50	1.48	19	53
Ammonia	NH_3	17.0	28	0.90	15	67

(B) PREPARATION OF DILUTE ACIDS AND BASES

Dil acetic acid (5N)	Dilute 285 ml of glacial acetic acid of 1.05 sp. gr. to one litre with distilled water.
Dil hydrochloric acid (5N)	Dilute 445 ml of conc. HCl of 1.18 sp. gr. to one litre with distilled water
Dil nitric acid	Dilute 315 ml of conc. HNO_3 of 1.12 sp. gr. to one litre with distilled water
Dil Sulphuric acid	Pour 140 ml of conc.H_2SO_4 with sp. gr. 1.84 slowly and with stirring into 500 ml of water. Cool and dilute to one litre with distilled water
Sodium hydroxide	Dissolve 200 g of solid NaOH in one litre of water.

APPENDIX VII

Density of Water at Different Temperatures

Temperature t (°C)	*Density* g/mL	*Temperature* t (°C)	*Density* g/mL
0	.99987	21	.99799
4	.99997	22	.99777
5	.99996	23	.99754
6	.99994	24	.99729
7	.99990	25	.99705
8	.99985	26	.99678
9	.99978	27	.99651
10	.99970	28	.99623
11	.99969	29	.99594
12	.99950	30	.99565
13	.99938	31	.99530
14	99925	33	.99504
15	.99910	33	.99471
16	.99895	34	.99440
17	.99878	35	.99408
18	.99860		
19	.99841		
20	.99821		

❑❑❑

APPENDIX VIII

Surface Tension of water at different temperatures (in dynes/cm)

Temperature	*Surface tension*	*Temperature*	*Surface tension*
0°C	*75.64*	*40°C*	*69.56*
5	74.92	45	68.00
10	74.22	50	67.91
15	73.49	60	66.18
18	73.05	70	64.42
20	72.75	80	62.61
25	71.97	90	60.75
30	71.18	100	58.85
35	70.38		

(a) Surface Tension of Liquids (dynes/cm)

Substance	0°C	10°C	15°C	20°C	25°C	30°C	40°C	50°C
Water	75.64	74.22	73.49	72.75	71.97	71.18	69.56	67.91
Ethyl alcohol	24.05	23.12	22.76	22.27	21.85	21.43	20.60	19.80
Ethyl acetate	26.5	-	24.36	23.9	-	22.55	-	20.2
Carbon tetra -chloride	29.4	28.00	27.36	26.77	26.14	25.33	24.41	23.14
Ethyl ether	-	-	17.62	17.01	16.50	15.95	-	13.47
Benzene	31.58	30.22	29.55	28.88	-	27.56	26.26	24.98

(a)

(b) Viscosities of Liquids (millipoise)

Substance	0°C	10°C	15°C	20°C	25°C	30°C	40°C	50°C
Aniline	122.0	64.5	53.3	42.7	36.4	31.1	23.6	18.6
Ethyl alcohol	17.72	14.66	13.0	12.00	10.78	10.03	8.34	7.02
Methyl alcohol	8.08	6.90	6.23	5.92	5.44	5.15	4.49	3.95
Benzene	9.00	7.57	6.96	6.47	5.96	5.61	4.92	4.36
Carbon tetra -chloride	13.51	11.38	10.40	9.75	9.03	8.48	7.46	6.62
Chloroform	6.99	6.25	5.96	5.63	-	5.10	4.64	4.24
Toluene	7.72	6.71	6.23	5.90	-	5.25	4.71	4.26
Water	17.865	13.04	11.37	10.02	8.9	8.01	6.54	5.477

(b)

APPENDIX X

Heat of Solution of Some Salts

Substance	Moles of H_2O per mole of salt	ΔH solution kcal mole	Substance	Moles of H_2O per moles of salt	ΔH solution kcal/ mole
NaCl	200	+ 1.28	$MgSO_47H_2O$	400	+ 23.80
KNO_3	200	+ 8.52	Na_2SO_4	400	− 0.55
NH_4Cl	200	+ 3.88	$Na_2SO_410H_2O$	400	+ 18.90
$BaCl_22H_2O$	400	+ 4.93	KCl	200	+ 4.20
$CuSO_4$ (anhydrous)	400	− 15.80	H_2SO_4(l)	200	− 17.91
$CuSO_45H_2O$	400	+ 2.75	KOH(s)	5	− 11.53
NH_4NO_2(s)	5	+ 4.24		-	-

❑❑❑

APPENDIX XI

Cooling Mixtures

Mixture	Temperature obtained
1. Finally crushed Ice	0 - 5°C
2. Common salt + crushed Ice	Below 0°C
(1 part common salt + 3 parts of Ice)	− 5 to − 18°C
3. $CaCl_2 \cdot 6H_2O$ (Crystalline) + Ice (5 parts of $CaCl_2 \cdot 6H_2O$ + 4 parts of Ice) [If Ice is not available the following mixtures can be used.]	−40° to − 50°C
4. NH_4Cl (1 part) + $NaNO_3$ (1 part) (2 parts)	−15°C to −30°C
5. NH_4Cl (3 parts) + water (10 parts)	13 to −15°C
6. $Na_2S_2O_3.5H_2O$ (11 parts) + water (10 parts)	11 to − 8°C

❑❑❑

APPENDIX XII

Drying Agents

Substance Used	Remarks
1. $CaSO_4$	Very efficient drying agent and is converted to its hydrate $2CaSO_4.H_2O$.
2. $MgSO_4$	Forms heptahydrate $MgSO_4.7H_2O$.
3. Na_2SO_4	Its capacity as drying agent is quite high but it possesses low action and efficiency.
4. $CaCl_2$	Fairly good capacity of dehydrating, cannot be used to dry amines, alcohols and phenols.
5. Metallic sodium	It is used in the form of wire for drying some organic solvents.

❑❑❑

Win Prizes!

ATTENTION READERS

We request you for your frank assessment regarding some of the aspects of the book given as under:

Experiments in Applied Chemistry

By

Dr. Sunita Rattan

Please fill up the given space in neat capital letters.

(i) What topics of your syllabus that are important from you examination point of view are not covered in the book ?

....

....

....

....

(ii) What are the chapters wherein the treatment of the subject matter is not systematic or organised or updated ?

....

....

....

....

(iii) Have you come across misprints/mistakes/factual inaccuracies in the book? Please specify the chapters and the page numbers.

....

....

....

....

(iv) What other books on the same subject have you found/heard better that the present book? Please specify in terms of price and quality.

....

....

....

....

(v) Further suggestions and comments for the improvement of the book:

....

....

....

....

(PS: If need be please attach a separate sheet to write your views.)

Other Details :

(i) Who recommended you the book ? (Please tick in the box near the option relevant to you.)

☐ Teachers ☐ Friends ☐ Bookseller

(ii) Name of the recommending teacher, his designation and address.

....

....

....

....

(iii) Name and address of the bookseller you purchased the book form:

....

....

....

....

(iv) Name and address of your institution (Please mention the University or Board, as the case may be)

....

....

....

....

(v) Your name and complete postal address:

....

....

....

....

(vi) Course you are enrolled in:

....

....

....

....

The best assessment will be awarded each month. The award will be in the form of our publications as decided by the Editorial Board.

Please mail the filled up coupon at your earliest to:

S.K. Kataria & Sons®

4760-61/23, Ansari Road, Daryaganj Delhi-110002
Ph. : 23243489, 23269324
Fax : 011-23243489
e-mail: katariabooks@yahoo.com
katson_sanjeev@yahoo.co.in
www.skkatariaandsons.com